Introductory Physics
A Mastery-Oriented Curriculum

Second Edition

NOVARE
SCIENCE & MATH

Austin, Texas
2016

Introductory Physics
A Mastery-Oriented Curriculum

John D. Mays

Second Edition

NOVARE
SCIENCE & MATH

Austin, Texas
2016

Published by Novare Science & Math

novarescienceandmath.com

Printed in the United States of America

ISBN: 978-0-9966771-8-9

Novare Science & Math is an imprint of Novare Science & Math LLC.

Cover design by Nada Orlic, http://nadaorlic.info/

NOVARE
SCIENCE & MATH

For a catalog of titles published by Novare Science & Math, visit novarescienceandmath.com.

To all those beautiful students who have helped me understand how to be a better teacher, and to how to help others do the same.

REVIEWER

This text was carefully reviewed for technical accuracy and clarity of expression by

Chris Mack   Adjunct Faculty, University of Texas at Austin
PhD, University of Texas at Austin, Chemical Engineering
MS, Electrical Engineering, University of Maryland
BS degrees in Physics, Electrical Engineering, Chemistry, Chemical Engineering, Rose-Hulman Institute of Technology

Any errors or ambiguities that remain are the responsibility of the author.

ACKNOWLEDGEMENTS

I wish to express my thanks to my companion-in-arms in teaching *Introductory Physics*, Cathy Waldo. No man could have a more cheerful and encouraging team teacher than she. Her ideas and our discussions together were indispensable in working out a curriculum that is both challenging and appropriate for grade-level students, and her buoyant spirit has cheered me through many a difficult day.

I wish also to thank my other colleagues in the Math-Science Department at Regents School of Austin, particularly Chris Corley and Dr. Christina Swan. Each of these has contributed in many ways to helping me discover better ways to teach.

Thanks to my daughter, Rebekah, for her editorial assistance in completing this volume.

Thanks, as usual, to my good friend Dr. Chris Mack, who has helped me in my teaching and writing in ways too numerous to name.

And finally, thanks to my brother, Jeffrey, for laboring with me in this writing and publishing project. I could not do this without him.

Pax Christi to all of these.

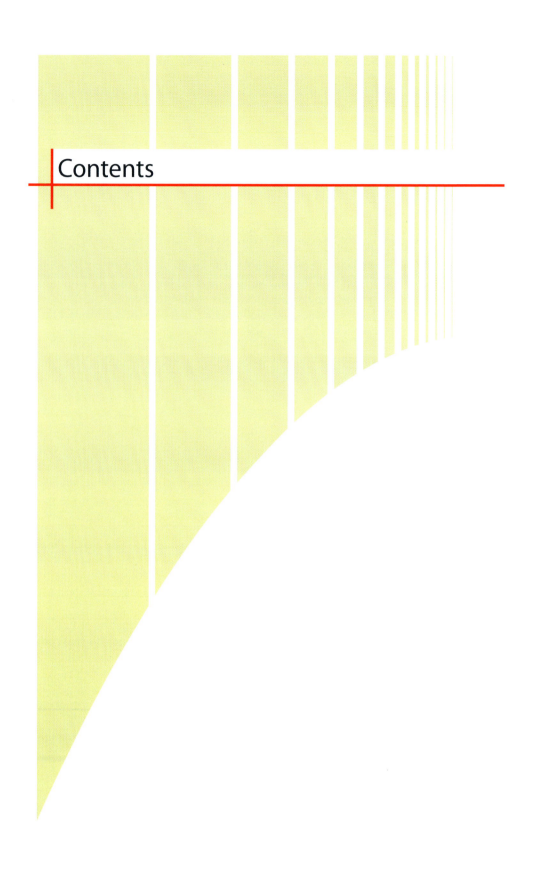

Contents

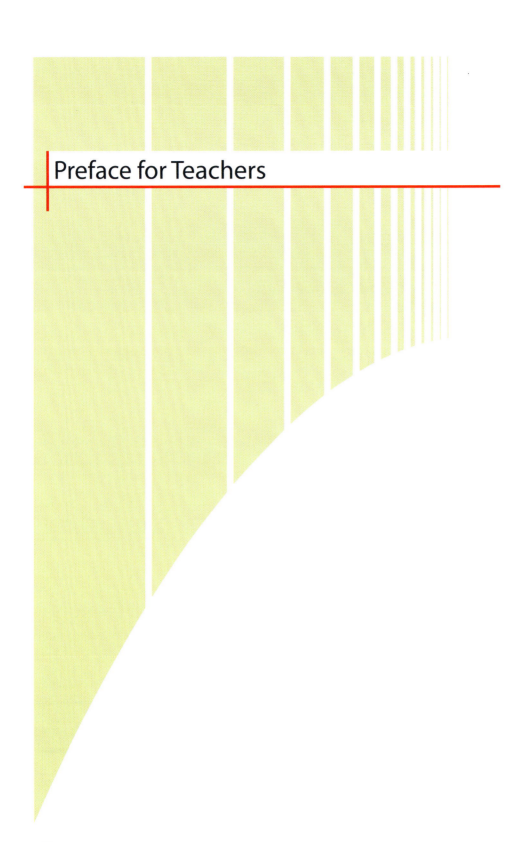

Preface for Teachers

Introductory Physics is a textbook designed for students who are performing at grade level in mathematics. For most users of this text, this means that students take the course in ninth grade while concurrently taking Algebra 1. Many schools prefer this "physics-first" science curriculum, which places introductory physics in ninth grade. By placing physics at the beginning of the high school science curriculum, students are already familiar with basic physical concepts such as matter, energy, radiation, and heat transfer when they enter biology and chemistry.

1.   A Different Textbook Philosophy

From the size of this text and the layout of its pages, it should be immediately apparent that the design philosophy represented here is significantly different from the thinking that governs science texts produced by other publishers. There are many specific differences that distinguish this text from others. Some of my comments in this Preface address these distinctive features. But there is a lot to say about our textbook philosophy, so for a full description I invite you to visit our website, novarescienceandmath.com.

Mastery, Integration, and Kingdom Perspective

I developed the philosophy and teaching methods for this course over a ten-year period, and they are described in detail in my book, *Teaching Science so that Students Learn Science.*[1] This philosophy may be summarized in terms of three key principles—Mastery, Integration, and Kingdom Perspective.

Mastery

The norm for classes in contemporary schools is what I call the *Cram-Pass-Forget Cycle*, in which students cram for tests, pass them, and then forget most of what they crammed in just a few weeks. Teachers across the nation know exactly what this looks like because they see it day after day. This cycle is a waste of time for teachers and students. Instead, students should *learn*, *master*, and *retain* what they have learned. Realizing this in the classroom requires both the teacher and the students to make significant changes in the ways they approach the tasks of teaching, testing, practicing, and studying. More on this below.

Integration

For a variety of complex reasons, effective science instruction requires that a number of content areas not usually represented adequately in curriculum materials be deeply embedded in the curriculum. These content areas include scientific process, basic epistemology, mathematics, scientific history, and English language usage. Typically, science classes do not place the necessary emphasis on these areas, and as a result students fail to have a coherent and effective learning experience. Science teachers need to think deeply about how their courses are addressing the need for these and other key areas of integration, and make adjustments to curriculum, teaching methods, assessments, and coordination between science courses.

---

1   *Teaching Science so that Students Learn Science*, by John D. Mays (2010). Published by Novare Science & Math and available at novarescienceandmath.com.

Kingdom Perspective

Science and mathematics provide us with unique ways of seeing God's creative presence in the world. Bringing biblical faithfulness to science classes is not accomplished by simply folding in a few Bible verses or prayers. In fact, much more is involved. Science and math teachers need to think broadly about how we fulfill Christ's mandate to love God with all our mind, how we teach our students effectively to engage issues, and how we perceive God's fingerprints in creation.

Learning vs. Not Learning

Typical high school science texts are 600 or 700 pages long, or even longer. I suppose this is because the authors believe that they can't leave anything out and that introductory courses should cover everything. The trouble is there is no way students can learn everything. On their part, teachers feel the obligation to "cover" what is in the text, or at least most of it, trusting that the author of the text must be a good judge of what course content for a year of science should look like. As a result, teachers often end up teaching through the text superficially, perpetuating the *Cram-Pass-Forget* Cycle.

This disastrous cycle, standard operating procedure in nearly every school in our nation, has brought us to the point of total exasperation. It is a very rare student in our day who masters what you teach him in a math or science class, remembers it, and can use it weeks or months down the road when it is needed. As a teacher, I consider this circumstance unacceptable. And thus the cornerstone of my work as a science and math teacher and writer has been to develop pedagogical methods and curriculum materials that lead students *en masse* to mastery and retention. I should emphasize that a teaching philosophy centered on mastery is not just for accelerated or honors-level students. Ordinary students benefit from mastery-based instruction just as much or more.

This text is designed to provide something completely different. Over the course of some 12 or 13 years teaching ninth-grade science, I focused on the pedagogy needed to bring students to mastery and retention, constantly studying and tweaking the curriculum to discover how much material students should reasonably master and retain in one year, and how to lead them to it. I discovered that a mastery-based course philosophy requires reigning in the curriculum and narrowing the topics covered down to a small enough set that true mastery becomes possible. Most high school freshmen, including the really bright ones, cannot master a 500+ page text in one year. If mastery is the teaching goal, as it is mine, deliberate choices must be made so that the course demands remain reasonable within the framework of other course programming.

But while carving the content down to size, I also discovered that mastery of fewer basics is vastly more powerful and satisfying than covering more topics superficially and forgetting most of them. In fact, a large percentage of my senior physics students still remember a great deal of what they learned from me as freshmen. As a result, we can move through the senior physics course—which includes all the standard topics—at a rapid pace. In short, the methods work.

2. Teaching With This Text

Skills and Prerequisites

As you will see when you look inside, the problems in the examples and problem sets almost amount to a celebration of the basic math skills that are used ubiquitously in science. These skills include using scientific notation, performing unit conversions, using the metric prefixes, determining significant digits, and solving equations for an unknown variable. Every one of these skills is basic and essential. Students study most of them in their pre-algebra courses. However, science classes in which the use of these skills is continuous—so that students truly master them—are few. The way to address this problem is to use these skills every day, week after week. Time and again I have brought classes of ordinary, grade-level students to true proficiency in the use of these skills.

In this text, I assume that students possess reasonable proficiency in the pre-algebra skills listed above, but teachers should expect to review these skills with their students in class. This text includes a fairly comprehensive tutorial on performing unit conversions, which most students have learned about but usually have not mastered by this point. Scientific notation is another skill that students have usually learned about but not mastered. Teachers should work with students to help them develop the mental math skills associated with adding or subtracting the powers of ten, but of particular importance is for students to know how to use the EE or EXP buttons on their calculators so that values in scientific notation can be entered into the calculator correctly. My motto in class is, if I am pushing buttons on my calculator the students are, too.

Optional Chapter Content

When I taught this course to ninth-grade, grade-level students, I taught the contents of this book and the five laboratory experiments, but without Chapters 8 (Pressure and Buoyancy) and 13 (Geometric Optics). If mastery is your goal, and I hope it is, then the other eleven chapters are plenty for grade-level freshmen to cover in a single course. My advice to teachers in similar circumstances would be to skip Chapters 8 and 13.

I added these two additional chapters after discussions with schools (and some home school parents) who have a non-vector physics course in their program for upper-level students. Grade-level juniors, for example, sometimes take such a course while concurrently enrolled in Algebra 2. These older students are capable of doing more in a year than freshmen are. Additionally, their more advanced status in mathematics means they can handle more algebra in solving problems. For these students, the extra two chapters will give them a broader exposure to basic topics in physics.

The mathematics involved in Chapters 8 and 13 is similar in difficulty to the math encountered in Chapters 5 (Momentum) and 11 (DC Circuits). For freshmen, the math in Chapters 5 and 11 is the most challenging in the text. Two chapters like this are enough for them. For older students, algebra is more familiar territory and the additional two chapters will give them a solid workout in algebra along with the benefit of the additional material.

Assignments, Homework, and the Weekly Workload

I do not recommend assigning very much homework addressing new material. Instead, I usually give students a fair amount of time to work on exercises in class. The major exception to this is the lab reports, which are completed entirely outside of class.

The reason outside assignments are kept to a minimum is that mastery requires regular review and practice. Since my goal for students is mastery, I help them achieve that goal by encouraging them to spend their study time at home rehearsing the material we covered in class to get it firmly in their memories, and working through review exercises to keep older material fresh. If they keep up with these tasks, they will spend two to three hours per week outside of class studying and rehearsing course material. Since I do not wish for the student workload to be any higher than this, I give students time in class to work on assignments addressing new material. Diligent students complete the majority of these exercises in class. Less efficient students inevitably end up completing some of their work at home.

Memory work is a significant part of every science course and this course is no different. It is impossible to think and converse about physics unless one not only understands the concepts, but also knows the major laws, equations, conversion factors, metric prefixes, and a few physical constants by heart. This is the reason I require students to memorize specific sets of information, as indicated in the Objectives Lists found at the beginning of each chapter. General cultural and scientific literacy also requires that students have a modest amount of historical information in their heads, so I have built in requirements to this effect in the course. Parents sometimes disagree with me about this, sometimes rather strongly. But having spent a significant portion of my adult life in graduate school, 14 years as a professional in the engineering world, and two decades as an educator, I am firmly persuaded that my point of view on this is sound.

But having established the need for some memorization, I hasten to add that there is no point in having students memorize reference data such as particle masses, element densities, or information from the Periodic Table of the Elements. Instead, students need to learn how to use resources like the periodic table or data tables and whenever possible I provide these for students to use on their quizzes and exercises. As a specific case in point, I require students to know the major conversion factors for working within the metric system and others for working within the U.S. Customary System of units (which most of them have known since they were children). But the only conversion factor I require freshmen to memorize for converting between these two systems is 1 inch = 2.54 centimeters. This one factor is used a lot and has the beauty of being exact. One can also get by with this conversion factor alone, even without any of the other SI to USCS length conversion factors. Other factors for converting miles to meters or gallons to liters can always be looked up when needed.

Experiments and Experimental Error

One of the conventional calculations in high school and college physics experiments is the so-called "experimental error." Experimental error is typically defined as the difference between the predicted value (which comes from scientific theory) and the experimental value, expressed as a percentage of the predicted value, or

$$\text{experimental error} = \frac{\left|\text{predicted or accepted value} - \text{experimental value}\right|}{\text{predicted or accepted value}} \times 100\%$$

Although the term "experimental error" is widely used, it is in my view a poor choice of words. When there is a mismatch between theory and experiment, the experiment may not be the source of the error. Often, it is the theory that is found wanting. This is how science advances.

It is, of course, true that at the introductory and intermediate level students are not generally engaged in research that uncovers weaknesses in scientific theories. At this level, the difference between prediction and experimental result may well be due entirely to "experimental error" arising from experimental limitations or inaccuracies. However, I prefer that students develop scientific habits of mind, and in the real world of scientific research in physics and engineering, the measurements are as accurate as the experimenters know how to make them and one does not know whether differences between mathematical prediction and experimental result are due to the mathematical model or error in the experiment.

I prefer to use the phrase *percent difference* to describe the value computed by the above equation. When quantitative results are compared to quantitative predictions or accepted values, students should compute the percent difference as

$$\text{percent difference} = \frac{|\text{predicted or accepted value} - \text{experimental value}|}{\text{predicted or accepted value}} \times 100\%$$

One more note. In the study of statistics, there is a calculation call the "percentage difference," in which the difference between two values is divided by their average. To avoid potential future confusion, you should note the distinction between the calculation we are using here and the one arising in statistics.

3.   Companion Resources

There are several important companion resources for instructors and students designed to be used alongside this text. These are described below and are available from Novare Science & Math at novarescienceandmath.com.

Teaching Science so that Students Learn Science

To achieve mastery with the basic math skills and all the physics content, students need to use what they learn every week. The weekly cumulative quiz regimen I use with *Introductory Physics* is a very important part of this process, and is one of the hallmarks of the course. The exercises, Weekly Review Guides, "daily questions," and other activities are all oriented toward enabling students to perform well on these quizzes, which account for the large majority of each student's grade. As I mentioned above, all these teaching methods are described in *Teaching Science so that Students Learn Science*. I commend this book as an essential companion volume to this text for anyone teaching the course.

Experiments Resources

Favorite Experiments for Physics and Physical Science describes in detail the background, apparatus, and practical considerations for the five experiments listed in the appendix of this volume and the other class demonstrations I use when teaching Introductory Physics. *Favorite Experiments* also describes all the experiments I use in my upper-level Physics course. Each presentation includes illustrative photographs. The background material students need for each laboratory experiment is included in the present volume, in Appendix C. But for the full details to assure that each experiment is a success, teachers will want to avail themselves of the detailed descriptions available in *Favorite Experiments*.

For those who seek only the information for the five experiments without the information about demonstrations and without information pertaining to upper-level physics, our small book *Experiments for Introductory Physics and ASPC* is available at significantly lower cost.

Teacher Resource CD

To facilitate the teaching methodology teachers should use with *Introductory Physics*, the following resources are available on the resource CD:

Course Overview This document outlines all the specifics of how I have taught this course, including the cumulative weekly quiz regimen, grading rubrics, and other details.

Quiz Bank This is a set of 28 quizzes for the weekly quiz. The files are in Microsoft Word to facilitate editing. PDF files of handwritten keys for the computations are included.

Final Exams These are two cumulative semester exams for fall and spring. The files are in Microsoft Word to facilitate editing. PDF files of handwritten keys for the computations are included.

Weekly Review Guide Bank As part of the mastery-oriented nature of this course, students are given a Weekly Review Guide each week beginning with week three. The Weekly Review Guides focus on rehearsal and review of previously-covered material so that it stays fresh. This set of 23 review guides is also formatted in Microsoft Word.

Course Schedule This sample lesson schedule is based on class meetings four days per week, and covers the entire text in one regular school year.

Sample Answers to Verbal Questions This document provides example answers to all the verbal questions in the text, as well as those on the quizzes and exams on the Resource CD.

The Student Lab Report Handbook

Copies of this book should be supplied to high school freshmen so they can use the book as a resource for science lab report writing year after year throughout high school and on into college. *The Handbook* presents virtually everything students need to know in order to write excellent lab reports.

Solutions Manual to Accompany Introductory Physics

This book contains fully written solutions for every computational problem in the text. The answers for the problems are included in the present volume. But for those who would like to have full solutions handy for reference, this manual fills that need.

4.   Revisions in the Second Edition

In this second edition of *Introductory Physics*, minor editorial revisions have been made throughout the text. However, teachers accustomed to the first edition should be aware that significant revisions to actual content have been made to the following topics:

- Chapter 1: The explanation of how we know truth has been revised and significantly expanded. Many new examples have been added to illustrate the distinction between truth claims and statements that we can actually know are true. The Cycle of Scientific Enterprise was revised to include Analysis and Review steps in the cycle.

- Chapter 2: An additional step regarding checking one's work for reasonableness was added to the Universal Problem Solving Strategy. The description of the Copernican Revolution has been significantly revised and expanded. In particular, Galileo's confrontation with authorities and trial has been revised to reflect the nuanced roles of church authorities, Church policy, and Galileo himself.

- Chapter 4: The definition of thermal energy has been revised.

- Chapter 6: The description of homogeneous mixtures and solutions has been revised to reflect the fact that these are now generally regarded as different names for the same type of substance. Plasma has been added to the descriptions of the basic phases of matter. Phase diagrams have been added to the discussion of phase transitions. The concepts of heat of fusion and heat of vaporization have been added to this discussion.

- Chapter 7: The definition of thermal energy has been revised.

- Chapter 9: The explanation of standing waves and resonance has been revised. The discussion of harmonics and timbre has been relocated to be included in the discussion of resonance.

- Glossary: A glossary of terms has been added.

- New Appendix: Appendix E on Making Accurate Measurements has been added.

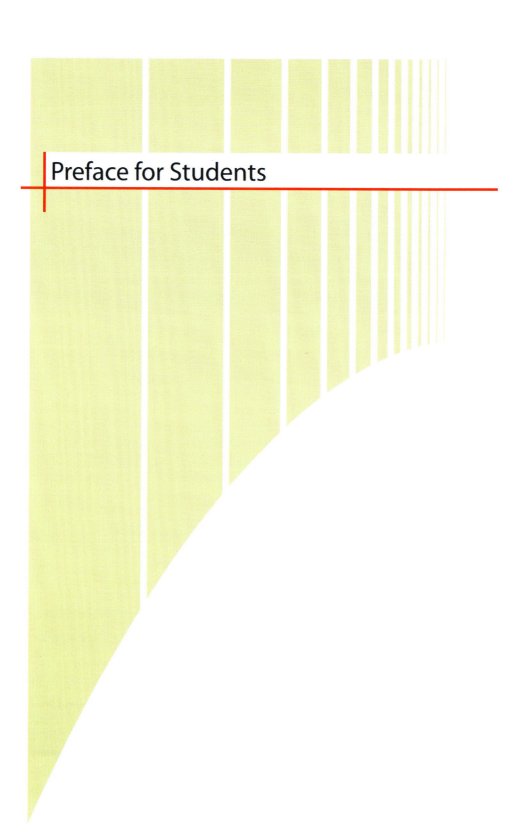

Preface for Students

This course is designed to challenge you, while bringing you to a solid level of mastery. Physics is a challenging subject. But this course is designed so that just about anyone with ordinary scientific and mathematical abilities can understand, and *master*, all the concepts and principles presented in this book. For you to succeed in *Introductory Physics*, there are a few essential things you—the student—should bring to the table.

First, you should have a sincere desire to learn. This is not the same as liking science. Nearly everyone likes science. We are all fascinated by octopuses and lasers, explosions and rockets. But since this course is specifically designed to enable the student to learn, master, and retain core knowledge and skills, you must desire to learn. And learning takes work, as does achievement in anything. If you play on a sports team or if you play a musical instrument, you probably really like doing it. But you also know that it takes a lot of practice to get good at it. Some of this practice is painful and much of it is exhausting. But because you love your sport or your instrument, you don't mind the pain, the fatigue, and the hours spent. The same things apply to this course of study.

Which brings us to the second thing: you need to develop the discipline to study and review the way you need to, as I will describe on the next page. Doing our work well is not really an option for Christians. Colossians 3:23 tells us, "Whatever you do, work heartily, as for the Lord and not for men." So I encourage you to view your studies as *the Lord's work*, something you should work at heartily for *Him*, and not something to ignore or put off. Regarding your work this way is part of being mature.

Third, you need to ask questions. Having taught courses like this one for many years, I can attest that students who are engaged in learning ask questions. If you never ask any, it is a sign to your teacher that there is trouble ahead for you. When reviewing new topics questions always arise. Ask them! When working through problem sets, there will be times when you do not understand something. Ask your instructor about it. This is an important two-way street. If you are engaged, questions will arise. You will get the answers you need by asking your instructor. And when you do, the instructor knows that since you are asking questions you are engaged with the material. The instructor also knows from your questions how well you are progressing, and this will enable him or her to bring things into the classroom that will help you and other students along.

Fourth, you need to be organized! If you are studying well, your notebook will be well organized. If it is not well organized, it is a sign that you are not doing the work that is required. When students claim that they are studying hard while their notebooks are messy and disorganized like the one in the photo (an actual student's notebook), it is clear that their claims are exaggerated. Effective study requires you to have separate sections in your course binder for notes, quizzes, review guides, homework papers, lab reports, and practice problems. These papers should all be filed in your binder, in order, so you can easily find and use them as you review. Unlike previous courses you may have had, this course is designed for *mastery*. There is a lot of review involved because we want you to remember the things you learn, rather than forgetting them a few weeks after completing the chapter.

Not a good sign.

Finally, you must apply yourself to all the exercises. In addition to the regular exercises assigned in each chapter, you will be given a Weekly Review Guide containing a list of exercises and review activities. The review guides tell you to make and rehearse flash cards, work problems, recite memory work, and other tasks. You must be diligent about accomplishing these tasks if you desire to succeed in *Introductory Physics*. I have outlined the essential components for a complete study strategy in the box below.

For years, students have said that *Introductory Physics* was one of the classes in which they learned the most. If you do the work assigned, including the review exercises, you will indeed learn a lot about this fascinating world God made. And you will have a great year, too!

A Solid Study Strategy

Your grade in Introductory Physics will be very strongly based on your performance on the weekly quizzes that occur throughout the year. These quizzes are cumulative, which means that once new material is covered in class you are responsible for it on quizzes all year long.

To be prepared for the weekly cumulative quizzes, you should establish a weekly study regimen encompassing each of the tasks listed below. You should spread out your review work so that you spend time with the material at least two or three separate days each week. Most students find that an hour spent two to three times per week is adequate for solid performance in the course.

These are the documents you must pay attention to and use in your weekly studies:

- Chapter Objectives Lists (11 total for the year, not counting the optional chapters, 8 and 13)
- Scientists List (see Appendix D)
- Conversion Factors and Constants (see Appendix A)
- Weekly Review Guides

1. Study the Objectives List for each new chapter carefully. Make it your policy that you will be able to do everything on the list (that is, for the objectives that have been covered so far in class) before quiz day each week.

2. Look over Objectives Lists from previous chapters regularly. Identify any item that you cannot do or cannot remember how to do and follow up on it.

3. Develop, maintain, and practice flash cards for each major category of information that you need to know. I recommend these four separate stacks of flash cards: 1) technical terms, laws, and equations; 2) scientists and experiments; 3) special lists to memorize (as indicated by the Objectives Lists); and 4) conversion factors, prefixes, and constants. Also, on cards for equations, indicate the units of measure for the variables involved and make saying those units part of your flash card practice routine.

4. Read every chapter in this text at least once, and preferably twice. Ideally, every time your instructor covers new material you should read the sections in this book corresponding to that material within 24 hours.

5. Go through the exercises described in the Weekly Review Guide every week.

Work each of the four review computations. The Review Guide prompts you to rehearse your flash cards, review older topics, and so on. Take the Weekly Review Guide seriously and do what it says.

6. Raise questions in class as often as you can. Asking questions and interacting with the instructor and the rest of the class is an effective way to help your brain engage, focus, and remember.

7. Go back and read the chapters in this book again when you are a month or two down the road. You will be amazed at how much easier it is to remember things when you have reread a chapter. (Besides, reading is more fun than rehearsing flash cards.)

8. When you are working on exercises involving computations, check your answers against the answer key. Every time you get an incorrect answer, dig in and stay with the problem until you identify your mistake and obtain the correct answer. If you can't figure out a problem after 10 or 15 minutes, raise the question in class.

9. Every time you lose significant points on a quiz, follow up and fill in the gaps in your learning. If you didn't understand something, raise the question with your instructor. If you forgot something, rehearse it more thoroughly until you have it down. If you failed to commit something to memory or didn't have it in your flash cards, then add it to the cards and commit it to memory. If you were not proficient enough at one or more of the computations, look up some similar problems from the exercises or from previous quizzes and practice them thoroughly. Always follow up before the next quiz. Remember, the quizzes are cumulative and the same questions come up again and again.

If you study for the course according to this study plan, you cannot help but be successful in the course and you will find that very satisfying. You will not only know a lot about physics, but you will have the satisfaction that comes from doing a job well. (By the way, you should apply this same strategy to your other classes. It works there, too!)

Introductory Physics
A Mastery-Oriented Curriculum

Second Edition

NOVARE
SCIENCE & MATH

CHAPTER 1
The Nature of Scientific Knowledge

From the 1920 report of Sir Arthur Eddington on the expedition to verify Albert Einstein's prediction of the bending of light around the sun during a solar eclipse:

In Plate 1 is given a half-tone reproduction of one of the negatives taken with the 4-inch lens at Sobral [Ceará, Brazil]. This shows the position of the stars, and, as far as possible in a reproduction of this kind, the character of the images, as there has been no retouching. A number of photographic prints have been made and applications for these from astronomers, who wish to assure themselves of the quality of the photographs, will be considered and as far as possible acceded to.

(The image above is a positive created from Eddington's negative.)

OBJECTIVES

After studying this chapter and completing the exercises, students will be able to do each of the following tasks, using supporting terms and principles as necessary:

1. Define science, theory, hypothesis, and scientific fact.
2. Explain the difference between truth and scientific facts and describe how we obtain knowledge of each.
3. Describe the difference between General Revelation and Special Revelation and relate these to our definition of truth.
4. Describe the "Cycle of Scientific Enterprise," including the relationships between facts, theories, hypotheses, and experiments.
5. Explain what a theory is and describe the two main characteristics of a theory.
6. Explain what is meant by the statement, "a theory is a model."
7. Explain the role and importance of theories in scientific research.
8. State and describe the steps of the "scientific method."
9. Define explanatory, response, and lurking variables in the context of an experiment.
10. Explain why experiments are designed to test only one explanatory variable at a time. Use the procedures the class followed in the Pendulum Experiment as a case in point.
11. Explain the purpose of the control group in an experiment.
12. Describe the possible implications of a negative experimental result. In other words, if the hypothesis is not confirmed, explain what this might imply about the experiment, the hypothesis, or the theory itself.

1.1 Modeling Knowledge

1.1.1 Kinds of Knowledge

There are many different kinds of knowledge. One kind of knowledge is *truth*. As Christians we are very concerned about truth because of its close relation to knowledge revealed to us by God. The facts and theories of science constitute a different kind of knowledge, and as students of the natural sciences we are also concerned about these.

Some people handle the distinction between the truths of the faith and scientific knowledge by referring to religious teachings as one kind of truth and scientific teaching as a different kind of truth. The problem here is that there are not different kinds of truth. There is only *one* truth, but there *are* different kinds of knowledge. Truth is one kind of knowledge, and scientific knowledge is a different kind of knowledge.

We are going to unpack this further over the next few pages, but here is a taste of where we are going. Scientific knowledge is not static. It is always changing as new discoveries are made. On the other hand, the core teachings of Christianity do not change. They are always true. We know this because God reveals them to us in his Word, which is true. This difference between scientific knowledge and knowledge from Scripture indicates to us that the knowledge we have from the Scriptures is a different kind of knowledge than what we learn from scientific investigations.

I have developed a model of knowledge that emphasizes the differences between what the Scriptures teach us and what scientific investigations teach us. This model is not perfect, nor is it exhaustive, but it is very useful, as all good models are. Our main goal in the next few sections is to develop this model of knowledge. The material in this chapter is crucial if you wish to have a proper understanding of what science is all about.

To understand science correctly, we need to understand what we mean by scientific knowledge. Unfortunately, there is much confusion among non-scientists about the nature of scientific knowledge and this confusion often leads to misunderstandings when we talk about scientific findings and scientific claims. This is nothing new. Misconceptions about scientific claims have plagued public discourse for thousands of years, and continue to do so to this day. This confusion is a severe problem, much written about within the scientific community in recent years.

To clear the air on this issue, it is necessary to examine what we mean by the term *truth*, as well as the different ways we discover truth. Then we must discuss the specific characteristics of scientific knowledge, including the key scientific terms *fact*, *theory*, and *hypothesis*.

1.1.2 What is Truth and How Do We Know It?

Epistemology, one of the major branches of philosophy, is the study of what we can know and how we know it. Both philosophers and theologians claim to have important insights on the issue of knowing truth, and because of the roles science and religion have played in our culture over the centuries, we need to look at what both philosophers and theologians have to say. The issue we need to treat briefly here is captured in this question: What is truth and how do we know it? In other words, what do we mean when we say something is *true*? And if we can agree on a definition for truth, how can we *know* whether something is true?

These are really complex questions, and philosophers and theologians have been working on them for thousands of years. But a few simple principles will be adequate for our purpose.

As for what truth is, my simple but practical definition is this:

> *Truth* may be defined as *the way things really are.*

Whatever reality is like, that is the truth. If there *really* is life on other planets, then it is true to say, "There is life on other planets."

The harder question is: How do we know the truth? According to most philosophers, there are two ways that we can know truth, and these involve either our senses or our use of reason. First, truths that are obvious to us just by looking around are said to be *evident*. It is evident that birds can fly. No proof is needed. So the proposition, "Birds can fly," conveys truth. Similarly, it is evident that humans can read books and that birds cannot. (Of course, when we speak of people knowing truth this way we are referring to people whose perceptive faculties are functioning normally.)

The second way philosophers say we can know truth is through the valid use of logic. Logical conclusions are typically derived from a sequence of logical statements called a *syllogism*, in which two or more statements (called *premises*) lead to a conclusion. For example, if we begin with the premises, "All dogs have four legs," and, "Buster is a dog," then it is a valid conclusion to state, "Buster has four legs." The truth of the conclusion of a logical syllogism definitely depends on the truth of the premises. The truth of the conclusion also depends on the syllogism having a valid structure. Some logical structures are not logically valid. (These invalid structures are called *logical fallacies*.) If the premises are true and the structure is valid, then the conclusion must be true.

So the philosophers provide us with two ways of knowing truth that most people agree upon—truths can be evident (according to our senses) or they can be proven (by valid use of reason from true premises).

Believers in some faith traditions—including Christianity—argue for a crucial third possibility for knowing truth, which is by revelation from supernatural agents such as God or angels. Jesus said, "I am the way, and the truth, and the life" (John 14:6). As Christians, we believe that Jesus was "God with us" and that all he said and did were revelations of truth to us from God the Father. Further, we believe that the Bible is inspired by God and reveals truth to us. We will come back to the ways God reveals truth to us at the end of this section.

Obviously, not everyone accepts the possibility of knowing truth by revelation. Specifically, those who do not believe in God do not accept the possibility of revelations from God. Additionally, there are some who accept the existence of a transcendent power or being, but do not accept the possibility of revelations of truth from that power. So this third way of knowing truth is embraced by many people, but certainly not by everyone.

Few people would deny that knowing truth is important. This is why we started our study by briefly exploring what truth is. But this is a book about science, and we need now to move to addressing a different question: what does *science* have to do with truth? The question is not as simple as it seems, as evidenced by the continuous disputes between religious and scientific communities stretching back over the past 700 years. To get at the relationship between science and truth, we will first look at the relationship between propositions and truth claims.

1.1.3 Propositions and Truth Claims

Not all that passes as valid knowledge can be regarded as *true*, which I defined in the previous section as "the way things really are." In many circumstances, we do not actually *know* the way things really are. People do, of course, often use propositions or statements with the intention of conveying truth. But with other kinds of statements, people intend to convey something else.

We will unpack this with a few example statements. Let's consider the following propositions:

1. I have two arms.

2. My wife and I have three children.

3. I worked out at the gym last week.

4. My car is at the repair shop.

5. Texas gained its independence from Mexico in 1836.

6. Atoms are composed of three fundamental particles—protons, neutrons, and electrons.

7. God made the world.

Among these seven statements are actually three different types of claims. From the discussion in the previous section you may already be able to spot two of them. But some of these statements do not fit into any of the categories we explored in our discussion of truth. We will discover some important aspects about these claims if we look at them one by one. So suppose for a moment that I, the writer, am the person asserting each of these statements as we examine the nature of the claim in each case.

I have two arms. This is true. I do have two arms, as is evident to everyone who sees me.

My wife and I have three children. This is true. To me it is just as evident as my two arms. I might also point out that it is true regardless of whether other people believe me when I say it. (Of course, someone could claim that I am delusional, but let's just keep it simple here and assume I am in normal possession of my faculties.) This bit about the statement being true regardless of others' acceptance of it comes up because of a slight difference here between the statement about children and the statement about arms. Anyone who looks at me will accept the truth that I have two arms. It will be evident, that is, obvious, to them. But the truth about my children is only really evident to a few people (my wife and I, and perhaps a few doctors and close family members). Nevertheless, the statement is true.

I worked out at the gym last week. This is also true; I did work out last week. The statement is evident to me because I clearly remember going there. Of course, people besides myself must depend on me to know it because they cannot know it directly for themselves unless they saw me there. Note that I cannot prove it is true. I can produce evidence, if needed, but the statement cannot be proven without appealing to premises that may or may not be true. Still, the statement is true.

My car is at the repair shop. Here is a statement that we cannot regard as a truth claim. It is merely a statement about where I understand my car to be at present, based on where I left it this morning and what the people at the shop told me they were going to do with it. For all I know, they may have taken my car joy riding and it may presently be flying along the back roads of the Texas hill country. I *can* say that the statement is correct as far as I know.

Texas gained its independence from Mexico in 1836. We Texans were all taught this in school and we believe it to be correct, but as with the previous statement we must stop short of calling this a truth claim. It is certainly a *historical fact*, based on a lot of historical evidence. The statement is correct as far as we know. But it is possible there is more to that story than we know at present (or will ever know) and none of those now living was there.

Atoms are composed of three fundamental particles—protons, neutrons, and electrons. This statement is, of course, a scientific fact. But like the previous two statements, this statement is not—surprise!—a truth claim. We simply do not know the truth about atoms. The truth about atoms is clearly not evident to our senses. We cannot guarantee the truth of any premises we might use to construct a logical proof about the insides of atoms, so proof is not able to lead us to the truth. And so far as I know, there are no supernatural agents who have revealed to us anything about atoms. So we have no access to knowing how atoms really are. What we do have are the data from many experiments, which may or may not tell the whole story. Atoms may have other components we don't know about yet. The best we can say about this statement is that *it is correct so far as we know* (that is, so far as the scientific community knows).

God made the world. This statement clearly is a truth claim, and we Christians joyfully believe it. But other people disagree on whether the statement is true. I include this example here because we will soon see what happens when scientific claims and religious truth claims get confused. I hope you are a Christian, but regardless of whether you are or not, the issue is important. We all need to learn to speak correctly about the different claims people make.

To summarize this section, some statements we make are evidently or obviously true. But for many statements, we must recognize that we don't know if they actually are true. The best we can say about these kinds of statements—and scientific facts are like this—is that they are correct so far as we know. Finally, there are metaphysical or religious statements about which people disagree; some claim they are true, some deny the same, and some say there is no way to know.

1.1.4 Truth and Scientific Claims

Let's think a bit further about the truth of reality, both natural and supernatural. I think most people agree that regardless of what different people think about God and nature, there is some actual truth or *reality* about nature and the supernatural. Regarding nature, there is some full reality about the way, say, atoms are structured, regardless of whether we currently understand that structure correctly. So far as we know, this reality does not shift or change from day to day, at least not since the early history of the universe. So the reality about atoms—the truth about atoms—does not change.

And regarding the supernatural, there is some reality about the supernatural realm, regardless of whether anyone knows what that is. Whatever these realities are, they are *truths*, and these truths do not change either.

Now, I have observed over the years that since (roughly) the beginning of the 20th century, careful scientists do not refer to scientific claims as truth claims. They do not profess to knowing the ultimate truth about how nature *really* is. For example, Niels Bohr, one of the great physicists of the 20th century, said, "It is wrong to think that the task of physics is to find out how nature *is*. Physics concerns what we can *say* about nature." Scientific claims are understood to be statements about *our best understanding* of the way things are. Most scientists believe that over time our scientific theories get closer and closer to the truth of the way things really are. But when they are speaking carefully, scientists do not claim that our present understanding of this or that is the truth about this or that.

1.1.5 Truth vs. Facts

Whatever the truth is about the way things are, that truth is presumably absolute and unchanging. If there is a God, then that's the way it is, period. And if matter is made of atoms as we think it is, then that is the truth about matter and it is always the truth. But what we call scientific facts, by their very nature, are not like this. Facts are subject to change, and sometimes do, as new information comes becomes known through ongoing scientific research. Our definitions for truth and for scientific facts need to take this difference into account. As we have seen, truth is the way things really are. By contrast, here is a definition for *scientific facts*:

> A scientific fact is a proposition that is supported by a great deal of evidence.
>
> Scientific facts are discovered by observation and experiment, and by making inferences from what we observe or from the results of our experiments.
>
> A scientific fact is *correct so far as we know*, but can change as new information becomes known.

So facts can change. Scientists do not put them forward as truth claims, but as propositions that are correct so far as we know. In other words, scientific facts are *provisional*. They

Examples of Changing Facts

In 2006, the planet Pluto was declared not to be a planet any more.

In the 17th century, the fact that the planets and moon all orbit the earth changed to the present fact that the planets all orbit the sun, and only the moon orbits the earth.

At present we know of only one kind of matter that causes gravitational fields. This is the matter made up of protons, electrons, and neutrons, which we will discuss in a later chapter. But scientists now think there may be another kind of matter contributing to the gravitational forces in the universe. They call it "dark matter" because apparently this kind of matter does not reflect or refract light the way ordinary matter does. (We will also study reflection and refraction later on.) For the existence of dark matter to become a scientific fact, a lot of evidence is required, evidence which is just beginning to emerge. If we are able to get enough evidence, then the facts about matter will change.

are always subject to revision in the future. As scientists make new scientific discoveries, they must sometimes revise facts that were formerly considered to be correct. But the truth about reality, whatever it is, is absolute and unchanging.

The distinction between truth and scientific facts is crucial for a correct understanding of the nature of scientific knowledge. Facts can change; truth does not.

1.1.6 Revelation of Truth

In Section 1.1.2, we examined the ways we can know truth. Here we need to say a bit more about what Christian theology says about revealed truth.

Christians believe that the supreme revelation of God to us was through Jesus Christ in the incarnation. Those who knew Jesus and those who heard Jesus teach were receiving direct revelation from God. Jesus said, "Whoever has seen me has seen the Father" (John 14:9).

Jesus no longer walks with us on the earth in a physical body (although we look forward to his return when he will again be with us). But Christians believe that when Jesus departed he sent his Holy Spirit to us, and today the Spirit guides us in the truth. According to traditional Christian theology, God continues to reveal truth to us through the Spirit in two ways: *Special Revelation* and *General Revelation*. Special Revelation is the term theologians use to describe truths God teaches us in the Bible, his Holy Word. General Revelation refers to truths God teaches us through the world he made. Sometimes theologians have described Special and General Revelation as the two "books" of God's revelation to us, the book of God's *Word* (the Bible) and the book of God's *Works* (nature). And it is crucial to note that the truths revealed in God's Word and those revealed in his Works *do not conflict*.

Truth is not discovered the same way scientific facts are. Truth is true for all people, all times, and all places. Truth never changes. Here are just a few examples of the many truths revealed in God's Word:

- Jesus is the divine Son of God (Matthew 16:16).
- All have sinned and fall short of what God requires (Romans 3:23).
- All people must die once and then face judgment (Hebrews 9:27).
- God is the creator of all that is (Colossians 1:16, Revelation 4:11).
- God loves us (John 3:16).

Each of these statements is true, and we know they are true because God has revealed them to us in his word. (The reasons for believing God's word are important for all of us to know and understand, but that is a subject for a different course of study.)

1.2   The Cycle of Scientific Enterprise

1.2.1 Science

Having established some basic principles about the distinction between scientific facts and truth, we are now ready to define *science* itself and examine what science is and how it works. Here is a definition:

> Science is the process of using experiment, observation, and logical thinking to build "mental models" of the natural world. These mental models are called *theories*.

We do not and cannot know the natural world perfectly or completely, so we construct models of how it works. We explain these models to one another with descriptions, diagrams, and mathematics. These models are our scientific theories. Theories never explain the world to us perfectly. To know the world perfectly, we would have to know the absolute truth about reality just as God knows it, which in this present age we do not. So theories always have their limits, but we hope they become more accurate and more complete over time, accounting for more and more physical phenomena (data, facts), and helping us to understand the natural world as a coherent whole.

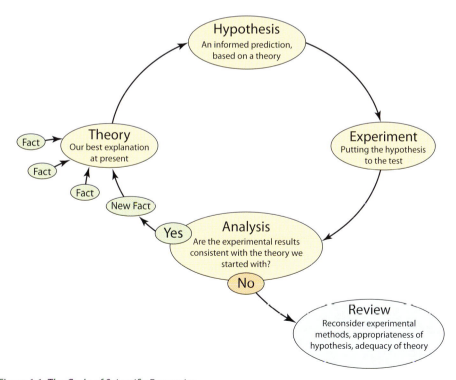

Figure 1.1. The Cycle of Scientific Enterprise.

Scientific knowledge is continuously changing and advancing through a cyclic process that I call the *Cycle of Scientific Enterprise*, represented in Figure 1.1. In the next few sections we will examine the individual parts of this cycle in detail.

1.2.2 Theories

Theories are the grandest thing in science. In fact, it is fair to say that theories are the *glory* of science, and developing good theories is what science is all about. Electromagnetic field theory, atomic theory, quantum theory, the general theory of relativity—these are all theories in physics that have had a profound effect on scientific progress and on the way we all live.[1]

Now, even though many people do not realize it, *all scientific knowledge is theoretically based.* Let me explain. A *theory* is a mental model or explanatory system that explains and relates together most or all of the facts (the data) in a certain sphere of knowledge. A theory is not a hunch or a guess or a wild idea. Theories are the mental structures we use to make sense of the data we have. We cannot understand any scientific data without a theory to organize it and explain it. This is why I wrote that all scientific knowledge is theoretically based. And for this reason, it is inappropriate and scientifically incorrect to scorn these explanatory systems as "merely a theory" or "just a theory." Theories are explanations that account for a lot of different facts. If a theory has stood the test of time, that means it has wide support within the scientific community.

It is popular in some circles to speak dismissively of certain scientific theories as if they represented some kind of untested speculation. It is simply incorrect—and very unhelpful—to speak this way. As students in high school science, one of the important things you need to understand is the nature of scientific knowledge, the purpose of theories, and the way scientific knowledge progresses. These are the issues this chapter is about.

All useful scientific theories must possess several characteristics. The two most important ones are:

- The theory accounts for and explains most or all of the related facts.

- The theory enables new hypotheses to be formed and tested.

Theories typically take decades or even centuries to gain credibility. If a theory gets replaced by a new, better theory, this also usually takes decades or even centuries to happen. No theory is ever "proved" or "disproved" and we should not speak of them in this way. We also should not speak of them as being "true" because, as we have seen, we do not use the word "truth" when speaking of scientific knowledge. Instead we speak of facts being correct so far as we know, or of current theories as representing our best understanding, or of theories being successful and useful models that lead to accurate predictions.

An experiment in which the hypothesis is confirmed is said to support the theory. After such an experiment, the theory is stronger but it is not proved. If a hypothesis is not confirmed by an experiment, the theory might be weakened but it is not disproved. Scientists require a great deal of experimental evidence before a new theory can be established as the best explanation for a body of data. This is why it takes so long for theories to become widely accepted. And since no theory ever explains everything perfectly, there are always phe-

---

1 The term *law* is just a historical (and obsolete) term for what we now call a theory.

nomena we know about that our best theories do not adequately explain. Of course, scientists continue their work in a certain field hoping eventually to have a theory that does explain all of the facts. But since no theory can explain everything perfectly, it is impossible for one experimental failure to bring down a theory. Just as it takes a lot of evidence to establish a theory, so it takes a large and growing body of conflicting evidence before scientists abandon an established theory.

Earlier, I wrote that theories are mental *models*. This statement needs a bit more explanation. A model is a representation of something, and models are designed for a purpose. You have probably seen a model of the organs in the human body in a science classroom or textbook. A model like this is a physical model and its purpose is to help people understand how the human body is put together. A mental model is not physical; it is an intellectual understanding, although we often use illustrations or physical models to help communicate to one another our mental ideas. But as in the example of the model of the human body, a theory is also a model. That is, a theory is a representation of how part of the world works. Frequently, our models take the form of mathematical equations that allow us to make numerical predictions and calculate the results of experiments. The more accurately a theory represents the way the world works, which we judge by forming new hypotheses and testing them with experiments, the better and more successful the theory is.

To summarize, a successful theory represents the natural world accurately. This means the model (theory) is useful because if a theory is an accurate representation, then it leads

Examples of Famous Theories

In the next chapter, you will learn a bit about Einstein's general theory of relativity, one of the most important theories in modern physics. Einstein's theory represents our best current understanding of how gravity works.

Another famous theory you will learn about later is the kinetic theory of gases, our present understanding of how molecules of gas too small to see are able to create pressure inside of a container.

Key Points About Theories

1. A theory is a way of modeling nature, enabling us to explain why things happen in the natural world from a scientific point of view.
2. A theory tries to account for and explain the known facts that relate to it.
3. Theories must enable us to make new predictions about the natural world so we can learn new facts.
4. Strong, successful theories are the glory and goal of scientific research.
5. A theory becomes stronger by producing successful predictions that are confirmed by experiment. A theory is gradually weakened when new experimental results repeatedly turn out to be inconsistent with the theory.
6. It is incorrect to speak dismissively of successful theories because theories are not just guesses.
7. We don't speak of theories as being proved or disproved. Instead, we speak of them in terms such as how successful they have been at making predictions and how accurate the predictions have been.

Figure 1.2. Key points about theories.

to accurate predictions about nature. When a theory repeatedly leads to predictions that are confirmed in scientific experiments, it is a strong, useful theory. The key points about theories are summarized in Figure 1.2.

1.2.3 Hypotheses

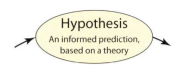

A *hypothesis* is a positively stated, informed prediction about what will happen in certain circumstances. We say a hypothesis is an *informed* prediction because when we form hypotheses we are not just speculating out of the blue. We are applying a certain theoretical understanding of the subject to the new situation before us and predicting what will happen or what we expect to find in the new situation based on the theory the hypothesis is coming from. Every scientific hypothesis is based on a particular theory.

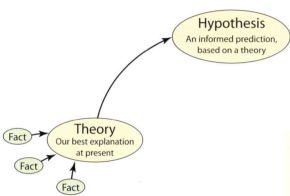

Often hypotheses are worded as *if-then* statements, such as, "If various forces are applied to a pickup truck, then the truck accelerates at a rate that is in direct proportion to the net force." Every scientific hypothesis is based on

a theory and it is the hypothesis that is directly tested by an experiment. If the experiment turns out the way the hypothesis predicts, the hypothesis is confirmed and the theory it came from is strengthened. Of course, the hypothesis may not be confirmed by the experiment. I will describe how scientists respond to this situation shortly.

Key Points About Hypotheses

1. A hypothesis is an informed prediction about what will happen in certain circumstances.

2. Every hypothesis is based on a particular theory.

3. Well-formed scientific hypotheses must be testable, which is what scientific experiments are designed to do.

Figure 1.3. Key points about hypotheses.

Examples of Famous Hypotheses

Einstein used his general theory of relativity to make an incredible prediction in 1917: that gravity causes light to bend as it travels through space. In the next chapter, you will read about the stunning result that occurred when this hypothesis was put to the test.

The year 2012 was a very important year for the standard theory in the world of subatomic particles, called the Standard Model. This theory led in the 1960s to the prediction that there are weird particles in nature, now called Higgs Bosons, which no one had ever detected. Until 2012, that is! An enormous machine that could detect these particles, called the Large Hadron Collider, was built in Switzerland and completed in 2008. In 2012, scientists announced that the Higgs Boson had been detected at last, a major victory for the Standard Model, and for Peter Higgs, the physicist who first proposed the particle that now bears his name.

The terms *theory* and *hypothesis* are often used interchangeably in common speech, but in science they mean different things. For this reason you should make note of the distinction.

One more point about hypotheses. A hypothesis that cannot be tested is not a scientific hypothesis. For example, horoscopes purport to predict the future with statements like, "You will meet someone important to your career in the coming weeks." Statements like this are so vague they are untestable and do not qualify as scientific hypotheses.

The key points about hypotheses are summarized in Figure 1.3.

1.2.4 Experiments

Experiments are tests of the predictions in hypotheses, under controlled conditions. Effective experiments are difficult to perform. Thus, for any experimental outcome to become regarded as a "fact" it must be replicated by several different experimental teams, often working in different labs around the world. Scientists have developed rigorous methods for conducting valid experiments. We will consider these briefly in Section 1.3.

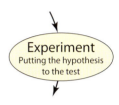

1.2.5 Analysis

In the Analysis phase of the Cycle of Scientific Enterprise, researchers must interpret the experimental results. The results of an experiment are essentially data, and data always have to be interpreted. The main goal of this analysis is to determine whether the original hypothesis has been confirmed. If it has, then the experiment has produced new facts that are consistent with the origi-

nal theory because the hypothesis was based on that theory. As a result, the support for the theory has increased—the theory was successful in generating a hypothesis that was confirmed by experiment. As a result of the experiment, our confidence in the theory as a useful model has increased and the theory is even more strongly supported than before.

1.2.6 Review

If the outcome of an experiment does not confirm the hypothesis, the researchers must consider all the possibilities for why this might have happened. Why didn't our theory, which is our best explanation of how things work, enable us to form a correct prediction? There are a number of possibilities, beginning with the experiment and going backwards around the cycle:

- The experiment may have been flawed. Scientists double check everything about the experiment, making sure all equipment is working properly, double checking the calculations, looking for unknown factors that may have inadvertently influenced the outcome, verifying that the measurement instruments are accurate enough and precise enough to do the job, and so on. They also wait for other experimental teams to try the experiment to see if they get the same results or different results, and then compare.

(Although, naturally, every scientific team likes to be the first one to complete an important new experiment.)

- The hypothesis may have been based on a incorrect understanding of the theory. Maybe the experimenters did not understand the theory well enough, and maybe the hypothesis is not a correct statement of what the theory says will happen.

- The values used in the calculation of the hypothesis' predictions may not have been accurate or precise enough, throwing off the hypothesis' predictions.

- Finally, if all else fails, and the hypothesis still cannot be confirmed by experiment, it is time to look again at the theory. Maybe the theory can be altered to account for this new fact. If the theory simply cannot account for the new fact, then the theory has a weakness, namely, there are facts it doesn't adequately account for. If enough of these weaknesses accumulate, then over a long period of time (like decades) the theory might eventually need to be replaced with a different theory, that is, another, better theory that does a better job of explaining all the facts we know. Of course, for this to happen someone would have to conceive of a new theory, which usually takes a great deal of scientific insight. And remember, it is also possible that the facts themselves can change.

1.3   The Scientific Method

1.3.1 Conducting Reliable Experiments

The so-called *scientific method* that you have been studying ever since about fourth grade is simply a way of conducting reliable experiments. Experiments are an important part of the *Cycle of Scientific Enterprise*, and so the scientific method is important to know. You probably remember studying the steps in the scientific method from prior courses, so they are listed in Table 1.1 without further comment.

We will be discussing variables and measurements a lot in this course, so we should take the opportunity here to identify some of the language researchers use during the experimental process. In a scientific experiment, the researchers have a question they are trying to answer (from the State the Problem step in the scientific method), and typically it is some kind of question about the way one physical quantity affects another one. So the researchers design an experiment in which one quantity can be manipulated (that is, deliberately varied in a controlled fashion) while the value of another quantity is monitored.

A simple example of this in everyday life that you can easily relate to is varying the amount of time you spend each week studying for your math class in order to see what effect the time spent has on the grades you earn. If you reduce the time you spend, will your grades go down? If you increase the time, will they go up? A precise answer depends on a lot

The Scientific Method	
1. State the problem.	5. Collect data.
2. Research the problem.	6. Analyze the data.
3. Form a hypothesis.	7. Form a conclusion.
4. Conduct an experiment.	8. Repeat the work.

Table 1.1. Steps in the scientific method.

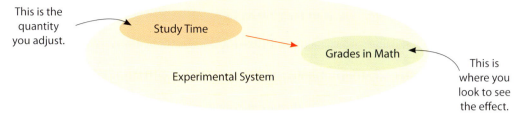

Figure 1.4. Study time and math grades in a simple experimental system.

of things, of course, including the person involved, but in general we would all agree that if a student varies the study time enough we would expect to see the grades vary as well. And in particular, we would expect more study time to result in higher grades. The way your study time and math grades relate together can be represented in a diagram such as Figure 1.4.

Now let us consider this same concept in the context of scientific experiments. An experiment typically involves some kind of complex system that the scientists are modeling. The system could be virtually anything in the natural world—a galaxy, a system of atoms, a mixture of chemicals, a protein, or a badger. The variables in the scientists' mathematical models of the system correspond to the physical quantities that can be manipulated or measured in the system. As I describe the different kinds of variables, refer to Figure 1.5.

1.3.2 Experimental Variables

When performing an experiment, the variable that is deliberately manipulated by the researchers is called the *explanatory variable*. As the explanatory variable is manipulated, the researchers monitor the effect this variation has on the *response variable*. In the example of study time versus math grade, the study time is the explanatory variable and the grade earned is the response variable.

Usually, a good experimental design will allow only one explanatory variable to be manipulated at a time so that the researchers can tell definitively what its effect is on the response variable. If more than one explanatory variable were changing during the course of the experiment, researchers may not be able to tell which one was causing the effect on the response variable.

A third kind of variable that plays a role in experiments is the *lurking variable*. A lurking variable is a variable that affects the response variable without the researchers being aware of it. This is undesirable, of course, because with unknown influences present the researchers may not be able to make a correct conclusion about the effect of the known

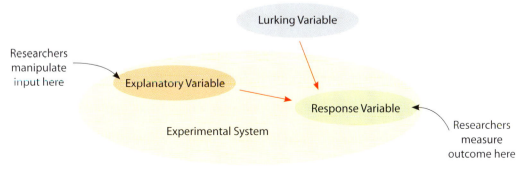

Figure 1.5. The variables in an experimental system.

explanatory variables on the response variable under study. So researchers have to study their experimental projects very carefully to minimize the possibility of lurking variables affecting their results.

Do You Know …   ## *What are double-blind experiments?*

The human mind is so powerful that if a person *believes* a new medication might help, the person's condition can sometimes improve even if the medication itself isn't doing a thing! This is amazing, but in medical research it means that the researchers can have a hard time determining whether a person was helped by the new medication, or by feeling positively about the medication, or even by the attention given to him or her by the doctor.

Pictured below is Lauren Wood, a clinician involved in vaccine research at the Center for Cancer Research, which is part of the National Cancer Institute. Just as with

every other scientific researcher, Dr. Wood's research is conducted according to methods that have been developed to ensure that people's beliefs about the research don't influence the outcome of the research.

The approach is to divide the patients who will participate in testing a new medication into two groups, control and experimental. The experimental group is given the new medication. The control group is given a *placebo*—a fake medication such as a sugar pill—that has no effect on the person's medical condition. Further, none of the patients know whether they are given the placebo or the real medication. This technique, called a *blind experiment*, allows the researchers to determine whether a new medication actually helps, as they compare the results of the control and experimental groups.

But there's more. It turns out that *the researchers themselves* can affect the results of the experiment if they know which patients are receiving a placebo and which ones are receiving the medication under study. How can this happen? Well, if the researchers know who is getting the real medication, they might subconsciously act more positively with them than with other patients. This might be because the researchers expect those getting the new medication to improve, and this expectation gets subconsciously communicated to the patients. The positive attitude might be perceived as more encouraging and patients might improve just because of the encouragement!

The way around this dilemma is to use a *double-blind experiment*. In a double-blind experiment, neither the patients nor the researchers know which patients are getting the placebo and which are getting the real treatment. A team of technicians is in the middle, administering the medication and keeping records of who received what. The researchers are not allowed to see the lists until the research results are finalized. The double-blind experiment is the standard protocol followed today for new medical research.

In our example about study time and math grades, there could be a number of lurking variables affecting the results of the experiment. Possible lurking variables include changes in the difficulty of the material from one chapter to the next and variations in the student's ability to concentrate due to fatigue from seasonal sports activities.

1.3.3 Experimental Controls

The last thing we will look at in this section is an important way researchers control an experiment to ensure the results are valid. You are probably aware that developing new medical treatments is one of the major goals of experimental research in the 21st century. Many experiments in the field of medical research are designed to test some new kind of treatment by comparing the results of the new treatment to those obtained using a conventional treatment or no treatment at all. This is the situation in medical research all the time for experiments testing new therapies, medications, or procedures.

Clinical trials are experiments conducted by researchers on people to test new therapies or medications. In experiments like these, the people (patients) involved in the study are divided into two groups—the *control group* and the *experimental group*. The control group receives no treatment or some kind of standard treatment. The experimental group receives the new treatment being tested. The results of the experimental group are assessed by comparing them to those of the control group.

Another example will help to clarify all these terms. Let's say researchers have developed a variety of fruit tree that they believe is more resistant to drought than other varieties. According to the researchers' *theoretical understanding* of how chemical reactions and water storage work in the biological systems of the plant, they *hypothesize* that the new variety of tree will be able to bear better fruit during drought conditions. To test this hypothesis by *experiment*, the scientists develop a group of the new trees. Then they place the trees in a test plot, along with other trees of other varieties, and see how they perform. Figure 1.6 shows a researcher working in an agricultural test plot. In our fruit tree example, the trees of the new variety are in the *experimental group* and the trees of the other varieties are in the *control group*.

The *response variable* is the quality of the plant's fruit. Researchers expect that under drought conditions the fruit of the new variety will be better than the fruit of the other varieties. The *explanatory variable* is the unique feature of the new variety that relates to the plant's use of water. The trees are exposed to drought conditions in the experiment. If the new variety produces higher quality fruit than the control group, then the hypothesis is confirmed, and the theory that led to the hypothesis has gained credibility through this success. One can imagine many different *lurking variables* that could affect the outcome of this experiment without the scientists' awareness. For example, the new variety trees could be planted in locations that receive different amounts of moisture or sun than the locations where the control group trees are, or, the nutrients in the soil in different locations might vary.

In a good experimental design, researchers seek to identify such factors and take measures to ensure that

Figure 1.6. An agricultural research assistant working in a test plot.

they do not affect the outcome of the experiment. They do this by making sure there are trees from both the experimental group and the control group in all the different conditions the trees will experience. This way, variations in sunlight, soil type, soil water content, elevation, exposure to wind, and other factors will be experienced equally by trees in both groups.

Chapter 1 Exercises

As you go through the chapters in this book, always answer the questions in complete sentences, using correct grammar and spelling.

Here is a tip that will help improve the quality of your written responses: avoid pronouns! Pronouns almost always make your responses vague or ambiguous. If you want to receive full credit for written responses, avoid them. (Oops. I mean, avoid pronouns!)

Study Questions

Answer the following questions with a few complete sentences.

1. Distinguish between theories and hypotheses.
2. Explain why a single experiment can never prove or disprove a theory.
3. Explain how an experiment can still provide valuable data even if the hypothesis under test is not confirmed.
4. Explain the difference between truth and facts and describe the sources of each.
5. State the two primary characteristics of a theory.
6. Does a theory need to account for all known facts? Why or why not?
7. It is common to hear people say, "I don't accept that; it's just a theory." What is the error in a comment like this?
8. Distinguish between facts and theories.
9. Distinguish between explanatory variables, response variables, and lurking variables.
10. Why do good experiments that seek to test some kind of new treatment or therapy include a control group?
11. Explain specifically how the procedure you followed in the Pendulum Experiment satisfies every step of the "scientific method."
12. This chapter argues that scientific facts should not be regarded as true. Someone might question this and ask, If they aren't true, then what are they good for? Develop a response to this question.
13. Explain what a model is and why theories are often described as models.
14. Consider an experiment that does not deliver the result the experimenters had expected. In other words, the result is negative because the hypothesis is not confirmed. There are many reasons why this might happen. Consider each of the following elements of the Cycle of Scientific Enterprise. For each one, describe how it might be the driving factor that results in the experiment's failure to con-

firm the hypothesis.

a. the experiment
b. the hypothesis
c. the theory

15. Identify the explanatory and response variables in the Pendulum Experiment, and identify two realistic possibilities for ways the results may have been influenced by lurking variables.

Do You Know ...   *How did Sir Humphry Davy become a hero?*

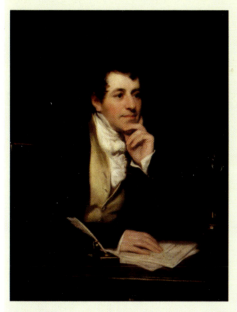

Sir Humphry Davy (1778–1829) was one of the leading experimenters and inventors in England in the early nineteenth century. He conducted many early experiments with gases; discovered sodium, potassium, and numerous other elements; and produced the first electric light from a carbon arc.

In the early nineteenth century, explosions in coal mines were frequent, resulting in much tragic loss of life. The explosions were caused by the miners' lamps igniting the methane gas found in the mines.

Davy became a national hero when he invented the Davy Safety Lamp (below). This lamp incorporated an iron mesh screen around the flame. The cooling

from the iron reduces the flame temperature so the flame does not pass through the mesh, and thus cannot cause an explosion. The Davy Lamp was produced in 1816 and was soon in wide use.

Davy's experimental work proceeded by reasoning from first principles (theory) to hypothesis and experiment. Davy stated, "The gratification of the love of knowledge is delightful to every refined mind; but a much higher motive is offered in indulging it, when that knowledge is felt to be practical power, and when that power may be applied to lessen the miseries or increase the comfort of our fellow-creatures."

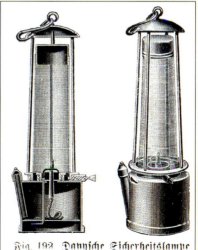

Fig. 192 Dännische Sicherheitslampe

CHAPTER 2
Motion

Orrery

Orreries, mechanical models of the solar system, were well-known teaching tools in the 18th century, often forming the centerpiece of lessons on astronomy. They demonstrated Copernicus' theory that the earth and other planets orbit the sun. This example, from around 1750, is smaller but otherwise similar to George II's grand orrery.

This photo of the orrery was taken in the British Museum in London.

OBJECTIVES

Memorize and learn how to use these equations:

$$d = vt \qquad\qquad a = \frac{v_f - v_i}{t}$$

After studying this chapter and completing the exercises, students will be able to do each of the following tasks, using supporting terms and principles as necessary:

1. Define and distinguish between velocity and acceleration.
2. Use scientific notation correctly with a scientific calculator.
3. Calculate distance, velocity, and acceleration using the correct equations, MKS and USCS units, unit conversions, and units of measure.
4. Use from memory the conversion factors, metric prefixes, and physical constants listed in Appendix A.
5. Explain the difference between accuracy and precision and apply these terms to questions about measurement.
6. Demonstrate correct understanding of precision by using the correct number of significant digits in calculations and rounding.
7. Describe the key features of the Ptolemaic model of the heavens, including all of the spheres and regions in the model.
8. State several additional features of the medieval model of the heavens and relate them to the theological views of the Christian authorities opposing Copernicanism.
9. Briefly describe the roles and major scientific models or discoveries of Copernicus, Tycho, Kepler, and Galileo in the Copernican Revolution. Also, describe the significant later contributions of Isaac Newton and Albert Einstein to our theories of motion and gravity.
10. Describe the theoretical shift that occurred in the Copernican Revolution and how Christian officials (both supporters and opponents) were involved.
11. State Kepler's first law of planetary motion.
12. Describe how the gravitational theories of Kepler, Newton, and Einstein illustrate the way the Cycle of Scientific Enterprise works.

2.1   Computations in Physics

In this chapter you will begin mastering the skill of applying mathematics to the study of physics. To do this well, you need to know a number of things about the way measurements are handled in scientific work. You also need to have a solid problem solving strategy you can depend on to help you solve problems correctly without becoming confused. These topics are addressed in this chapter.

2.1.1 The Metric System

Units of measure are crucial in science. Science is about making measurements and a measurement without its units of measure is a meaningless number. For this reason, your answers to computations in scientific calculations must *always* show the units of measure.

The two major unit systems you should know about are the SI (from the French *Système international d'unités*), typically known in the United States as the metric system, and

Unit	Symbol	Quantity
meter	m	length
kilogram	kg	mass
second	s	time
ampere	A	electric current
kelvin	K	temperature
candela	Cd	luminous intensity
mole	mol	amount of substance

Table 2.1. The seven base units in the SI unit system.

the USCS (U.S. Customary System). You have probably studied these systems before and should already be familiar with some of the SI units and prefixes, so our treatment here will be brief.

If you think about it, you would probably agree that the USCS is cumbersome. One problem is that there are many different units of measure for every kind of physical quantity. For example, just for measuring length or distance we have the inch, foot, yard, and mile. The USCS is also full of random numbers like 3, 12, and 5,280, and there is no inherent connection between units for different types of quantities.

By contrast, the SI system is simple and has many advantages. There is usually only one basic unit for each kind of quantity, such as the meter for measuring length. Instead of having many unrelated units of measure for measuring quantities of different sizes, fractional and multiple prefixes based on powers of ten are used with the units to accommodate various sizes of measurements.

A second advantage is that since quantities with different prefixes are related by some power of ten, unit conversions can often be performed mentally. To convert 4,555 ounces into gallons, we first have to look up the conversion from ounces to gallons (which is hard to remember), and then use a calculator to perform the conversion. But to convert 40,555 cubic centimeters into cubic meters is simple—simply divide by 1,000,000 and you have 0.040555 m³. (If you are not clear on the reason for dividing by 1,000,000, just hold on until we get to the end of Section 2.1.3.)

Another SI advantage is that the units for different types of quantities relate to one another in some way. Unlike the gallon and the foot, which have nothing to do with each other, the liter (a volume) relates to the centimeter (a length): 1L = 1,000 cubic centimeters.[1] For all these reasons, the USCS is not used much in scientific work. The SI system is the international standard and it is important to know it well.

In the SI unit system, there are seven *base units*, listed in Table 2.1. (In this text we will use only the first five of them.) There are also many additional units of measure, known as *derived units*. All the derived units are formed by various combinations of base units. To illustrate, I will show a few examples of derived units that you will learn about and use in this book. Note, however, that we won't be working much with the messy fractions; they are simply shown to illustrate how base units are combined to form derived units.

- the newton (N) is the SI unit for measuring force: $1\,N = 1\dfrac{kg \cdot m}{s^2}$

- the joule (J) is the SI unit for measuring energy: $1\,J = 1\dfrac{kg \cdot m^2}{s^2}$

- the watt (W) is the SI unit for measuring power: $1\,W = 1\dfrac{kg \cdot m^2}{s^3}$

---

1 The liter is not actually an official SI unit of measure, but it is used all the time anyway in scientific work.

Using the SI system entails knowing the units of measure—base and derived—and the prefixes that are applied to the units to form fractional units (such as the centimeter) and multiple units (such as the kilometer). The complete list of metric prefixes is shown in Appendix A in Table A.1. The short list of prefixes you need to know by memory for use in this course is in Table A.2. Note that even though the kilogram is a base unit, prefixes are not added to the kilogram. Instead prefixes are added to the gram to form units such as the milligram and microgram.

2.1.2 MKS Units

A handy subset of the SI system is the so-called *MKS system*. The MKS system uses only base units—such the *meter*, *kilogram*, and *second* (hence, "MKS") as units for mass, length, and time—along with other units derived from the base units. The mass, length, and time units, and the symbols and variables used with them, are listed in Table 2.2.

Variable	Variable Symbol	Unit	Unit Symbol
length	d (distance) L (length) h (height) r (radius), etc.	meter	m
mass	m	kilogram	kg
time	t	second	s

Table 2.2. The three base units in the MKS system.

Dealing with different systems of units can become quite confusing. But the wonderful thing about sticking to the MKS system is that *any calculation performed with MKS units produces a result in MKS units*. This is why the MKS system is so handy. The MKS system dominates calculations in physics and we will use it almost all the time in this course.

To convert the units of measure given in problems into MKS units, you must know the conversion factors listed in Appendix A in Tables A.2, A.3, and A.4. Table A.5 lists several common unit conversions that you are not required to memorize but should have handy when working problem assignments.

2.1.3 Converting Units of Measure

One of the most basic skills scientists and engineers use is re-expressing quantities into equivalent quantities with different units of measure. These calculations are called *unit conversions* and mastery of this skill is essential for any student in high school science. You have studied unit conversions in your math classes for the past few years. But this skill is so important in science that we are going take the time in this section to review in detail how to perform unit conversions.

Let's begin with the basic principle of how this works. First, you know that multiplying any value by unity (one) leaves its value unchanged. Second, you also know that in any fraction if the numerator and denominator are equivalent, the value of the fraction is *unity*, which means *one*. A "conversion factor" is simply a fractional expression in which the numerator and denominator are equivalent ways of writing the same physical quantity. This means a conversion factor is just a special way of writing unity (one). Third, we know that when multiplying fractions, factors that appear in both the numerator and denominator may be "cancelled out." So when performing common unit conversions, what we are doing is repeatedly multiplying our given quantity by unity so that cancellations alter the units of measure until they are expressed the way we wish. Since all we are doing is multiplying by one, the value of our original quantity is unchanged; it simply looks different because it is expressed with different units of measure.

Do You Know ...

The definitions of the base units are fascinating and they all have interesting stories behind them. The official definition of the second is based on the waves of light emitted by cesium atoms. The meter is defined as the distance light travels in a specific tiny fraction of a second (1/299,792,458 of a second). The kilogram is the only base unit that is still defined by a man-made physical object (an *artifact*). It is also the only base unit that uses a metric prefix. The official kilogram is a golf-ball sized platinum cylinder kept in a vault in Paris, France. There are a number of copies of the official kilogram stored in different countries. One of these replicas is shown to the right. In 2014, officials decided to explore new possibilities for defining the kilogram that use only natural constants.

How is the kilogram defined?

Let me elaborate a bit more on the idea of unity I mentioned above, using one common conversion factor as an example. School kids all learn that there are 5,280 feet in one mile, which means 5,280 ft = 1 mi. One mile and 5,280 feet are equivalent ways of writing the same length. If we place these two expressions into a fraction, the numerator and denominator are equivalent, so the value of the fraction is unity, regardless of the way we write it. The equation 5,280 ft = 1 mi can be written in a conversion factor two different ways, and the fraction equals unity either way:

$$\frac{5280 \text{ ft}}{1 \text{ mi}} = \frac{1 \text{ mi}}{5280 \text{ ft}} = 1$$

So if you have a measurement such as 43,000 feet that you wish to re-express in miles, the conversion calculation is written this way:

$$43,000 \text{ ft} \cdot \frac{1 \text{ mi}}{5280 \text{ ft}} = 8.1 \text{ mi}$$

There are two important comments to make here. First, since any conversion factor can be written two ways (depending on which quantity is placed in the numerator), how do we know which way to write the conversion factor? Well, we know from algebra that when we have quantities in the numerator of a fraction that are multiplied, and quantities in the denominator of the fraction that are multiplied, any quantities that appear in both the numerator and denominator cancel. Most units of measure are mathematically treated as multiplied quantities that can be cancelled out.[2] In the example above, we desire that "feet" in the given quantity (which is in the numerator) cancels out, so the conversion factor is written with feet in the denominator and miles in the numerator.

Second, if you perform the calculation above, the result written on your calculator screen is 8.143939394. So why didn't I write down all those digits in my result? Why did I round my answer off to simply 8.1 miles? The answer to that question has to do with the

---

2   An example of a unit not treated this way is the degree.

significant digits in the value 43,000 ft that we started with. We will address the issue of significant digits later in this chapter, but in the examples that follow I always write the results with the correct number of significant digits for the values involved in the problem.

There are several important techniques you must use to help you perform unit conversions correctly. I will illustrate them below with examples. You should rework each of the examples on your own paper as practice to make sure you can do them correctly. As a reminder, the conversion factors used in the examples below are all listed in Appendix A. You should study Appendix A to see which ones you must know by memory and which ones will be provided to you on quizzes.

1. Use only horizontal bars in your unit fractions. Never use slant bars.

In printed materials one often sees values written with a slant fraction bar in the units, as in the value 35 m/s. Although writing the units this way is fine for a printed document, you should not write values this way when you are performing unit conversions. This is because it is easy to get confused and not notice that one of the units is in the denominator in such an expression (s, or seconds, in my example), and the conversion factors used must take this into account.

▼ Example 2.1

Convert 57.66 mi/hr into m/s.

Writing the given quantity with a horizontal bar makes it clear that "hour" is in the denominator. This helps you to write the hour-to-seconds factor correctly.

$$57.66 \ \frac{\text{mi}}{\text{hr}} \cdot \frac{1609 \ \text{m}}{\text{mi}} \cdot \frac{1 \ \text{hr}}{3600 \ \text{s}} = 25.77 \ \frac{\text{m}}{\text{s}}$$

Now that you have your result, you may write it as 25.77 m/s if you wish, but do not use slant fraction bars in the units when you are working out the unit conversion.

▲

2. The term "per" implies a fraction.

Some units of measure are commonly written with a "p" for "per," such as mph for miles per hour or gps for gallons per second. Change these expressions to fractions with horizontal bars when you work out the unit conversion.

▼ Example 2.2

Convert 472.15 gps to L/hr.

When you write down the given quantity change the gps to gal/s, and write these units with a horizontal bar:

$$472.15 \ \frac{\text{gal}}{\text{s}} \cdot \frac{3.786 \ \text{L}}{1 \ \text{gal}} \cdot \frac{3600 \ \text{s}}{1 \ \text{hr}} = 6,435,000 \ \frac{\text{L}}{\text{hr}}$$

▲

3. Use the ⊠ and ⊟ keys correctly when entering values into your calculator.

When dealing with several numerator terms and several denominator terms, multiply all the numerator terms together first, hitting the ⊠ key between each, then hit the ⊟ key and enter all the denominator terms, hitting the ⊟ key between each. This way you do not need to write down intermediate results and you do not need to use any parentheses.

▼ Example 2.3

Convert 43.17 mm/hr into km/yr.

The setup with all the conversion factors is as follows:

$$43.17 \ \frac{mm}{hr} \cdot \frac{1 \ m}{1000 \ mm} \cdot \frac{1 \ km}{1000 \ m} \cdot \frac{24 \ hr}{1 \ day} \cdot \frac{365 \ day}{1 \ yr} = 0.378 \ \frac{km}{yr}$$

To execute this calculation in your calculator, enter the values and operations in this sequence:

$$43.17 \times 24 \times 365 \div 1000 \div 1000 =$$

▲

4. When converting units for area and volume such as cm² or m³, use the appropriate length conversion factor twice for areas or three times for volumes.

The unit "cm²" for an area means the same thing as "cm × cm." Likewise, "m³" means "m × m × m." So when you use a length conversion factor such as 100 cm = 1 m or 1 in = 2.54 cm, you must use it twice to get squared units (areas) or three times to get cubed units (volumes).

▼ Example 2.4

Convert 3,550 cm³ to m³.

$$3550 \ cm^3 \cdot \frac{1 \ m}{100 \ cm} \cdot \frac{1 \ m}{100 \ cm} \cdot \frac{1 \ m}{100 \ cm} = 0.00355 \ m^3$$

▲

Notice in Example 2.4 that the unit cm occurs three times in the denominator, giving us cm³ when they are all multiplied together. This cm³ term in the denominator cancels with the cm³ term in the numerator. And since the m unit occurs three times in the numerator, they multiply together to give us m³ for the units in our result. Notice also that the denominator is 100·100·100 = 1,000,000. This is why I wrote earlier that to convert from cm³ to m³ we can just divide by 1,000,000. Don't make the common (and silly) mistake of dividing by 100!

This issue only arises when you have a unit raised to a power, such as when using a length unit to represent an area or a volume. When using a conversion factor such as 3.786 L = 1 gal, the units of measure are written using units that are strictly volumetric (liters and gallons), and are not obtained from lengths the way in², ft², cm³, and m³ are. Another common unit that uses a power is acceleration, which has units of m/s² in the MKS unit system.

 Example 2.5

Convert 5.85 mi/hr² into MKS units.

$$5.85 \ \frac{mi}{hr^2} \cdot \frac{1609 \ m}{1 \ mi} \cdot \frac{1 \ hr}{3600 \ s} \cdot \frac{1 \ hr}{3600 \ s} = 0.000726 \ \frac{m}{s^2}$$

With this example you see that since the "hour" unit is squared in the given quantity, the conversion factor converting hours to seconds must appear twice in the conversion calculation.

▲

2.1.4 Accuracy and Precision

The terms *accuracy* and *precision* refer to the limitations inherent in making measurements. Science is all about investigating nature and to do that we must make measurements. Accuracy relates to error, which is the difference between a measured value and the true value. The lower the error is in a measurement, the better the accuracy. Error can be caused by a number of different factors, including human mistakes, malfunctioning equipment, incorrectly calibrated instruments, or unknown factors that influence a measurement without the knowledge of the experimenter. All measurements contain error because (alas!) perfection is simply not a thing we have access to in this world.

Precision refers to the resolution or degree of "fine-ness" in a measurement. The limit to the precision that can be obtained in a measurement is ultimately dependent on the instrument being used to make the measurement. If you want greater precision, you must use a more precise instrument. The precision of a measurement is indicated by the number of *significant digits* (or significant figures) included when the measurement is written down (see next section).

Figure 2.1 is a photograph of a machinist's rule and an architect's scale set side by side. Since the marks on the two scales line up consistently, these two scales are equally accurate. But the machinist's rule (on top) is more precise. The architect's scale is marked in 1/16-inch increments, but the machinist's rule is marked in 1/64-inch increments.

It is important that you are able to distinguish between accuracy and precision. Here is an example to illustrate the difference. Let's say Shana and Marius each buy digital thermometers for their homes. The thermometer Shana buys cost $10 and measures to the nearest 1°F. Marius pays $40 and gets one that reads to the nearest 0.1°F. Shana reads the directions and properly installs the sensor for her new thermometer in the shade. Marius doesn't read the directions and mounts his sensor in the direct sunlight, which causes a significant error in the

Figure 2.1. The accuracy of these two scales is the same, but the machinist's rule on the top is more precise.

measurement for much of the day. The result will be that Shana has lower-precision, higher-accuracy measurements!

2.1.5 Significant Digits

The precision in any measurement is indicated by the number of *significant digits* it contains. Thus, the number of digits we write in any measurement we deal with in science is very important. The number of digits is meaningful because it shows the precision that was present in the instrument used to make the measurement.

Let's say you are working a computational exercise in a science book. The problem tells you that a person drives a distance of 110 miles at an average speed of 55 miles per hour and wants you to calculate how long the trip takes. The correct answer to this problem *will be different* from the correct answer to a similar problem with given values of 110.0 miles and 55.0 miles per hour. And if the given values are 110.0 miles and 55.00 miles per hour, the correct answer is different yet again. Mathematically, of course, all three answers are the same. If you drive 110 miles at 55 miles per hour, the trip takes two hours. But scientifically, the correct answers to these three problems are different: 2.0 hours, 2.00 hours, and 2.000 hours, respectively. The difference between these cases is in the precision indicated by the given data, which are *measurements*. (Even though this is just a made-up problem in a book and not an actual measurement someone made in an experiment, the given data are still measurements. There is no way to talk about distances or speeds without talking about measurements, even if the measurements are only imaginary or hypothetical.)

When you perform a calculation with physical quantities (measurements), you cannot simply write down all the digits shown by your calculator. The precision inherent in the measurements used in a computation governs the precision in any result you calculate from those measurements. And since the precision in a measurement is indicated by the number of significant digits, data and calculations must be written with the correct numbers of significant digits. To do this, you need to know how to count significant digits and you must use the correct number of significant digits in all your calculations and experimental data.

Correctly counting significant digits involves four different cases:

1. Rules for determining how many significant digits there are in a given measurement.

2. Rules for writing down the correct number of significant digits in a measurement you are making and recording.

3. Rules for computations you perform with measurements—multiplication and division.

4. Rules for computations you perform with measurements—addition and subtraction.

In this course, we will not use the rules for addition and subtraction, so we will leave those for a future course (probably chemistry). We will now address the first three cases, in order.

Case 1   We begin with the rule for determining how many significant digits there are in a given measurement value. The rule is as follows:

> The number of significant digits (or figures) in a number is found by counting all the digits from left to right beginning with the first nonzero digit on the left. When no decimal is present, trailing zeros are not considered significant.

Let's apply this rule to several example values to see how it works:

15,679   This value has five significant digits.

21.0005   This value has six significant digits.

37,000   This value has only two significant digits because when there is no decimal trailing zeros are not significant. Notice that the word *significant* here is a reference to the *precision* of the measurement, which in this case is rounded to the nearest thousand. The zeros in this value are certainly *important*, but they are not *significant* in the context of precision.

0.0105   This value has three significant digits because we start counting with the first nonzero digit on the left.

0.001350   This value has four significant digits. Trailing zeros count when there is a decimal.

The significant digit rules enable us to tell the difference between two measurements like 13.05 m and 13.0500 m. Mathematically, of course, these values are equivalent. But they are different in what they tell us about the process of how the measurements were made. The first measurement has four significant digits. The second measurement is more precise. It has six significant digits and was made with a more precise instrument.

Now, just in case you are bothered by the zeros at the end of 37,000 that are not significant, here is one more way to think about significant digits that may help. The precision in a measurement depends on the instrument used to make the measurement. If we express the measurement in different units, this should not change the precision. A measurement of 37,000 grams is equivalent to 37 kilograms. Whether we express this value in grams or kilograms, it still has two significant digits.

Case 2   The second case addresses the rules that apply when you record a measurement yourself, rather than reading a measurement someone else has made. When you take measurements yourself, as you do in laboratory experiments, you need to know the rules for which digits are significant in the reading you are taking on the measurement instrument. The rule for taking measurements depends on whether the instrument you are using is a digital instrument or an analog instrument. Here are the rules for these two possibilities:

Rule 1 for digital instruments

For the digital instruments commonly found in high school or undergraduate science labs, assume all the digits in the reading are significant except leading zeros.

Rule 2 for analog instruments

The significant digits in a measurement include all the digits known with certainty, plus one digit at the end that must be estimated between the finest marks on the scale of your instrument.

The first of these rules is illustrated in Figure 2.2. The reading on the left has leading zeros, which do not count as significant. Thus, the first reading has three significant digits. The second reading also has three significant digits. The third reading has five significant digits.

0042.0 **42.0** **42.000** **42,000**

Figure 2.2. With digital instruments, all digits are significant except leading zeros. Thus, the numbers of significant digits in these readings are, from left to right, three, three, five, and five.

The fourth reading also has five significant digits because with a digital display, the only zeros that don't count are the leading zeros. Trailing zeros are significant with a digital instrument. However, when you write this measurement down, you must write it in a way that shows those zeros to be significant. The way to do this is by using scientific notation. Thus, the right-hand value in Figure 2.2 must be written as 4.2000×10^4.

Dealing with digital instruments is actually more involved than the simple rule above implies, but the issues involved go beyond what we typically deal with in introductory or intermediate science classes. So, simply take your readings and assume that all the digits in the reading except leading zeros are significant.

Now let's look at some examples illustrating the rule for analog instruments. Figure 2.3 shows a machinist's rule being used to measure the length in millimeters (mm) of a brass block. We know the first two digits of the length with certainty; the block is clearly between 31 mm and 32 mm long. We have to estimate the third significant digit. The scale on the rule is marked in increments of 0.5 mm. Comparing the edge of the block with these marks, I would estimate the next digit to be a 6, giving a measurement of 31.6 mm. Others might estimate the last digit to be 5 or 7; these small differences in the last digit are unavoidable because the last digit is estimated. Whatever you estimate the last digit to be, two digits of this measurement are known with certainty, the third digit is estimated, and the measurement has three significant digits.

The photograph in Figure 2.4 shows a measurement in milliliters (mL) being taken with a piece of apparatus called a *buret*—a long glass tube used for measuring liquid volumes. Notice in this figure that when measuring liquid volume the surface of the liquid curls up at the edge of the cylinder. This curved surface is called a *meniscus*. The liquid measurement must be made at the bottom of the meniscus for most liquids, including water. The scale on the buret shown is marked in increments of 0.1 mL. This means we estimate to the nearest 0.01 mL. To one person, the bottom of the meniscus (the black curve) may appear to be just below 2.2 mL, so that person would call this measurement 2.21 mL. To someone else, it may seem that the bottom of the meniscus is right on 2.2, in which case that person would call the reading 2.20 mL. Either way, the reading has three significant digits and the last digit is estimated to be either 1 or 0.

As a third example, Figure 2.5 shows a liquid volume measurement being taken with a piece of apparatus called a *graduated cylinder*. (You will use graduated cylinders in an experiment we perform later on in this course.) The scale on the graduated cylinder shown is marked in increments of 1 mL. In the photo, the entire meniscus appears silvery in color with a black curve at the bottom. For the liquid shown in the figure, we know

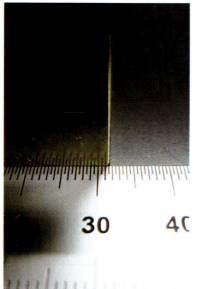

Figure 2.3. Reading the significant digits with a machinist's rule.

the first two digits of the volume measurement with certainty because the reading at the bottom of the meniscus is clearly between 82 mL and 83 mL. We have to estimate the third digit, and I would estimate the black line to be at 40% of the distance between 82 and 83, giving a reading of 82.4 mL.

It is important for you to keep the significant digits rules in mind when you are taking measurements and entering data for your lab reports. The data in your lab journal and the values you use in your calculations and report must correctly reflect the use of the significant digits rules as they apply to the actual instruments you use to take your measurements. Note also the helpful fact that when a measurement is written in scientific notation, the digits written in the stem (the numerals in front of the power of 10) *are* the significant digits.

Figure 2.4. Reading the significant digits on a buret.

Case 3 The third case of rules for significant digits applies to the calculations (multiplication and division) you perform with measurements. The main idea behind the rule for multiplying and dividing is that the precision you report in your result cannot be higher than the precision that is in the measurements to start with. The precision in a measurement depends on the instrument used to make the measurement, nothing else. Multiplying and dividing things cannot improve that precision, and thus your results can be no more precise than the measurements that go into the calculations. In fact, your result can be no more precise than the *least precise value* used in the calculation. The least precise value is, so to speak, the "weak link" in the chain, and a chain is no stronger than its weakest link.

There are two rules for combining the measured values into calculated values, including any unit conversions that must be performed. Here are the two rules for using significant digits in our calculations in this course:

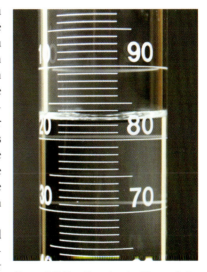

Figure 2.5. Reading the significant digits on a graduated cylinder.

Rule 1

Count the significant digits in each of the values you use in a calculation, including the conversion factors you use. (Exact conversion factors are not considered.) Determine how many significant digits there are in the least precise of these values. The result of your calculation must have this same number of significant digits.

Rule 1 is the rule for multiplying and dividing, which is what most of our calculations entail. (As I mentioned previously, there is another rule for adding and subtracting that you will learn when you take chemistry.)

> **Rule 2**
>
> When performing a multi-step calculation, you must keep at least one extra digit during intermediate calculations and round off to the final number of significant digits you need at the very end. This practice ensures that small round-off errors don't add up during the calculation. This extra digit rule also applies to unit conversions performed as part of the computation.

As I present example problems in the coming chapters, I frequently refer to these rules and show how they apply to the example at hand. As you take your quizzes, your instructor might give you a few weeks to practice and master the correct use of significant digits without penalizing you for mistakes. But get this skill down as soon as you can because soon you must use significant digits correctly in your computations to avoid losing points on your quiz scores.

2.1.6 Scientific Notation

You have probably studied scientific notation before. However, in this course you must master it, including the use of the special key found on scientific calculators for working with values in scientific notation. Mastery of scientific notation is important because working with values in scientific notation is a basic and common occurrence in scientific work. We review the basic principles next.

Mathematical Principles   Scientific notation is a way of expressing very large or very small numbers without all the zeros, unless the zeroes are *significant*. This is of enormous benefit when one is dealing with a value such as 0.0000000000001 cm (the approximate diameter of an atomic nucleus). The basic idea will be clear from a few examples.

Let's say we have the value 3,750,000. This number is the same as 3.75 million, which can be written as $3.75 \times 1,000,000$. Now, 1,000,000 itself can be written as 10^6 (which means one followed by six zeros), so our original number can be expressed equivalently as 3.75×10^6. This expression is in scientific notation. The numerals in front, the stem, are always written as one digit followed by a decimal and the other digits. The multiplied 10 raised to a power has the effect of moving the decimal over as many places as necessary to recreate our original number.

As a second example, the current population of earth is about 7,290,000,000, or 7.29 billion. One billion has nine zeros, so it can be written as 10^9. So we can express the population of earth in scientific notation as 7.29×10^9.

When dealing with extremely small numbers such as 0.000000016, the process is the same, except the power on the 10 is negative. The easiest way to think of it is to count how many places the decimal in the value must be moved over to get 1.6. To get 1.6, the decimal has to be moved to the right eight places, so we write our original value in scientific notation as 1.6×10^{-8}.

Using Scientific Notation with a Scientific Calculator   All scientific calculators have a key for entering values in scientific notation. This key is labeled EE or EXP on most calculators, but others use a different label.[3] It is *very* common for those new to scientific calculators to use this key incorrectly and obtain incorrect results. So read carefully as I outline the general procedure.

The whole point of using the $\boxed{\text{EE}}$ key is to make keying in the value as quick and error-free as possible. When using the scientific notation key to enter a value, you do not press the $\boxed{\times}$ key, nor do you enter the 10. The scientific calculator is designed to reduce all this key entry, and the potential for error, by use of the scientific notation key. You only enter the stem of the value and the power on the ten and let the calculator do the rest.

Here's how. To enter a value, simply enter the digits and decimal in the stem of the number, then hit the $\boxed{\text{EE}}$ key, then enter the power on the ten. The value is now entered and you may do with it as you wish. As an example, to multiply the value 7.29×10^9 by 25 using a standard scientific calculator, the sequence of key strokes is as follows:

7.29 $\boxed{\text{EE}}$ 9 $\boxed{\times}$ 25 $\boxed{=}$

Notice that between the stem and the power the only key pushed is the $\boxed{\text{EE}}$ key.

When entering values in scientific notation with negative powers on the 10, the $\boxed{+/-}$ key is used before the power to make the power negative. Thus, to divide 1.6×10^{-8} by 36.17, the sequence of key strokes is:

1.6 $\boxed{\text{EE}}$ $\boxed{+/-}$ 8 $\boxed{\div}$ 36.17 $\boxed{=}$

Again, neither the "10" nor the "×" sign that comes before it is keyed in. The $\boxed{\text{EE}}$ key has these built in.

Students sometimes wonder why it is incorrect to use the $\boxed{10^x}$ key for scientific notation. To execute 7.29×10^9 times 25, they are tempted to enter the following:

7.29 $\boxed{\times}$ $\boxed{10^x}$ 9 $\boxed{\times}$ 25 $\boxed{=}$

The answer is that sometimes this works, and sometimes it doesn't, and calculator users must use key entries that *always* work. The scientific notation key ($\boxed{\text{EE}}$) keeps a value in scientific notation all together as one number. That is, when the $\boxed{\text{EE}}$ key is used, then to the calculator 7.29×10^9 is not two numbers, it is a single numerical value. But when the $\boxed{\times}$ key is manually inserted, the calculator treats the numbers separated by the $\boxed{\times}$ key as two separate values. This causes the calculator to render an *incorrect* answer for a calculation such as

$$\frac{3.0 \times 10^6}{1.5 \times 10^6}$$

The denominator of this expression is exactly half of the numerator, so the value of this fraction is obviously 2. But when using the $\boxed{10^x}$ key, the 1.5 and the 10^6 in the denominator are separated and treated as separate values. The calculator then performs the following calculation:

$$\frac{3.0 \times 10^6}{1.5} \times 10^6$$

---

3 One infuriating model uses the extremely unfortunate label $\boxed{\times 10^x}$ which looks a *lot* like $\boxed{10^x}$, a different key with a completely different function.

This comes out to 2,000,000,000,000 (2×10^{12}), which is not the same as 2!

The bottom line is that the $\boxed{EE}$ key, however it may be labeled, is the correct key to use for scientific notation.

2.1.7 Problem Solving Methods

Organizing problems on your paper in a reliable and orderly fashion is an essential practice. Physics problems can get very complex, and proper solution practices can often make the difference between getting most or all of the points for a problem and getting few or none. Each time you start a new problem, you must set it up and follow the steps according to the outline presented in the box on pages 36 and 37, entitled *Universal Problem Solving Method*. It is very important that you always show all your work. Do not give in to the temptation to skips steps or take shortcuts. Develop correct habits for problem solving and stick with them!

2.2   Motion

In this course, we address two types of *motion*: motion at a constant *velocity*, when an object is not accelerating, and motion with a *uniform acceleration*. Defining these terms is a lot simpler if we stick to motion in one dimension, that is, motion in a straight line. So in this course, this is what we will do.

2.2.1 Velocity

Figure 2.6. A car traveling with the cruise control on is an example of an object moving with constant velocity.

When thinking about motion, one of the first things we must consider is how fast an object is moving. The common word for how fast an object is moving is *speed*. A similar term is the word *velocity*. For the purposes of this course, you may treat these two terms as synonyms. The difference is technical. Technically, the term velocity means not only *how fast* an object is moving, but also in what *direction*. The term speed refers only to how fast an object is moving. But since we are only going to consider motion in one direction at a time, we can use the terms *speed* and *velocity* interchangeably.

An important type of motion is motion at a constant velocity, like a car with the cruise control on (Figure 2.6). At a constant velocity, the velocity of an object is defined as the distance the object travels in a certain period of time. Expressed mathematically, the velocity, *v*, of an object is calculated as

$$v = \frac{d}{t}$$

The velocity is calculated by dividing the distance the object travels, *d*, by the amount of time, *t*, it takes to travel that distance. So, if you walk 5.0 miles in 2.0 hours, your velocity is *v* = (5.0 miles)/(2.0 hours), or 2.5 miles per hour.

Notice that for a given length of time, if an object covers a greater distance it is moving with a higher velocity. In other words, the velocity is proportional to the distance traveled in a certain length of time. When performing calculations using the SI System of units, distances are measured in meters and times are measured in seconds. This means the units for a velocity are meters per second, or m/s.

The relationship between velocity, distance, and time for motion at a constant velocity is shown graphically in Figure 2.7. Travel time is shown on the horizontal axis and distance traveled is shown on the vertical axis. The steeper curve[4] shows distances and times for an object moving at 2 m/s. At a time of one second, the distance traveled is two meters because the object is moving at two meters per second (2 m/s). After two seconds at this speed, the object has moved four meters: (4 m)/(2 s) = 2 m/s. And after three seconds, the object has moved six meters: (6 m)/(3 s) = 2 m/s.

The right-hand curve in Figure 2.7 represents an object traveling at the much slower velocity of 0.5 m/s. At this speed, the graph shows that an object travels two meters in four seconds, four meters in eight seconds, and so on.

To see this algebraically, look again at the velocity equation above. If we multiply both sides of this equation by the time, t, and cancel, we have

$d = vt$

This is the same equation, just written in a different form. It still applies to objects moving at a constant velocity. Written this way, t is the independent variable, d is the dependent variable, and v serves as the slope of the line relating d to t. With this form of the velocity equation, we can calculate how far an object travels in a given amount of time, assuming the object is moving at a constant velocity.

Now we will work a couple of example problems, following the problem-solving method described on pages 36–37. And remember, all the unit conversion factors you need are listed in Appendix A.

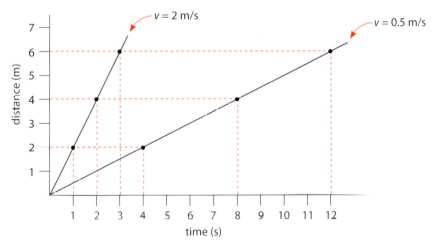

Figure 2.7. A plot of distance versus time for an object moving at constant velocity. Two different velocity cases are shown.

4   Note that when discussing graphs, the lines or curves on the graph are all referred to as *curves*, whether they are curved or straight.

Universal Problem Solving Method

Solid Steps to Reliable Problem Solving

In *Introductory Physics*, you will learn how to use math to solve scientific problems. Developing a sound and reliable method for approaching problems is very important. The problem solving method shown below is used in scientific work everywhere. You must follow every step closely and show all your work.

1. Write down the given quantities at the left side of your paper. Include the variable quantities given in the problem statement and the variable you must solve for. Make a mental note of the precision in each given quantity.
2. For each given quantity that is not already in MKS units, work immediately to the right of it to convert the units of measure into MKS units. To help prevent mistakes, always use horizontal fraction bars in your units and unit conversion factors. Write the results of these unit conversions with one extra digit of precision over what you need in your final result.
3. Write the standard form of the equation you will use to solve the problem.
4. If necessary, use algebra to get the variable you are solving for alone on the left side of the equation. Never put values into the equation until this step is done.
5. Write the equation again with the values in it, using only MKS units, and compute the result.
6. If you are asked to state the answer in non-MKS units, perform the final unit conversion now.
7. Write the result with the correct number of significant digits and the correct units of measure.
8. Check your work.
9. Make sure your result is reasonable.

Example Problem

If you want a complete and happy life, do 'em just like this!

A car is traveling at 35.0 mph. The driver then accelerates uniformly at a rate of 0.15 m/s² for 2 minutes and 10.0 seconds. Determine the final velocity of the car in mph.

Step 1 Write down the given information in a column down the left side of your page, using horizontal lines for the fraction bars in the units of measure.

$$v_i = 35.0 \ \frac{mi}{hr}$$

$$a = 0.15 \ \frac{m}{s^2}$$

$$t = 2 \ min \ 10.0 \ s$$

$$v_f = ?$$

Step 2   Perform the needed unit conversions, writing the conversion factors to the right of the given quantities you wrote in the previous step.

$$v_i = 35.0 \ \frac{mi}{hr} \cdot \frac{1609 \ m}{mi} \cdot \frac{1 \ hr}{3600 \ s} = 15.6 \ \frac{m}{s}$$

$$a = 0.15 \ \frac{m}{s^2}$$

$$t = 2 \ min \ 10.0 \ s - 130.0 \ s$$

$$v_f = ?$$

Step 3   Write the equation you will use in its standard form.

$$a = \frac{v_f - v_i}{t}$$

Step 4   Perform the algebra necessary to get the unknown you are solving for alone on the left side of the equation.

$$a = \frac{v_f - v_i}{t}$$

$$at = v_f - v_i$$

$$v_f = v_i + at$$

Step 5   Using only values in MKS units, insert the values and compute the result.

$$v_f = v_i + at = 15.6 \ \frac{m}{s} + 0.15 \ \frac{m}{s^2} \cdot 130.0 \ s = 35.1 \ \frac{m}{s}$$

Step 6   Convert to non-MKS units, if required in the problem.

$$v_f = 35.1 \ \frac{m}{s} \cdot \frac{1 \ mi}{1609 \ m} \cdot \frac{3600 \ s}{1 \ hr} = 78.5 \ \frac{mi}{hr}$$

Step 7   Write the result with correct significant digits and units of measure.

$$v_f = 79 \ mph$$

Step 8   Check over your work, looking for errors.

Step 9   Make sure your result is reasonable. First, check to see if your result makes sense. The example above is about an accelerating car, so the final velocity we calculate should be a velocity a car can have. A result like 14,000 mph is obviously incorrect. (And remember that nothing can travel faster than the speed of light, so make sure your results are reasonable in this way as well.) Second, if possible, estimate the answer from the given information and compare your estimate to your result. In step 6 above, we see that 3600/1609 is about 2, and 2·35.1 is about 70. Thus our result of 79 mph makes sense.

(Optional Step 10: Revel in the satisfaction of knowing that once you get this down you can work physics problems perfectly nearly every time!)

▼ Example 2.6

Sound travels 1,120 ft/s in air. How much time does it take to hear the crack of a gun fired 1,695.5 m away?

First, write down the given information and perform the required unit conversions so that all given values are in MKS units. Check to see how many significant digits your result must have and do the unit conversions with one extra significant digit. The given speed of sound has three significant digits, so we perform our unit conversions with four digits.

$$v = 1120 \ \frac{\text{ft}}{\text{s}} \cdot \frac{0.3048 \ \text{m}}{\text{ft}} = 341.4 \ \frac{\text{m}}{\text{s}}$$

$$d = 1695.5 \ \text{m}$$

$$t = ?$$

Next, write the appropriate equation to use.

$$v = \frac{d}{t}$$

Perform any necessary algebra, insert the values in MKS units, and compute the result.

$$v = \frac{d}{t}$$

$$t = \frac{d}{v} = \frac{1695.5 \ \text{m}}{341.4 \ \frac{\text{m}}{\text{s}}} = 4.966 \ \text{s}$$

Next, round the result so that it has the correct number of significant digits. In the velocity unit conversion and in the calculated result, I used four significant digits. The given velocity has three significant digits and the given distance has five significant digits. Thus, our result must be reported with three significant digits, but all intermediate calculations must use one extra digit. This is why I used four digits. But now we are finished and our result must be rounded to three significant digits because the least precise measurement in the problem has three significant digits. Rounding our result accordingly, we have

$$t = 4.97 \ \text{s}$$

The final step is to check the result for reasonableness. The result should be roughly the same as 1500/300 or 2000/400, both of which equal 5. Thus, our result makes sense.

2.2.2 Acceleration

An object's velocity is a measure of how fast it is going; it is not a measure of whether its velocity is changing. The quantity we use to measure if a velocity is changing, and if so, how fast it is changing, is the *acceleration*. If an object's velocity is changing, the object is accelerating, and the value of the acceleration is the rate at which the velocity is changing.

The equation we use to calculate uniform acceleration in terms of an initial velocity v_i and a final velocity v_f is

$$a = \frac{v_f - v_i}{t}$$

where a is the acceleration (m/s^2), t is the time spent accelerating (s), and v_i and v_f are the initial and final velocities, respectively, (m/s).

Did you notice that the MKS units for acceleration are meters per second *squared* (m/s^2)? These units often drive students crazy, and we need to pause here and discuss what this means so you can sleep peacefully tonight. I wrote just above that the acceleration is the *rate* at which the velocity is changing. The acceleration simply means that the velocity is increasing by so many meters per second, per second. Now, "per" indicates a fraction, and if a velocity is changing so many meters per second, per second, we write these units in a fraction this way and simplify the expression:

$$\frac{\frac{m}{s}}{s} = \frac{\frac{m}{s}}{\frac{s}{1}} = \frac{m}{s} \cdot \frac{1}{s} = \frac{m}{s^2}$$

Because the acceleration equation results in negative accelerations when the initial velocity is greater than the final velocity, you can see that a negative value for acceleration means the object is slowing down. In future physics courses, you may learn more sophisticated interpretations for what a negative acceleration means, but in this course you are safe associating negative accelerations with decreasing velocity. In common speech, people sometimes use the term "deceleration" when an object is slowing down, but mathematically we just say the acceleration is negative.

Before we work through some examples, let's look at a graphical depiction of uniform acceleration the same way we did with velocity. Figure 2.8 shows two different acceleration curves, representing two different acceleration values. For the curve on the right, after 1 s

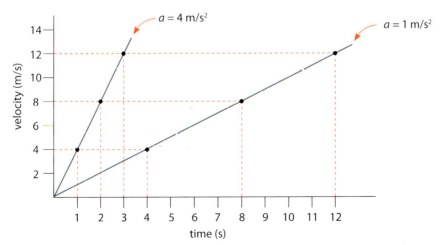

Figure 2.8. A plot of velocity versus time for an object accelerating uniformly. Two different acceleration cases are shown.

the object is going 1 m/s. After 2 s, the object is going 2 m/s. After 12 s, the object is going 12 m/s. You can take the velocity that corresponds to any length of time (by finding where their lines intersect on the curve) and calculate the acceleration by dividing the velocity by the time to get $a = 1$ m/s^2. The other curve has a higher acceleration, 4 m/s^2. An acceleration of 4 m/s^2 means the velocity is increasing by 4 m/s every second. Accordingly, after 2 s the velocity is 8 m/s, and after 3 s, the velocity is 12 m/s. No matter what point you select on that curve, $v/t = 4$ m/s^2.

We must be very careful to distinguish between velocity (m/s) and acceleration (m/s^2). Acceleration is a measure of how fast an object's velocity is changing. To see the difference, note that an object can be at rest ($v = 0$) and accelerating *at the same instant*.

Now, although you may not see this at first, it is important for you to think this through and understand how this counter-intuitive situation can come about. Here are two examples. The instant an object starts from rest, such as when the driver hits the gas while sitting at a traffic light, the object is simultaneously at rest and accelerating. This is because if an object at rest is to ever begin moving, its velocity must *change* from zero to something else. In other words, the object must accelerate. Of course, this situation only holds for an instant; the velocity instantly begins changing and does not stay zero.

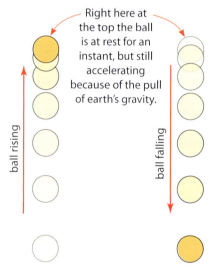

Right here at the top the ball is at rest for an instant, but still accelerating because of the pull of earth's gravity.

ball rising

ball falling

Figure 2.9. A rising and falling ball helps illustrate the difference between velocity and acceleration.

Perhaps my point will be easier to see with this second example. As depicted in Figure 2.9, when a ball is thrown straight up and reaches its highest point, it stops for an instant as it starts to come back down. At its highest point, the ball is simultaneously at rest and accelerating due to the force of gravity pulling it down. As before, this situation only holds for a single instant.

The point of these two examples is to help you understand the difference between the two variables we are discussing, velocity and acceleration. If an object is moving at all, then it has a velocity that is not zero. The object may or may not be accelerating. But acceleration is about whether the velocity itself is changing. If the velocity is constant, then the acceleration is zero. If the object is speeding up or slowing down, then the acceleration is not zero.

And now for another example problem, this time using the acceleration equation.

▼ Example 2.7

A truck is moving with a velocity of 42 mph (miles per hour) when the driver hits the brakes and brings the truck to a stop. The total time required to stop the truck is 8.75 s. Determine the acceleration of the truck, assuming the acceleration is uniform.

Begin by writing the givens and performing the unit conversions.

$$v_i = 42 \; \frac{\text{mi}}{\text{hr}} \cdot \frac{1609 \text{ m}}{\text{mi}} \cdot \frac{1 \text{ hr}}{3600 \text{ s}} = 18.8 \; \frac{\text{m}}{\text{s}}$$

$v_f = 0$

$t = 8.75$ s

$a = ?$

Now write the equation and complete the problem.

$$a = \frac{v_f - v_i}{t} = \frac{0 - 18.8 \; \frac{\text{m}}{\text{s}}}{8.75 \text{ s}} = -2.15 \; \frac{\text{m}}{\text{s}^2}$$

The initial velocity has two significant digits, so I did the calculations with three significant digits until the end. Now we round off to two digits giving

$$a = -2.2 \; \frac{\text{m}}{\text{s}^2}$$

If you keep all the digits in your calculator throughout the calculation and round to two digits at the end, you have −2.1 m/s². This answer is fine, too. Remember, the last digit of a measurement or computation always contains some uncertainty, so it is reasonable to expect small variations in the last significant digit. A check of our work shows the result should be about −20/10, which is −2. Thus the result makes sense.

One more point on this example: Notice that the calculated acceleration value came out negative. This was because the final velocity was lower than the initial velocity. Thus we see that a negative acceleration means the vehicle is slowing down.

If you haven't yet read the example problem in the yellow Universal Problem Solving Method box, you should read it now to see a slightly more difficult example using this same equation.

2.3   Planetary Motion and the Copernican Revolution

2.3.1 Science History and the Science of Motion

People have been fascinated with the heavens since ancient times. God's people love to quote Psalm 19:

The heavens declare the glory of God, and the sky above proclaims his handiwork.
Day to day pours out speech, and night to night reveals knowledge.

The psalmist tells us that the glory of the stars and other heavenly bodies reveals the glory of their creator, our God. This means they convey truth to us, the truth we call General Revelation.

The study of motion has always been associated with the motion of the heavenly bodies we see in the sky, so it is particularly fitting in this chapter on motion for us to review the history of views about the solar system and the rest of the universe, referred to as "the heavens" by those in ancient times. As we will see, the particular episode known as the Copernican Revolution was a pivotal moment in that history and was the setting for the emergence of our contemporary understanding of scientific epistemology—what knowledge is and how we know what we know.

As you recall, in Chapter 1 we addressed the Cycle of Scientific Enterprise and examined the way science works. You learned that science is an ongoing process of modeling nature—at least that is the way we understand science now. We now understand that scientists use theories as models of the way nature works, and over time theories change and evolve as scientists learn more. Sometimes scientists find that a theory is so far off the mark that they have to toss it out completely and replace it with a different one.

The present general understanding among scientists that science is a process of modelling nature took hold around the beginning of the 20th century. The ideas that led to this understanding began to emerge at the time of the Copernican Revolution in the 16th and 17th centuries. But since natural philosophy was then entering new territory, there was a period of difficult struggle that involved both theologians and philosophers.

There are a lot of misconceptions about what happened at that time. The conflict in Galileo's day is often regarded as a fight between faith and science and these misconceptions have led many people in today's world to the position that faith is dead and only science gives us real knowledge. But that depiction is not even close to what really happened and that belief about science is not even close to the truth. The real issue with Galileo was about epistemology. The so-called "faith versus science" debate rages today as much as ever, so it is worth spending some time to understand that crucial period in scientific history.

2.3.2 Aristotle

The study of astronomy and astrology dates back to the ancient Babylonians, but we will pick up the story with the ancient Greeks and the Greek philosopher Aristotle in the 4th-century BC (Figure 2.10). Aristotle was a highly influential philosopher who wrote a lot about philosophy, physics, biology, and other fields of learning. Back then, science was called *natural philosophy* and there was really no distinction between scientists and philosophers.

That time was also many centuries before experiments became part of scientific research. Natural philosophy did involve making observations about the world, but the conclusions reached by ancient philosophers like Aristotle were based simply on observation and philosophical thought. It was still about 2,000 years before natural philosophers realized that the way things appear to our ordinary senses might not be the way they actually are and that to understand more about the world required scientific experiments. For example, if you just walk outside and quietly look around you notice that the earth does not appear to be in motion; it feels solid and at rest. The sun, planets, and stars appear to move across the sky each day. In fact, watching a sunrise gives the distinct impression that the sun is moving up and then across the sky. Today, we understand things differently, but that is the result of the revolution we are about to explore and the experimental science that emerged at that time.

Figure 2.10. Greek philosopher Aristotle (384–322 BC).

Aristotle's ideas were grounded in the concept of *telos*—a Greek term meaning purpose, goal, or end. Aristotle believed that each thing that exists has its own *telos*, an idea that we can heartily embrace today as Christians who believe that God made the world with specific purposes in mind.

Aristotle observed the serene beauty of the stars, the planets, the sun, and moon as they appear majestically to rotate around the earth day after day. He also noticed that nothing in the heavens ever seems to change. Other than the motions of the heavenly bodies, everything in the heavens seems to be pure and eternal. On earth, of course, Aristotle was surrounded by change: decay, corruption, birth, and death are all around. Animals and plants live and die, forests grow and burn, rivers flow and flood, storms come and go. These observations led Aristotle to conclude that change and corruption occur only on the earth. He wrote that imperfection and change of any kind occur only on the earth, while the heavens are pure and unchanging. Aristotle taught that the heavenly bodies—planets, stars, sun, and moon—are eternal and perfect. Further, he said that their motions must be in perfect circles since the circle is the purest and most perfect geometric shape. He conceived of the sun, moon, and planets as inhabiting celestial spheres, centered on the earth, one inside the other—an exquisite *geocentric* (earth-centered) system.

Aristotle was a tremendous moral philosopher whose ideas still have a profound influence on us today. Back in ancient times, he was regarded so highly that questioning his ideas was virtually unthinkable. Thus, his views about the heavenly motions became the basis for all further work on understanding the motions of the heavenly bodies.

2.3.3 Ptolemy

In the second century AD, the famous Alexandrian astronomer Ptolemy (Figure 2.11) worked out a detailed mathematical system based on Aristotle's ideas. (By the way, the "P" in Ptolemy is silent.) As with all ancient astronomers, Ptolemy's goal was to be able to make predictions about the movements of the planets and stars, along with other astronomical events such as eclipses, because these events were widely used as omens signifying important events on earth.

Ptolemy started with Aristotle's basic ideas and developed a complex mathematical system—a model—that was quite effective in making the desired predictions. There were other astronomers around that time who developed different systems, but Ptolemy's system became the most widely accepted understanding of the heavens for over a thousand years.

Figure 2.11. Alexandrian astronomer Ptolemy (c. 100–170 AD).

2.3.4 The Ptolemaic Model

The basic structure of Ptolemy's geocentric model of the heavens is depicted in Figure 2.12. As with Aristotle, there are seven heavenly bodies, each inhabiting a *sphere* centered on the earth. Each of the heavenly bodies is also itself a perfect sphere.

The contents of the spheres are summarized in Table 2.3. The first seven spheres contain the five planets (not including the earth), the sun, and the moon. Sphere 8 contains the so-called *Firmament*, the fixed layer of stars. The stars do not move relative to each

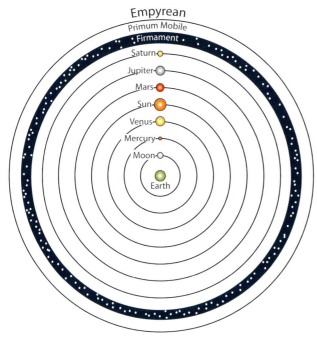

Figure 2.12. The Ptolemaic model of the heavens.

other; their positions are fixed and they rotate as a body in the eighth sphere each day. Within the firmament, the stars are arranged according to the *zodiac*, a belt of twelve constellations around the earth. The term zodiac derives from the Latin and Greek terms meaning "circle of animals," and is so named because many of the constellations in the zodiac represent animals.

The ninth sphere contains the *Primum Mobile*, which is Latin for "prime mover" (or "first mover"). The *Primum Mobile* is the sphere set into motion by God or the gods. As the *Primum Mobile* turns, it pulls all the other spheres with it, making them rotate as well. Outside the ninth sphere is the so-called *Empyrean*, the dwelling place of God or the gods.

Figure 2.12 shows the basic structure of Ptolemy's model, but there is a great deal more to the model than shown there. This is because all seven of the heavenly bodies appear to move around in the nighttime sky against the background of the fixed stars. If the all the heavenly bodies simply moved in their spheres around the earth together once each day, there would be no way to account for why the planets' positions change relative to the stars.

Sphere 1	Moon
Sphere 2	Mercury
Sphere 3	Venus
Sphere 4	Sun
Sphere 5	Mars
Sphere 6	Jupiter
Sphere 7	Saturn
Sphere 8	The Firmament. This region consists of the stars arranged in their constellations according to the zodiac.
Sphere 9	The *Primum Mobile*. This Latin name means "first mover." This sphere rotates around the earth every 24 hours and drags all the other spheres with it, making them all move.
Beyond	The Empyrean. This is the region beyond the spheres. The Empyrean is the abode of God, or the gods.

Table 2.3. Contents of the spheres in the Ptolemaic model.

Ptolemy accounted for the changes by a system of *epicycles*. An epicycle is a circular planetary orbit with its center moving in a separate circular path, as depicted in Figure 2.13. As the center of an epicycle moves along its path in the sphere, the planet in the epicycle rotates about the center of the epicycle, as if the epicycle were a wheel rolling around a path centered on the earth.

To help you understand why epicycles are necessary in Ptolemy's model, we will discuss them in more detail in the next section. A planet moving in an epicycle moves in a path similar to a person riding in a "tea cup ride" at an amusement park, like the one picture in Figure 2.14. To account for the complex motions of the heavenly bodies, Ptolemy's model contained some 80 different epicycles. Some of the planets were located in a epicycle riding on the rim of another epicycle, which in turn moved in the sphere around the earth. Ptolemy's system was mathematically very complex, but its genius was that it worked pretty well! The main features of Ptolemy's model are summarized in the box below.

Among the different astronomers of the ancient world there were those who held to variations on this basic model. For example, some astronomers reckoned that Mercury and Venus orbited the sun while the other heavenly bodies orbited the earth. But the basic Ptolemaic model is as described in the box.

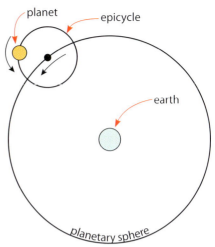

Figure 2.13. A planet moving in a path defined by an epicycle around the earth.

Figure 2.14. The people in the cups spin in a circle while the cup moves in a larger circle, motion like that of a planet moving on an epicycle.

The Main Principles in Ptolemy's Celestial Model

1. There are seven heavenly bodies.
2. All the heavenly bodies move in circular orbital regions called spheres. In the model, there are nine spheres plus the region beyond the spheres, with contents as listed in Table 2.3.
3. All the heavenly bodies are perfectly spherical.
4. All the spheres are centered on the earth, so this system is a *geocentric* system.
5. Corruption and change only exist on earth. All other places in the universe, including all the heavenly bodies and stars, are perfect and unchanging.
6. All the spheres containing the heavenly bodies and all the stars in the Firmament rotate completely around the earth every 24 hours.
7. Epicycles are used to explain the motion of the planets relative to the stars.

2.3.5 The Ancient Understanding of the Heavens

We will soon address the new ideas that began unfolding when Nicolaus Copernicus introduced his new *heliocentric* (sun-centered) model of the heavens. But before pressing on, let's pause to consider a couple of things about the way the motion of the planets in the night sky appears to observers on earth. This will make it easier to understand why Ptolemy's system became so widely accepted.

Stationary Earth First, as I mentioned above, the earth does not seem to be moving. To you and me, who grew up in a time when everyone knows that the earth and other planets orbit the sun, it seems obvious that day and night are caused by the earth's rotation on its axis. We have heard about this all our lives. But stop and consider that if all we had to go on was our simple observations, it does *appear* that everything is orbiting around the earth while the earth sits still: the sun and moon rise each day, track across the sky, and set, and the planets and stars all do the same thing. Also, it doesn't feel at all like earth is rotating. We all know that anytime we spin in a circle, like people on a merry-go-round, we have to hold on to keep from falling off. We also feel the wind in our hair. Again, if we have something with us on the merry-go-round that is tall and flexible, like a sapling, it does not remain vertical when it is moving in a circular fashion like this. Instead, it bends over because of the acceleration pulling it in its circular motion.

Figure 2.15. Greek mathematician and geographer Eratosthenes (c. 276–194 BC).

Now, the ancients knew about the large size of the earth—the Greek mathematician and geographer Eratosthenes (Figure 2.15) made a very accurate estimate of the earth's circumference—a bit under 25,000 miles—as far back as 240 BC. If a sphere that size spins in a circle once a day, the people on its surface move very fast (over 1,000 miles per hour on the equator). For this to be the case, it seemed that we would be hanging on for dear life! The trees would be laying down and we would constantly feel winds that make a hurricane seem like a calm summer day!

For all these reasons, it did not seem reasonable to believe that the apparent motion of the heavenly bodies across the sky every day was due to the earth's rotation. These arguments seemed obvious to nearly everyone before 1500, and to everyone except a few cutting-edge astronomers right up to the end of the 17th century. Only a crazy person imagined that the earth spins, and people used these arguments all the way up to the time of Galileo to prove that the earth was not orbiting the sun and spinning around once a day. Back then these were persuasive arguments.

Forward and Retrograde Motion The second item to consider here has to do with the apparent motion of the planets in the sky against the background of the stars. If you go out and look at, say, Mars each night and make a note of its location against the stars, you see that it is in a slightly different place each night. The planet gradually works its way along in a pathway against the starry background night after night. If you track the planet for several months or a year, it moves quite far. As mentioned above, Ptolemy used epicycles to account for this *forward motion* of planets against the background of fixed stars.

Going back to watching Mars, if you follow the planet's progress long enough, you see that there are periods of time lasting several weeks when the nightly progress of the planet reverses course. Mars appears to be backing up! This apparent backing up is called *retrograde motion*. Ptolemy used epicycles to account for this, too.

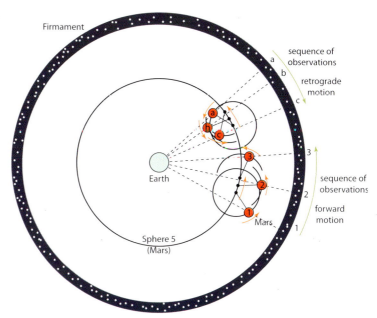

Figure 2.16. Using epicycles to explain the forward and retrograde motion of heavenly bodies against the background of fixed stars.

Figure 2.16 is a diagram showing how epicycles are used in the geocentric system to account for the planetary motions—both forward and retrograde—against the background of the fixed stars. Mars is shown in red moving on an epicycle, while the center of the epicycle moves around the earth. The dashed lines are the lines of sight from earth to Mars, and the letters and numbers outside the firmament show the locations where Mars appears among the stars at different times.

The lower right part of the diagram shows Mars in three locations (labeled 1, 2, and 3) over the course of a few weeks. Compared to the back-

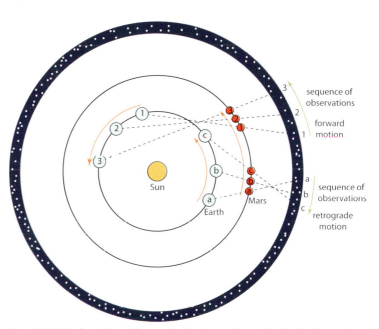

Figure 2.17. Explanation of forward and retrograde motion in the Copernican system.

ground of fixed stars, Mars exhibits forward motion during a sequence of nighttime observations.

The upper right part of the figure shows Mars' locations during a different sequence of observations (a, b, and c) some months later. Mars is now on the other side of its epicycle. The center of the epicycle continues to move in the same direction in its sphere around earth. But since Mars is on the near side of its epicycle, the sequence of observations of its location against the starry background—a, b, and c—maps along the starry background in the opposite direction. This apparent motion of Mars in the opposite direction is retrograde motion.

While we are on the subject, we may as well look at how forward and retrograde motion are explained in a *heliocentric* system—a system in which the planets orbit the sun. The system introduced by Copernicus is a heliocentric system. Assuming that the earth moves faster in its orbit than Mars (which is correct), the explanation is straightforward. As shown in the upper part of Figure 2.17, when the earth and Mars are on opposite side of their orbits, the observations of Mars' location against the stars exhibit forward motion. But when the earth and Mars are on the same side of the sun, as in the center-right part of the figure, the earth's greater velocity makes Mars' position against the stars exhibit retrograde motion.

To summarize, none of the planets actually reverses course in its orbit, and neither the geocentric nor heliocentric models depict planets as reversing direction. But depending on the system, the presence of epicycles and the relative locations of earth and a planet can combine to produce the appearance of forward or retrograde motion of the planet against the fixed background of the stars.

2.3.6 The Ptolemaic Model and Theology

We will soon continue our history of the science of planetary motion by reviewing the momentous events of the 16th and 17th centuries. Between Ptolemy and Copernicus were 1,300 years of theology and philosophy. During this long period of history, a strong tradition emerged among many theologians that the Ptolemaic model of the heavens aligned very well with certain passages in the Bible. This circumstance led theologians in this tradition to assume that such passages were to be interpreted as literal descriptions of the motions of the heavenly bodies. Here are a few examples of passages that seem to describe the earth as motionless, with the sun and stars going around the earth:

He set the earth on its foundations, so that it should never be moved (Psalm 104:5).

He made the moon to mark the seasons; the sun knows its time for setting (Psalm 104:19).

[The sun's] rising is from the end of the heavens, and its circuit to the end of them (Psalm 19:6).

The sun rises and the sun goes down, and hastens to the place where it rises (Ecclesiastes 1:5).

Additionally, other features in the Ptolemaic model (derived from Aristotle) seemed to line up with biblical symbolism. For example:

- Seven is the biblical number symbolizing perfection, so it made sense that God's creation contains seven heavenly bodies.

- Circles are the most perfect shape, regarded as divine from the times of the ancient Greeks, so the spherical bodies inhabiting spheres in which they move seemed to reflect the perfection of their Creator.

- Corruption was thought to exist only on earth, and it seemed this was obviously because of the curse that resulted from the Fall of man.

The result of such teaching was that many theologians assumed that the biblical passages and doctrines described above, along with the Ptolemaic model of the heavens, were literal descriptions of the true nature of reality. To these theologians, anyone who had different ideas about the heavens—such as, for example, the idea that the earth moved and orbited the sun—should be censored and prevented from spreading teachings they felt were unbiblical.

Although widespread, this tradition of associating the Ptolemaic model with the Bible was by no means universal. Many theologians took a completely different position, including the great theologian and philosopher Augustine, a bishop in northern Africa in the 4th and 5th centuries AD. An insightful and relevant passage from Augustine is found in his book *On the Literal Meaning of Genesis*:

> Usually, even a non-Christian knows something about the earth, the heavens, and the other elements of this world, about the motion and orbit of the stars and even their sizes and relative positions, about the predictable eclipses of the sun and moon, the cycles of the years and the seasons, about the kinds of animals, shrubs, stones, and so forth, and this knowledge he holds to as being certain from reason and experience. Now it is a disgraceful and dangerous thing for an infidel to hear a Christian, presumably giving the meaning of Holy Scripture, talking nonsense on these topics, and we should take all means to prevent such an embarrassing situation, in which people show up vast ignorance in a Christian and laugh it to scorn. The shame is not so much that an ignorant individual is derided, but that people outside the household of faith think our sacred writers held such opinions, and, to the great loss of those for whose salvation we toil, the writers of our Scripture are criticized and rejected as unlearned men.

As we open the curtain now on the rest of our story, it is key to remember that many church theologians were strong supporters of those engaged in natural philosophy. The Roman Catholic Church—which figures prominently in these events—had a long tradition of supporting intellectual inquiry, including natural philosophy, and many of the individual theologians in the church were admirers of the scientists we will encounter.

2.3.7 Copernicus and Tycho

Nicolaus Copernicus (Figure 2.18), a Polish astronomer, first proposed a detailed, mathematical heliocentric model of the heavens, with the earth rotating on its axis, all the planets moving in circular orbits around the sun, and the moon orbiting the earth.

Copernicus' system was about as accurate—and about as complex—as the Ptolemaic system. Copernicus' model still used circular orbits and because of this he still had to use epicycles to make the model accurate. Still, the model is an arrangement that is a lot closer to today's understanding than the Ptolemaic model is.

As mentioned in the accompanying box, Copernicus dedicated his famous work *On the Revolutions of the Heavenly Spheres* to Pope Paul III. This dedication indicates that the Roman Catholic church itself was not opposed to Copernicus' ideas. Nevertheless, Copernicus knew there were scholars in the Church who were strongly opposed to the suggestion that the earth moved. Being a sensitive and godly man, he didn't want to cause trouble so he published his work privately to his close friends in 1514. Just before Copernicus' death in

Figure 2.18. Polish astronomer Nicolaus Copernicus (1473–1543).

1543, his student and admirer, mathematician and astronomer Georg Joachim Rheticus, persuaded Copernicus to publish the work. Rheticus delivered the manuscript to the printer and brought proofs back to Copernicus to review. Rheticus was not continuously present with the printer, and during his absence a theologian named Andreas Osiander added an unsigned "note to the reader" to the front of Copernicus' book stating that the heliocentric ideas were *hypotheses* (although *theory* is the better term, since we are talking about a *model*) that were useful for the purpose of performing computations and not descriptions of actual reality. Because of this note, people generally thought that it expressed Copernicus' own viewpoint. However, Rheticus was outraged by the addition and marked it out with a red crayon in the copies he sent to people. Copernicus did not live to see the final printed version of his book, but Rheticus' reaction to Osiander's note suggests that Copernicus regarded his model as more than merely an imaginary convenience that made computations easier.

Nicolaus Copernicus gave us a beautiful description of our Creator, one that is often quoted. In the preface to his book *On the Revolutions of Heavenly Spheres* he dedicated the book to Pope Paul III. Copernicus wrote:

"I can reckon easily enough, Most Holy Father, that as soon as certain people learn that in these books of mine which I have written about the revolutions of the spheres of the world I attribute certain motions to the terrestrial globe, they will immediately shout to have me and my opinions hooted off the stage."

Copernicus went on to review the shortcomings of the work of other astronomers, and then justified his own work:

"Accordingly, when I had meditated upon this lack of certitude in the traditional mathematics concerning the composition of movements of the spheres of the world, I began to be annoyed that the philosophers, who in other respects had made a very careful scrutiny of the least details of the world, had discovered no sure scheme for the movements of the machinery of the world, which has been built for us by the Best and Most Orderly Workman of all."

—from Nicolaus Copernicus, *On the Revolutions of Heavenly Spheres* (1543)

Tycho Brahe (Figure 2.19), was a Danish nobleman and astronomer. Tycho[5] built a magnificent observatory called the Uraniborg on an island Denmark ruled at the time. This observatory is depicted in Figure 2.20.

Tycho was a passionate and hotheaded guy, as evidenced by the fact he had the bridge of his nose cut off in a duel. (You can see his prosthesis in Figure 2.19 if you look closely.) Even though Tycho's Uraniborg must have been the most palatial observatory in the world, he had a falling out with the new King of Denmark and decided to leave. In 1597, Tycho moved to Prague in Bohemia (the modern day Czech Republic) and became Imperial Mathematician for Rudolph II, King of Bohemia and Holy Roman Emperor there. Tycho spent his life cataloging astronomical data for over 1,000 stars (with cleverly contrived instruments, but only a primitive telescope). His work was published much later (1627) by Johannes Kepler in a new star catalog that identified the positions of these stars with unprecedented accuracy.

Figure 2.19. Danish astronomer Tycho Brahe (1546–1601).

Tycho witnessed and recorded two astronomical events that became historically very important. First, in 1563 he observed a *conjunction* between Jupiter and Saturn. A conjunction, illustrated in Figure 2.21, occurs when two planets are in a straight line with the earth so that from earth they appear to be in the same place in the sky. Tycho predicted the date for this conjunction using Copernicus' new heliocentric model. The prediction was close (this is good) but was still off by a few days (not so good). The error indicated that there was still something lacking in Copernicus' model. (There was: the orbits are not circular as Copernicus assumed.) Second, in 1572 Tycho observed what he called a "nova" (which is Latin for *new*; today we would call it a supernova) and proved that it was a new star.

This discovery rocked the Renaissance world because it was strong evidence that the stars are not perfect and unchanging as Aristotle had thought and as the Ptolemaic model of the heavens declared.

Although familiar with Copernicus' model, Tycho was a proud advocate of his own model, in which the sun and moon orbit the earth and the other planets orbit the sun, which in turn orbits the earth. His model did have the advantage of maintaining a stationary earth, which allowed Tycho to avoid controversy with those who

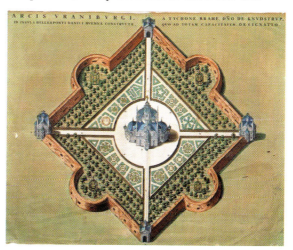

Figure 2.20. Tycho's Danish observatory, the Uraniborg.

---

5   I know it is appropriate to refer to historical figures by their last names, but most references in the literature refer to Tycho; historians rarely call him Brahe. I love the name Tycho, so I also call him that.

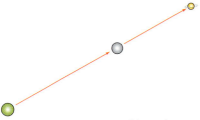

Figure 2.21. The alignment of three planets, called a conjunction.

insisted that the Bible taught that the earth did not move.

Tycho also had a good technical reason for rejecting Copernicus' model. If the earth moves in an orbit, then earth's location is different in the summer from its location in the winter. This means the relative positions of the stars should be slightly different at these different times of the year, an effect called *stellar parallax*. (As an analogy, imagine yourself looking at the trees in a forest. If you take a few steps to one side, the positions of the trees relative to each other in your new location are different.) At that time, no stellar parallax had been observed, and Tycho knew that this meant that either the earth was stationary or the stars were incredibly far away. Copernicus had accepted the great distance of the stars but Tycho did not, and famously wondered, "What purpose would all that emptiness serve?" In fact, stellar parallax was not observed until 1838, when telescopes were finally up to the task. The discovery of stellar parallax in 1838 was the *first actual evidence* that Copernicus was right. It helps to keep this in mind when we get to the controversy surrounding Galileo.

2.3.8 Kepler and the Laws of Planetary Motion

Johannes Kepler (Figure 2.18), a German astronomer and mathematician, was invited in 1600 to join the research staff at Tycho's observatory in Prague and became the Imperial Mathematician there the following year, after Tycho's death. Kepler had access to Tycho's massive body of research data and used it to develop his famous three *laws of planetary motion*, the first two of which were published in 1609. He discovered the third law a few years later and published it in 1619.

Kepler was a godly man and took his faith very seriously, even though he was caught in the middle during the Counter-Reformation, a time of serious disagreement between Roman Catholics and Protestants. Kepler was also an amazing scientist who believed that he had been called to glorify God through his discoveries. In addition to his astronomical discoveries, he made important discoveries in geometry and optics, he figured out some of the major principles of gravity later synthesized by Isaac Newton, and he was the first to hypothesize that the sun exerted a force on the earth.

Figure 2.18. German astronomer and mathematician Johannes Kepler (1571 1630).

I want to show you the three beautiful laws of planetary motion Kepler discovered. For your memory work you may focus on remembering only the first one. But I want you to see some things about the way the solar system is designed, and the One who designed it, so I am going to describe all three of Kepler's Laws.

Kepler's first law of planetary motion is as follows:

First Law	Each of the planetary orbits is an ellipse, with the sun at one focus.

A planet in an elliptical orbit is depicted in Figure 2.19. You may not have studied *ellipses* yet in math, so I will describe them. An ellipse is a geometric figure shaped like this: ⬭. An ellipse is similar to a circle, except that instead of having a single point locating the center, an ellipse has two points on either side of the center called *foci* that define its shape. (The term *foci* is plural, and pronounced FOH-sigh; the singular is *focus.*) Out in space, each planet travels on a path defined by a geometrical ellipse. The planetary orbits all have one focus located at the same place in space and this is where the sun is. Think how incredible it is that Kepler figured this out! He was a monster mathematician (no calculator!) and an extremely careful scientist, and the fact that scientists had understood the orbits to be circular for two thousand years did not get in his way. To me, this is simply amazing.

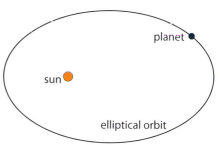

Figure 2.19. A planet in an elliptical orbit around the sun (Kepler's First Law).

Kepler's second law is not hard to understand. It is in the next box.

Second Law	A line drawn from the sun to any planet sweeps across the same area in space in any given period of time.

The second law is depicted in Figure 2.20. The idea is that for a given period of time, say, a month or a week or whatever, the shaded region in the figure will have the same area, regardless of where the planet is in its orbit. Now, since the sun is off-center, this law implies that the planets travel faster when they are closer to the sun and slower when they are farther away. Keep thinking about how

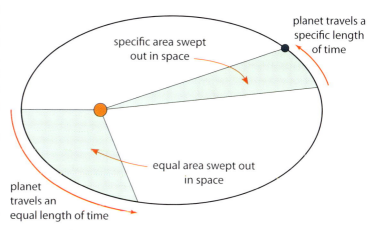

Figure 2.20. Equal areas are swept in space for equal periods of time (Kepler's Second Law).

stunning it is that a guy without a calculator or any modern computer could figure this out, all from the observational data that Tycho had assembled.

Kepler's third law is definitely more mathematically complex than the first two. This law is shown in the next box.

> **Third Law**   The orbits of any two planets are related as follows:
>
> $$\left(\frac{T_1}{T_2}\right)^2 = \left(\frac{R_1}{R_2}\right)^3$$
>
> where T_1 and T_2 are the planets' orbital periods, and R_1 and R_2 are their mean distances from the sun.

You might be relieved to learn that we will not be doing any computations with this law. However, Kepler's third law of planetary motion is a stunning example of the mathematical modeling that physicists do all the time, so I want to comment on it here for a bit.

The third law is quite accurate. The equation can be expressed in a way that shows that the orbital period, T, for any planet as a function of the planet's mean distance from the sun, R. In equation form, this expression of the third law can be written as

$$T = kR^{3/2}$$

In this equation, k is simply a constant that depends on the units used for T and R. I am not planning to go crazy with the math here, and I know you may be freaking out wondering what it means to raise a variable like R to the 3/2 power. Right now it doesn't matter. You will learn all that when you get to Algebra 2. I just want to show how simple Kepler's

Johannes Kepler viewed his discoveries of the mathematical order of nature as amazing revelations given to him by God. Some of the things Kepler worked on were very strange, such as his attempt to develop a theory of the spheres associated with the five regular Platonic solids and the mathematics of musical ratios developed by the Greeks. Although those ideas were abandoned, Kepler had the courage to look carefully at the astronomical data and this led him to his discovery of the laws of planetary motion.

Read the prayer Kepler wrote at the end of his book *Harmonies of the World*:

O Thou Who dost by the light of nature promote in us the desire for the light of grace, that by its means Thou mayest transport us into the light of glory, I give thanks to Thee, O Lord Creator, Who hast delighted me with Thy makings and in the works of Thy hands have I exulted. Behold! now, I have completed the work of my profession, having employed as much power of mind as Thou didst give to me; to the men who are going to read those demonstrations I have made manifest the glory of Thy works, as much of its infinity as the narrows of my intellect could apprehend. My mind has been given over to philosophizing most correctly: if there is anything unworthy of Thy designs brought forth by me — a worm born and nourished in a wallowing place of sins—breathe into me also that which Thou dost wish men to know, that I may make the correction: If I have been allured into rashness by the wonderful beauty of Thy works, or if I have loved my own glory among men, while I am advancing in the work destined for Thy glory, be gentle and merciful and pardon me; and finally design graciously to effect that these demonstrations give way to Thy glory and the salvation of souls and nowhere be an obstacle to that.

—from Johannes Kepler, *Harmonies of the World* (1619)

third law really is. Think about it. This simple equation accurately relates the period of any planet's orbit to that planet's mean distance from the sun.

Now I don't know about you, but when I see an equation that is as amazing and as simple as this, it sets me thinking. First, Kepler's work as a scientist is first class. He figured this out from data collected in the era before calculators and before computers. This was only three years after Shakespeare died!

Second, this equation says something deep about the universe we live in. The universe can be modeled with simple mathematics that can be understood by high school kids! How do you think this could be possible? Is it possible that a randomly evolving universe that occurred by chance, with no plan, could exhibit this kind of deep mathematical structure? I do not believe it is and I am not alone. Many great scientists—even non-Christian scientists—have called attention to the beautiful mathematical structure that appears everywhere in nature and have called it either a great mystery or evidence of God's handiwork. The fact that our solar system has the kind of beautiful and simple mathematical structure represented by Kepler's third law is strong evidence for an intelligent creator. This is not to say that Kepler's third law is itself the truth about nature. It is quite accurate, but as we will see below, claiming that it is the *truth* is an overstatement. But the fact that nature can be accurately modeled with mathematics by humans—even if we don't know the exact truth of nature itself—is because nature exhibits an order and regularity that can only be explained by the hand of "the Best and Most Orderly Workman of all."

2.4.3 Galileo

Galileo Galilei (Figure 2.21) worked at the university at Padua, Italy, and later as chief mathematician and philosopher for the ruling Medici family in Florence, Italy. Galileo's work in astronomy represents the climax of the Copernican Revolution. He made significant improvements to the telescope and used the telescope to see the craters on the moon and sunspots, which provided additional evidence that the heavens were not perfect and unchanging as Aristotle and Ptolemy had maintained. In 1610, he used the telescope to discover four of the moons around Jupiter, which was clearly in conflict with the idea that there had to be exactly seven heavenly bodies. He was fully on board with all the new science of the Copernican model, but, oddly, he never did accept Kepler's discovery that the planets' orbits were elliptical rather than circular. Galileo published his early astronomical discoveries in 1610 in a book called *The Starry Messenger*.

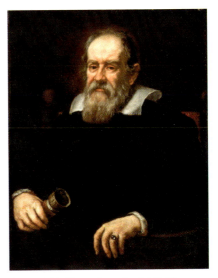

Most people know that Galileo was put on trial in 1633 by the Holy Office of the Inquisition established by Roman Catholic Church. However, the reasons for that trial are widely and seriously misunderstood. The real story is rather illuminating, and I will explain it here as briefly as possible.

Galileo is famous for this remark: "Philosophy is written in this grand book—I mean the universe—which stands continually open to our gaze, but it cannot be understood unless one first learns to comprehend the language in which it is written. It is written in the language of mathematics." This beautiful state-

Figure 2.21. Florentine scientist Galileo Galilei (1564–1642).

ment calls attention to the mathematical structure of creation, which, as we saw above, is strong evidence for a wise creator behind the existence of the universe. However, Galileo erred in taking his statement too far, claiming that the mathematics he used in his astronomical discoveries was the *truth* about nature. Galileo felt that his work *proved* beyond question that Copernicus was right. Galileo, along with other scientists over the next three centuries, still had to learn the limits associated with human modeling of nature.

In contrast to Galileo's attitude toward his work is the attitude of his friend Cardinal Roberto Bellarmine, an important Church official. Bellarmine was a great admirer of Galileo's work, but Bellarmine's thinking was very much along the lines of our discussion in the previous chapter: scientific inquiry leads to theories which are *models* and cannot be regarded as truths; models are provisional and subject to change. In this attitude of an important Church official, we recognize a remarkable early statement of the attitude toward scientific knowledge that today is held as the correct way to think about scientific theories. Bellarmine cautioned Galileo that no natural science could make claims as to truth and urged him to present his ideas as everyone thought Copernicus had done—as *hypotheses* rather than as *truths*.

There were definitely mistakes on both sides of the conflict that led eventually to Galileo's trial by the Holy Office. On Galileo's part, the mistake was in pushing his ideas too forcefully with the claim that they were true. Galileo's position implied that the theologians who claimed that the Bible lined up with the Ptolemaic system were wrong in their interpretation of the Bible. Galileo wrote an important letter at that time explaining that the theologians needed to reconsider their interpretations of Scripture in light of what the scientific evidence was showing. Galileo was correct in his views about interpreting Scripture, but his claims pertaining to the truth of his discoveries went too far.

To the theologians and church officials who held to the Ptolemaic view, having their views called into question was equivalent to calling the Bible itself into question. This was their mistake: they did not yet understand that the Bible has to be interpreted just as observations of nature have to be interpreted, and they were not yet ready to reconsider their views about the heavens and their interpretations of Scripture. It is interesting to note that in 1992, Pope John Paul II gave an address in which he commended Galileo's comments on the necessity of interpreting Scripture!

Recall that at this time, there was as yet no physical evidence that the earth was moving and rotating on an axis. As I mentioned previously, evidence for earth's motion around

the sun came in 1838 with the discovery of stellar parallax. Evidence for the rotation of the earth came a bit later in 1851 with the invention of Foucault's pendulum, like the one shown in Figure 2.22. The rotation of the earth causes a minute change of direction in each swing of a very long, massive pendulum. If the earth were not rotating, the pendulum would swing steadily back and forth in the same direction.

Now, briefly, here is the sequence of events that led to Galileo's trial. Rumors got around that Galileo had been secretly examined by the Holy Office and forced to abjure (renounce) his views about Coper-

Figure 2.22. A Foucault Pendulum in the Panthéon in Paris.

nicus. To help Galileo fight these annoying rumors, Cardinal Bellarmine wrote Galileo a letter in 1616 stating that the rumors were false. The letter went on to say that though Galileo had not been taken before the Holy Office, he had been told not to defend or teach as true the system of Copernicus. Then in 1632, Galileo published another major work on astronomy in which he did in fact uphold the system of Copernicus against the system of Ptolemy. The Pope at the time, Urban VIII, was also a friend and admirer of Galileo, but when he heard that Galileo had published such a book after having specifically been told not to, he was extremely upset and had no choice but to have Galileo examined by the Holy Office. This was Galileo's famous 1633 trial.

The details leading to Galileo's trial are very complex, but the controversy boils down to two the issues I have emphasized. First, Galileo pushed his scientific claims too far, claiming truth for a scientific theory which could not be regarded as more than a model of nature. Second, he published a book in defiance of an injunction against doing so. Galileo was a pious and godly man. There is good evidence that he never did actually intend to fall afoul of the injunction. But when the Holy Office persuaded him that he had, he was immediately ready to confess his actions and abjure them. This he did. Galileo was never tortured, but it was necessary that he be punished in some way. His friend Pope Urban VIII made it as easy on Galileo as he could by confining him to "house arrest" and prohibited him from further publishing. He lived for a few months in Rome in the palace of one of the cardinals, and then was allowed to return to his home in Florence were he lived in house arrest for the last eight years of his life.

In addition to his work in astronomy, Galileo developed ground-breaking ideas in physics over the course of 30 years of work. These ideas were published after his trial in what would be his final book.[6] Before Galileo, scientists had always accepted Aristotle's physics, which held that a force was needed to keep an object moving. Galileo broke with this 2,000-year-old idea and hypothesized that force was needed to *change* motion but not to *sustain* motion as Aristotle had taught. Galileo was the first to formulate the idea of a friction force that caused objects to slow down. By conducting his own experiments, Galileo also discovered that all falling objects accelerate at the same rate (the acceleration of gravity, 9.80 m/s^2), which is mathematically very close to Isaac Newton's second law of motion (our topic in the next chapter). Galileo's studies in physics thrust forward the Scientific Revolution and set the stage for the work of Isaac Newton, where the Scientific Revolution reached its climax.

The saga of the Copernican Revolution ends more or less with Galileo. Within 50 years of Galileo's death the heliocentric model of the planetary orbits was well established. But while we are studying the planets and gravity, the whole story just isn't complete unless we mention two more key figures in the history of science.

2.3.9 Newton, Einstein, and Gravitational Theory

Sir Isaac Newton (Figure 2.23) is perhaps the most celebrated mathematician and scientist of all time. He was English, as his title implies, and he was truly phenomenal. He held a famous professorship in mathematics at Cambridge University. He developed calculus. He developed the famous laws of motion, which we will examine. He developed an entire theory of optics and light. He formulated the first quantitative law of gravity called the *law*

---

6   Since he was forbidden to publish through the Catholic Church, the book was published by a Protestant publisher in Holland.

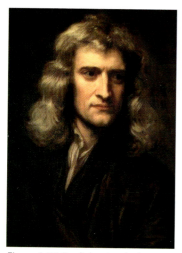

Figure 2.23. English scientist Isaac Newton (1643–1727).

of universal gravitation. His massive work on motion, gravity, and the planets, *Principia Mathematica*, was published in 1687. This work is one of the most important publications in the history of science.

We will not perform computations with Newton's law of universal gravitation, and you do not need to memorize the equation for it. But let's look at it here briefly. The law is usually written as

$$F = G \frac{m_1 m_2}{d^2}$$

where G is a constant, m_1 and m_2 are the masses of any two objects (such as the sun and a planet), and d is the distance between the centers of the two objects.

Newton theorized that every object in the universe pulls on every other object in the universe, which is why his law is called the law of *universal* gravitation. We now understand that he was correct. *Everything* in the universe pulls on *everything* else. I have no idea how Newton figured this out. The equation above gives the force of gravitational attraction between any two objects in the universe. Amazingly, this equation is quite accurate, too! Notice from the equation that Newton's model depends

Do You Know ...   ## *Who built the first monster telescope?*

William Herschel was a German astronomer who moved to England when he was a young man. He was a major contributor to pushing the technology of the reflecting telescope to new limits, and spent vast amounts of time casting and polishing his own mirrors. He constructed the largest telescopes ever built at the time.

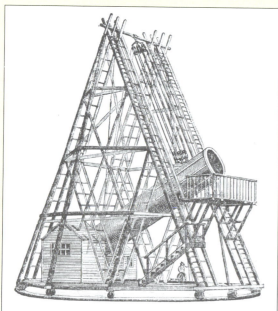

In 1781, Herschel discovered the planet Uranus. Herschel's sister Caroline was an important astronomer herself. She worked closely with her brother. Herschel gave her a telescope of her own and with it she discovered many new comets, for which she became recognized.

Herschel's monster 40-foot telescope, shown to the left, had a primary mirror over four feet in diameter. In 1789, on the first night of using the new telescope, Herschel discovered a new moon of Saturn. He discovered another new moon about a month later.

on each object having mass because the force of gravity has both masses in it multiplied together. Newton's model implies that if either mass is zero, the force of gravitational attraction is zero.

While we are here looking at Isaac Newton, we should pause and consider the relationship between his physical theories (including law of universal gravitation and his laws of motion) and Kepler's mathematical theory of planetary motion. It turns out that Kepler's discovery about the elliptical orbits and the relationship between the period and mean radius of the orbit can be directly derived from Newton's theories, and Newton does derive them in *Principia Mathematica*. But Newton's equations apply much more generally than Kepler's do. As we will see in the next chapter, Newton's laws apply to all objects in motion—planets, baseballs, rockets—while Kepler's laws apply to the special case of the planets' orbits. If we consider this in light of my comments

Figure 2.24. German physicist Albert Einstein (1879–1955).

in Chapter 1 about the way theories work, we see that Newton's laws explain everything Kepler's laws explain, and more. This places Newton's theory about motion and gravity above Kepler's, so Newton's theories took over as the most widely-accepted theoretical model explaining gravity and motion in general. However, even though Newton's laws ruled the scientific world for nearly 230 years, they do not tell the whole story.

This is where the German physicist Albert Einstein (Figure 2.24) comes in with his *general theory of relativity*, published in 1915. Einstein's theory explains gravity in terms of the curvature of space (or more accurately, *spacetime*) around a massive object, such as the sun or a planet. This spacetime curvature is represented visually in Figure 2.25. Fascinatingly, since Einstein's theory is about curving space, the theory predicts that even phenomena without mass, such as rays of light, are affected by gravity. Einstein noticed this and made the stunning prediction in 1917 that starlight bends as it travels through space when it passes near a massive object such as the sun. He formed this hypothesis, including the amount light bends, based on his general theory of relativity, which was based *completely on mathematics*. What do you think about that? It practically leaves me speechless.

Einstein became instantly world famous in 1919 when his prediction was confirmed. To test this hypothesis, Einstein proposed photographing the stars we see near the sun during a solar eclipse. This has to be done during an eclipse because looking at the sky while the sun is nearby means it is broad daylight and we aren't able to see the stars. Einstein predicted that the apparent position of the stars shifts a tiny amount relative to where they appear when the sun is not near the path of the starlight. British scientist Sir Arthur Eddington commissioned two teams of photographers to photograph the stars during the solar eclipse of 1919. After analyzing their photographic plates (one of which is shown on the opening page of Chapter 1), they found the starlight shifted by exactly the amount Einstein said it would. Talk about sudden fame—Einstein became the instant

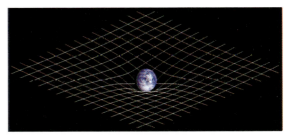

Figure 2.25. A visual representation of the curvature of spacetime around the earth.

global rock star of physics when this happened! (And his puppy dog eyes contributed even more to his popularity!)

Just as Kepler's laws were superseded by Newton's laws and can be derived from Newton's laws, Newton's law of universal gravitation was superseded by Einstein's general theory of relativity and can be derived from general relativity. Einstein believed that his own theories would some day be superseded by an even more all-encompassing theory, but so far (after 101 years) that has not happened. The general theory of relativity remains today the reigning champion theory of gravity, our best understanding of how gravity works, and one of the most important theories in 20th- and 21st-century physics.

Chapter 2 Exercises

Unit Conversions

It is time for you to get busy learning the metric prefixes and unit conversion factors in Appendix A. Perform the following unit conversions, showing all your work in detail. (Showing just the answers is not adequate; show all the conversion factors involved in the conversion for each problem.) Check your work against the answer key on the next page. Where possible, express your results in both standard notation and scientific notation, using the correct number of significant digits. For the first 20 problems, use the standard method of multiplying conversion factors. The last problem requires an extra step that I think you can figure out.

	Convert this Quantity	Into these Units
1	1,750 meters (m)	feet (ft)
2	3.54 grams (g)	kilograms (kg)
3	41.11 milliliters (mL)	liters (L)
4	7×10^8 m (radius of the sun)	miles (mi)
5	1.5499×10^{-12} millimeters (mm)	m
6	750 cubic centimeters (cm^3 or cc) (size of the engine in my old motorcycle)	m^3
7	2.9979×10^8 meters/second (m/s) (speed of light)	ft/s
8	168 hours (hr) (one week)	s
9	5,570 kilograms/cubic meter (kg/m^3) (average density of the earth)	g/cm^3
10	45 gallons per second (gps) (flow rate of Mississippi River at the source)	m^3/minute (m^3/min)
11	600,000 cubic feet/second (ft^3/s) (flow rate of Mississippi River at New Orleans)	liters/hour (L/hr)
12	5,200 mL (volume of blood in a typical man's body)	m^3

	Convert this Quantity	Into these Units
13	5.65×10^2 mm^2 (area of a postage stamp)	square inches (in^2)
14	32.16 ft/s^2 (acceleration of gravity, or one "g")	m/s^2
15	5,001 μg/s	kg/min
16	4.771 g/mL	kg/m^3
17	13.6 g/cm^3 (density of mercury)	mg/m^3
18	93,000,000 mi (distance from earth to the sun)	cm
19	65 miles per hour (mph)	m/s
20	633 nanometers (nm) (wavelength of light from a red laser)	in
21	5.015% of the speed of light	mph

Answers

(A dash indicates that it is either silly or incorrect to write the answer that way, so I didn't: silly because there are simply too many zeros, or no zeros at all; incorrect because we are unable to express the result that way and still show the correct number of significant digits.)

	Standard Notation	Scientific Notation
1	5,740 ft	5.74×10^3 ft
2	0.00354 kg	3.54×10^{-3} kg
3	0.04111 L	4.111×10^{-2} L
4	400,000 mi	4×10^5 mi
5	–	1.5499×10^{-15} m
6	0.00075 m^3	7.5×10^{-4} m^3
7	983,600,000 ft/s	9.836×10^8 ft/s
8	605,000 s	6.05×10^5 s
9	5.57 g/cm^3	–
10	–	1.0×10^1 m^3/min
11	60,000,000,000 L/hr	6×10^{10} L/hr
12	0.0052 m^3	5.2×10^{-3} m^3
13	0.876 in^2	8.76×10^{-1} in^2
14	9.802 m/s^2	–
15	0.0003001 kg/min	3.001×10^{-4} kg/min
16	4,771 kg/m^3	4.771×10^3 kg/m^3

	Standard Notation	Scientific Notation
17	13,600,000,000 mg/m³	1.36×10^{10} mg/m³
18	–	1.5×10^{13} cm
19	29 m/s	2.9×10^{1} m/s
20	0.0000249 in	2.49×10^{-5} in
21	33,700,000 mph	3.37×10^{7} mph

Motion Exercises

1. A train travels 25.1 miles in 0.50 hr. Calculate the velocity of the train.

2. Convert your answer from the previous problem to km/hr.

3. How far can you walk in 4.25 hours if you keep up a steady pace of 5.0000 km/hr? State your answer in km.

4. For the previous problem, how far is this in miles?

5. On the German autobahn there is no speed limit and in good weather many cars travel at velocities exceeding 150.0 mi/hr. How fast is this in km/hr?

6. Referring again to the previous question, how long does it take a car at this velocity to travel 10.0 miles? State your answer in minutes.

7. An object travels 3.0 km at a constant velocity in 1 hr 20.0 min. Calculate the object's velocity and state your answer in m/s.

8. A car starts from rest and accelerates to 45 mi/hr in 36 s. Calculate the car's acceleration and state your answer in m/s².

9. A rocket traveling at 31 m/s fires its retro-rockets, generating a negative acceleration (it is slowing down). The rockets are fired for 17 s and afterwards the rocket is traveling at 22 m/s. What is the rocket's acceleration?

10. A person is sitting in a car watching a traffic light. The light is 14.5 m away. When the light changes color, how long does it take the new color of light to travel to the driver so that he can see it? State your answer in nanoseconds. (The speed of light in a vacuum or air, *c*, is one of the physical constants listed in Appendix A that you need to know.)

11. A proton is uniformly accelerated from rest to 80.0% of the speed of light in 18 hours, 6 minutes, 45 seconds. What is the acceleration of the proton?

12. A space ship travels 8.96×10^{9} km at 3.45×10^{5} m/s. How long does this trip take? Convert your answer from seconds to days.

13. An electron experiences an acceleration of 5.556×10^{6} cm/s² for a period of 45 ms. If the electron is initially at rest, what is its final velocity?

14. A space ship is traveling at a velocity of 4.005×10^{3} m/s when it switches on its rockets. The rockets accelerate the ship at 23.1 m/s² for a period of 13.5 s. What is the final velocity of the rocket?

15. A more precise value for *c* (the speed of light) than the value given in Appendix A is 2.9979×10^{8} m/s. Use this value for this problem. On a particular day the earth

is 1.4965 × 10⁸ km from the sun. If on this day a solar flare suddenly occurs on the sun, how long does it take an observer on the earth to see it? State your answer in minutes.

Answers

1. 22 m/s	2. 79 km/hr	3. 21.3 km	4. 13.2 mi
5. 241.4 km/hr	6. 4.00 min	7. 0.63 m/s	8. 0.56 m/s²
9. −0.53 m/s²	10. 48.3 ns	11. 3,680 m/s²	12. 301 days
13. 2.5 × 10³ m/s	14. 4.32 × 10³ m/s	15. 8.3197 min	

Ptolemaic Model and Copernican Revolution Study Questions

1. Make a list of all the regions in the Ptolemaic Model in their correct order. (There are 10 of them and the first nine are called spheres.) For each of the last three regions write a brief description of the meaning of the name.

2. Describe why some theologians in the 16th century were strongly opposed to Copernicus' heliocentric theory.

3. State six features of the Ptolemaic model other than the spheres.

4. Describe Copernicus' model of the heavens.

5. What are some of the "proofs" people used in arguing that there is no way that the earth rotates on an axis?

6. For what reason did Copernicus decide to keep his theory private?

7. Write a description of the two key observations Tycho made (including dates) that challenged the Ptolemaic system.

8. Briefly describe the cosmological model put forward by Tycho.

9. State Kepler's first law of planetary motion.

10. This is a bit difficult, but explain retrograde motion and epicycles as well as you can in a few sentences.

11. Explain the two main mistakes individuals made that led to Galileo's trial.

12. Explain the actual cause of Galileo's trial and the results of that trial.

13. Describe why Pope John Paul II commended Galileo in 1992.

14. Distinguish between Newton's and Einstein's theories of gravitation. According to each of these two geniuses, what is the cause of gravity and what are the effects of gravity?

15. The theories reviewed in this chapter suggest that the universe possesses a very deep mathematical structure. What does this structure indicate about where the universe came from?

16. Describe some of Kepler's scientific achievements, aside from his laws of planetary motion.

CHAPTER 3
Newton's Laws of Motion

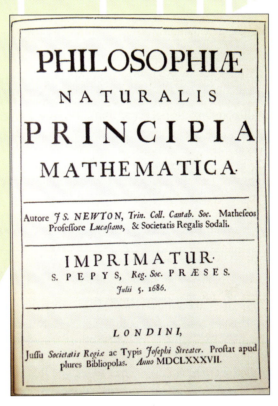

Principia Mathematica

This is a photograph of the original title page of Isaac Newton's Principia Mathematica. *This historically important work presents Isaac Newton's three laws of motion.*

OBJECTIVES

Memorize and learn how to use these equations:

$$a = \frac{F}{m} \qquad\qquad F_w = mg$$

After studying this chapter and completing the exercises, students will be able to do each of the following tasks, using supporting terms and principles as necessary:

1. Define and distinguish between matter, inertia, mass, force, and weight.
2. State Newton's laws of motion.
3. Calculate the weight of an object given its mass and vice versa.
4. Perform calculations using Newton's second law of motion.
5. Give several examples of applications of the laws of motion that illustrate their meaning.
6. Explain why Newton's first law of motion is called the law of inertia.
7. Use Newton's laws of motion to explain how a rocket works.
8. Apply Newton's laws of motion to application questions, explaining the motion of an object in terms of Newton's laws.

3.1   Matter, Inertia, and Mass

To study Sir Isaac Newton's three laws of motion, one needs to have a clear understanding of three closely related and somewhat confusing terms, *matter*, *inertia*, and *mass*. *Matter* is simply the term we use for anything made of atoms or parts of atoms. Examples of matter are all around us and include things like bananas, protons, carbon atoms, and planets. Examples of things that are *not* matter include light, radio waves, Beethoven's third symphony, justice, and the Holy Spirit. These are all real things, but they are not material objects and thus they have no mass.

To set the stage for the next of these three terms (inertia), consider that all matter possesses certain properties. For example, one property or quality you already know about is that matter takes up space. We *quantify* the amount of space that a material object occupies with the variable *volume*, and our measurements of the volume of an object are stated in certain units of measure. In the MKS system, the units for volume are cubic meters (m^3).

Now let's consider another property all matter possesses—*inertia*. The effect of this property is that a material object always resists changes to its *state of motion*. Saying an object has a particular state of motion is equivalent to saying that it has a specific velocity or that it is at rest (with a velocity of zero.)

Put another way, the effect of inertia is that in order to change the motion of an object a force is required. If an object is at rest and no net force is acting on it, it remains at rest. The same applies to an object moving with a constant velocity; if there is no net force acting on it, it continues moving with a constant velocity. In the previous chapter, we saw that a changing velocity is called acceleration. Now you can see that *to say an object's state of motion is changing is equivalent to saying its velocity is changing, which is also equivalent to saying that the object is accelerating*. In summary, a net force on an object is required to make it accelerate. No net force, no acceleration. Without a net force on an object, the object's inertia makes it continue doing whatever it is presently doing.

Finally, *mass* is the variable we use to specify a *quantity of matter*. In the MKS system, the unit of measure for mass is the kilogram (kg). As we saw in the previous chapter, the kilogram is one of the seven base units of measure in the SI system of units (as are the meter and the second for distance and time).

Students are often confused by the distinction between inertia and mass. The best way to keep them straight is to think of inertia as a *quality* of matter and mass as the *quantity* of a specific portion of matter. Inertia is one of the qualities or properties all matter possesses. Mass is a quantitative variable that specifies an amount of matter, a quantity of matter.

Now that you have a handle on matter, inertia, and mass, there is one more term we need to discuss before getting to Newton's three famous laws of motion. This is the term *force*, or as you will see me write over and over, *net force*. A force is simply a push or a pull. The concept of force is fairly easy to understand and Newton's laws of motion tell us very specific things about the way forces affect objects in nature.

But to understand the laws we really need to have an understanding of *net* force. Consider the forces presently acting on me as I sit here at my computer writing this book. Because of the gravitational attraction of the earth, the earth is pulling me toward its center, which is *down* from my point of view. If this were the only force on me right now I would be accelerating downward, that is, falling. Luckily for me, there is another force pushing up on me from the chair I am sitting in. So although there are these two forces acting on me at present, they happen to be equal and directed in opposite directions, so they cancel each other out. This means there is no *net force* on me at the moment. A net force is a force that is not balanced out or cancelled out by some other force.

3.2   Newton's Laws of Motion

3.2.1 The Three Laws of Motion

Isaac Newton's three *laws of motion* may be the most famous scientific statements ever made. Accordingly, we must learn them. Here is the first one.

> **First Law**   An object at rest remains at rest and an object in motion continues moving in a straight line at a constant speed, unless it is compelled to change that state by forces acting on it.

The first law applies when there is no net force acting on an object and describes what objects do in such circumstances. As mentioned before, objects behave the way this law describes because of their inertia. For this reason the first law is sometimes called the *law of inertia*.

In this book so far, I have avoided using the term *speed*. Instead, I have mostly used the term *velocity* for our discussion about motion. But the classic statement of Newton's first law uses the term speed (Newton called it "uniform motion"), so I kept it in there. The difference between the terms speed and velocity is a bit technical, but for those who are wondering I will address this issue briefly here. In the previous chapter, we saw that velocity is the rate at which an object's distance is changing, but I also said we were only going to consider motion in a straight line. The term velocity actually involves not only *how fast* an object is moving, which is what we have discussed, but also the *direction* in which the object is moving. The term speed denotes only the "how fast" part of an object's motion, not the direction part. For our purposes, since we are concentrating on motion in a straight line, the terms speed and velocity are essentially synonyms.

Here is Newton's second law.

Second Law   The acceleration of an object is proportional to the force acting on it, or

$$a = \frac{F}{m}$$

where a is the acceleration of the object (m/s²), F is the net force on the object in newtons (N), and m is the object's mass (kg).

I must now inform you that by writing the second law as $a = F/m$, I have boldly written it in a form different from the way it is written in nearly every physics book in the world, which is $F = ma$. If you ask nearly anyone to state Newton's second law, he or she says $F = ma$. Some people reading this might wonder if I have taken leave of my senses to restate one of the most famous equations in physics! No I haven't. I have a good reason for stating the law this way—this is closer to the way Isaac Newton stated it![1] I don't know why everyone started saying $F = ma$. Mathematically, the equation $F = ma$ implies that force is dependent on acceleration. This is because when writing algebraic equations we customarily call the variable on the left the *dependent variable*, and we think of it as depending on the values we put in for the variables on the right, which are called *independent variables*.

When we write $F = ma$, F is in the position of the dependent variable in the equation. But we do not usually think of objects this way. When we think of objects accelerating, we usually think of the acceleration as being dependent on the force acting on the object. If a certain net force is applied to an object, the object accelerates in proportion to the force. This is the way we think, in most circumstances, and this is the way Newton said it. So this is the way I'm saying it.

The second law applies when there is a net force present and says what the object does as a result. As we saw previously, when a net force is present on an object, the object changes its state of motion. In other words, the object accelerates, and according to Newton's second law, the acceleration is directly proportional to the force. This means if we double or triple the net force on an object, we double or triple its acceleration.

The MKS unit of measure for force is the newton (N), in honor of Sir Isaac Newton's contributions to science. This is the first derived unit we have encountered in this course. As you know, we are using the MKS system of units—meters, kilograms, and seconds. Most of the other units of measure that we will encounter are derived from combinations of these three main units. Since force is mass times acceleration (that is, $F = ma$), the unit for force is the product of the units for mass and acceleration, or

$$1\ \mathrm{N} = 1\ \mathrm{kg} \cdot \frac{\mathrm{m}}{\mathrm{s}^2}$$

---

1   Newton's actual statement was, "The change in motion is proportional to the motive force impressed." Newton's definition for *motion* is equivalent to what we now call *momentum*. We will take up this topic in Chapter 5.

(Compare this to the units shown at the bottom of page 22.)

Now for the third law, which requires some wordsmithing. The traditional way to state the third law is, "For every action there is an equal and opposite reaction." Newton himself wrote this law as, "To every action there is always opposed an equal reaction." The problem with these statements is that the words *action* and *reaction* in the traditional statement of Newton's third law do not mean anything like what we mean today when we use those words. Newton meant these terms to refer to *forces*. They are not, in the way we use the terms today, actions, reactions, events, or processes. So a better way to state the law now is: For every force, there is an equal and opposite push-back force. That is, if object A pushes to the right on object B, object B pushes to the left with the same strength on object A. Of course, it is good for us to know about the traditional language. So I have developed a hybrid version of the third law that students should use. Here it is:

| Third Law | For every action force, there is an equal and opposite reaction force. |

The third law describes the way objects always act on *one another* when forces are present. If an object experiences a force—any force—a second object is experiencing one too, identical in strength but in the opposite direction. These two forces are the so-called "action-reaction pairs" the third law is famous for. It is not possible for one object to push on another object without the second object pushing back in the opposite direction on the first object with the same amount of force. The third law applies all the time, whether the forces result in the acceleration of one or both of the objects involved, or the objects remain motionless as in the example of sitting in my chair at my computer.

As another example, consider again Newton's law of universal gravitation that we bumped into briefly near the end of Chapter 2. The law of universal gravitation says that every object in the universe pulls on every other object. If objects are large enough and close enough together, as in the case of a planet and the sun, these forces are very large and keep the planet in its orbit. Without a large attractive force, the planet's huge inertia would cause it to fly away in space at a constant speed in a straight line, as Newton's first law says.

As illustrated in Figure 3.1, a planet going around the sun is attracted to the sun with an amount of force given by the equation in Newton's law of universal gravitation, which can be written as

$$F = G\frac{m_1 m_2}{r^2}$$

(In the last chapter I used the variable d in the denominator of this equation. Here I used r because it is common to use the variable r instead of d when orbital types of motion are involved.) But let's consider this situation from the sun's point of view. Newton's third law of motion says that for every force there is an equal and opposite reaction force, another force pushing back with the same amount of force but in the opposite direction. This means that the sun is attracted to the planet with the exact same amount of force, pointing in exactly the opposite direction.

The result of these forces for the planet is that planet stays in orbit. But the result for the sun is that it wobbles in place as the planet goes around. Now, since the sun is so much more massive than any of the planets, it doesn't move very far. Additionally, the sun is being tugged on by all of the planets at once, and since they are spread out around the sun in their different orbits, their forces on the sun tend to balance each other out to some extent. Nevertheless, the sun does wobble because of all the planets pulling on it. In fact, all the planets wobble too because they are all pulling on each other. (Observing these wobbles in the past enabled astronomers to discover the planets Uranus and Neptune, which could not be seen with the telescopes from Galileo's era.) On top of that, planets with moons, like the earth and Jupiter, wobble even more because of their moons pulling on them! All these forces due to the planets and moons and sun pulling on each other produce an interacting system of forces that is so complex no one has ever been able to develop equations to model it that can be solved exactly. Instead, we model the planetary motions using approximations and the aid of sophisticated modern computers!

We will return soon to discussing the laws of motion and how to apply them. But before we do, let's get you started on using the two new equations that we are introducing in this chapter. We begin with some examples using the second law.

▼ Example 3.1

A force of 8.61 nN is applied to a proton with a mass of 1.673×10^{-24} g. Determine the acceleration of the proton.

As always, begin by writing down the given information and converting all of the quantities into MKS units.

$$F = 8.61 \text{ nN} \cdot \frac{1 \text{ N}}{10^9 \text{ nN}} = 8.61 \times 10^{-9} \text{ N}$$

$$m = 1.673 \times 10^{-24} \text{ g} \cdot \frac{1 \text{ kg}}{1000 \text{ g}} = 1.673 \times 10^{-27} \text{ kg}$$

$$a = ?$$

Now write the equation, insert the values in MKS units, and compute the result.

$$a = \frac{F}{m} = \frac{8.61 \times 10^{-9} \text{ N}}{1.673 \times 10^{-27} \text{ kg}} = 5.146 \times 10^{18} \frac{\text{m}}{\text{s}^2}$$

Since the given force, with three significant digits, has the least precision among the given quantities, we round this result to three significant digits to get

$$a = 5.15 \times 10^{18} \frac{\text{m}}{\text{s}^2}$$

▲

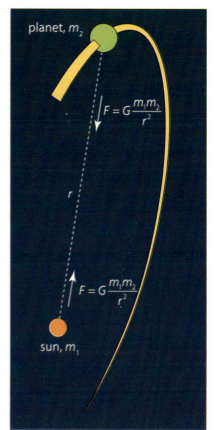

Figure 3.1. Newton's third law applied to a planet and the sun.

Calculations with Newton's second law are pretty easy because the equation itself is simple. The problems we can solve can actually become a lot more interesting if we combine these kinds of problems with those we learned last chapter to make a two-part problem. The reason we can do this is that we now know two different equations that contain the acceleration. These two-part problems arise when we use one of the equations to compute the acceleration, and then use the other equation with the acceleration we found to compute something else. The next example should make this clear.

▼ Example 3.2

A bullet from a rifle is traveling at 2,250 ft/s when it lodges in a tree, coming to rest 45 ms after impact. If the mass of the bullet is 65.5 g, determine the force of friction in the tree that stop the bullet.

This is a two-part problem. We must first use the velocity and time information to determine the acceleration. Once we have that value, we use it in Newton's second law to determine the force that produces it.

Begin by writing down the givens for the acceleration calculation and performing the unit conversions.

$$v_i = 2250 \ \frac{\text{ft}}{\text{s}} \cdot \frac{0.3048 \ \text{m}}{\text{ft}} = 685.8 \ \frac{\text{m}}{\text{s}}$$

$$v_f = 0$$

$$t = 45 \ \text{ms} \cdot \frac{1 \ \text{s}}{1000 \ \text{ms}} = 0.045 \ \text{s}$$

$$a = ?$$

Now write the equation, put in the MKS values, and compute the first result, keeping one extra digit of precision.

$$a = \frac{v_f - v_i}{t} = \frac{0 - 685.8 \ \frac{\text{m}}{\text{s}}}{0.045 \ \text{s}} = -15,240 \ \frac{\text{m}}{\text{s}^2}$$

In this intermediate result, I kept an extra significant digit (two extra digits actually) over what I need at the end. Notice that the acceleration is negative, meaning that the bullet slows down. (This is pretty obvious from the problem, isn't it?)

Now we deal with the force calculation using the second law, beginning with the givens and the unit conversions to MKS units.

$$m = 65.5 \ \text{g} \cdot \frac{1 \ \text{kg}}{1000 \ \text{g}} = 0.0655 \ \text{kg}$$

$$F = ?$$

Now write down the second law, do the algebra, and perform the computation.

$$a = \frac{F}{m}$$

$$F = ma = 0.0655 \text{ kg} \cdot \left(-15,240 \ \frac{m}{s^2}\right) = -998 \text{ N}$$

I have two comments to make here. First, the negative sign in front of the force simply means that the force is opposing the motion of the object, and thus is slowing it down. This is what friction always does. So write the negative sign if you wish, but since the problem is about a friction force, we know which direction friction forces point (against the motion), so the negative sign is not really necessary. Suit yourself and don't worry about it.

Second, we need to round this result to two significant digits, because the time given in the problem only has two significant digits. However, rounding this value gives 1,000 N, a value indicating only one significant digit. So we must express our result in scientific notation. In a value without a decimal, scientific notation is the only way to show that trailing zeros are significant. Thus, with two significant digits our result is

$$F = 1.0 \times 10^3 \text{ N}$$

Another way to display the correct number of significant digits is to convert the answer to kilonewtons (kN).

$$F = 998 \text{ N} \cdot \frac{1 \text{ kN}}{1000 \text{ N}} = 0.998 \text{ kN}$$

Rounding this result to two significant digits we have

$$F = 1.0 \text{ kN}$$

3.2.2 Actions and Reactions

When discussing the third law, we must emphasize that the actions-reaction pairs in the law are pushes or pulls, not events or processes. A generic statement of how action-reaction pairs of forces work is, "A pushes B and B pushes back equally and oppositely on A." As an illustration of how students often misunderstand this, consider the following scenario, illustrated in Figure 3.2 on the next page. A glass of milk is sitting on a table. A boy walks by and accidently bumps the table, causing the glass of milk to spill. It is not uncommon for students to say that the action is the boy bumping the table and the reaction is the milk spilling, but this is *not* a correct way to identify the forces involved. Remember, a "reaction" is a force; "spilling milk" is not a force. The correct way of describing the action-reaction pair is "the boy bumps the table and the table bumps the boy."

So why does the milk spill if it is not involved in the action-reaction pair? When the boy bumps the table, the table accelerates briefly just as the second law states. The glass is sitting on the table, so friction between the table and the bottom of the glass causes the bottom portion of the glass to accelerate with the table. But the inertia of the upper part of the glass makes the upper part of the glass resist change to its state of motion. If the bottom of

The Incident
A boy bumps into a table.
Black arrows represent the action-
reaction force pair.

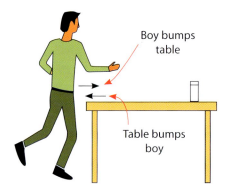

Boy bumps
table

Table bumps
boy

The Results
Results of the boy bumping the table:
table accelerates to right.

Results of the table bumping the boy:
boy feels the bump on his leg.

Inertia of upper portion of glass
makes it try to remain at rest
(Newton's first law)

Table and lower portion of
glass accelerate to the right
(Newton's second law)

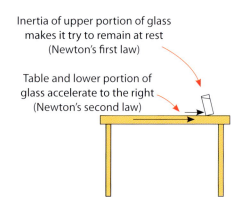

Figure 3.2. Force analysis of the boy bumping the table and the table bumping the boy (top). Several different outcomes result from these bumps (bottom).

the glass accelerates while the top part stays put, the glass tips over. In this illustration, all three of Newton's laws of motion are involved.

3.2.3 Showing Units of Measure in Computations

Now that I have worked a number of example problems in this text, I hope you have noticed that I always write the units with every quantity I write, every time I write one. To students your age, this usually seems tedious and unnecessary and many students resist doing it. But I want you to write them. Every time you write down a value of any kind, in any problem, you must write the units that go with it. Every time.

There are several reasons for this, actually, and we are not just being unnecessarily tedious when we insist on it. First, just as with the overall problem-solving method that you are learning, successfully solving problems in physics depends a lot on having an orderly and reliable approach. You must develop proper habits and using the problem-solving method presented in this book is one of those proper habits you must develop. Writing down the units is another one. Over the years, I have literally seen hundreds of points lost because of students failing to develop these proper habits. Without the proper habits, it is very easy to make silly mistakes.

A second reason is that physics problems get more complicated the farther you go in the subject. Keeping track of the units of measure you are working with can be a life saver in more complex problems, so now is the time to get in the habit of always writing them down. Third, by always writing down the units you learn the units more thoroughly. Students sometimes forget what the units are for force or energy or power. Students don't forget as easily when they write them down every time they work a problem. And finally, by writing down the units of measure you have a constant reminder in front of you to help you do the right thing. When you insert the velocity of an object into an equation, writing in a value in m/s is a little confirmation to you that you are doing the right thing, rather than, say, writing in the object's acceleration by mistake and not noticing your error.

The bottom line is always write down the units. Always write them down. Always write them. *Always*.

3.2.4 Weight

The term *weight* simply means the force with which gravity is pulling on an object. Weight is the force on an object due to gravity. The equation to calculate the weight of an object is

$$F_w = mg$$

where F_w is the weight of the object (N) and g is the acceleration due to gravity, which for earth is approximately equal to 9.80 m/s^2 (at sea level). We use this equation any time we have an object's mass and need its weight, or vice versa, which makes it quite important. The constant for g, the acceleration due to gravity on earth, is one of the constants listed in Appendix A that you must memorize.

As you recall, Galileo discovered that all falling objects accelerate at the same rate. This means that when released from rest, all objects experience the same acceleration in free fall. We designate this acceleration with the letter g. On earth at sea level, the average value of this acceleration due to gravity is $g = 9.80$ m/s^2. On the surface of earth, any object that falls freely accelerates with an acceleration of 9.80 m/s^2.

Now, although it may take some getting used to, notice that the weight equation above is just a special case of Newton's second law using g as the acceleration. The weight equation is a case in which it does make sense to write the second law as $F = ma$, with force as the dependent variable. An object's weight depends on the gravitational field it is in.

When the astronauts were walking on the moon back in the late 1960s, they weighed only 1/6 of their earth weight. This is because the moon's gravitational attraction on a person walking on the surface

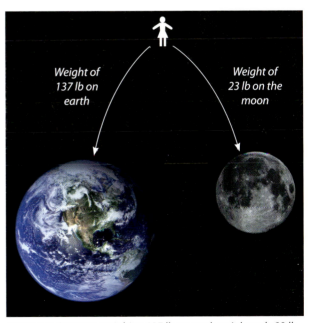

Figure 3.3. A person weighing 137 lb on earth weighs only 23 lb on the moon.

is only 1/6 as great as the earth's. However, the astronauts' masses were unchanged because mass is directly related to the atoms in an object. This difference in gravitational attraction is depicted in Figure 3.3 (in which I decided to imagine a female astronaut). The astronauts' weights depend on the acceleration of gravity where they are. The same equation for weight applies on the moon or other planets as well, but the value of g is different on the moon or other planets.

▼ Example 3.3

A certain athlete weighs 225.00 lb. Determine the athlete's mass.

As always, begin by writing down the given information and converting the units to MKS units. This problem requires the use of a conversion factor we have not seen before. This conversion factor is listed in Appendix A, and is one you do not need to memorize.

$$F_w = 225.00 \text{ lb} \cdot \frac{4.45 \text{ N}}{1 \text{ lb}} = 1001.3 \text{ N}$$

$$m = ?$$

Continue with writing down the equation, doing the algebra, and computing the result.

$$F_w = mg$$

$$m = \frac{F_w}{g} = \frac{1001.3 \text{ N}}{9.80 \ \frac{m}{s^2}} = 102.2 \text{ kg}$$

Since the value of g we are using has three significant digits, this determines the precision of our result, which must also have three significant digits. Rounding accordingly, we have

$$m = 102 \text{ kg}$$

3.2.5 Applying Newton's Laws of Motion

If you really *understand* Newton's laws of motion, you are able to apply them to different circumstances, explaining what happened in terms of one of the laws. I will illustrate this with a few examples, but before I do, here are some guidelines to follow. First, ask yourself if the situation is about an object that has a net force on it and is accelerating or if it is about an object that does not have a net force on it and is maintaining its state of motion. If there is no net force and no acceleration, the first law is the one to apply. In any scenario that depicts objects at rest and remaining at rest, or moving at a constant velocity and continuing to do so, the first law applies. Do not apply the first law to any scenario which depicts the state of motion as changing. In short, if the velocity is constant, the first law applies.

If the velocity is not constant, the second law applies. If any object is depicted as accelerating, apply the second law, which says that the acceleration of the object is directly proportional to the net force applied to it.

If the scenario refers to two objects pushing or pulling on each other, regardless of whether they are moving, apply the third law. Remember that the action-reaction pair is always a pair of forces, not anything else like actions, processes, or events. So you must always apply this law by saying something like, "A pushes B and B pushes back equally on A." Do not make the mistake of using events, movements, or actions as the action force or the reaction force. Remember, when Newton used the term *action* in his third law, the term did not have the same meaning that it has today. Newton was referring to *forces*, that is, *pushes* or *pulls*.

Please take the time to read the box on the next pages called *Thinking About Newton's Laws of Motion*. Then study the following examples of applying Newton's laws of motion to specific scenarios involving motion and forces.

▼ Example 3.4

Scenario	An asteroid speeding through space comes close to a star. When it does, it begins to speed up and turn in its course. Use one of Newton's laws of motion to comment on what is happening. (Note that since a force is required to change the direction of a moving object, turning is an example of acceleration.)
Explanation	The asteroid is accelerating because of the attraction of the star's gravitational field. Since there is a force (gravity) on the asteroid, its velocity is changing. That is, it is speeding up and changing direction. The acceleration of the asteroid is according to Newton's second law, $a = F/m$. The acceleration of the asteroid is in proportion to the strength of the gravitational attraction of the star.

▼ Example 3.5

Scenario	While driving, a man wearing glasses falls asleep at the wheel and runs into a tree. The man is wearing his seat belt and is unharmed but his glasses fly right off his face. Use one of Newton's laws of motion to comment on what happens in this scenario.
Explanation	The glasses are moving with the same velocity as the man and the car. While the man and the car experience forces that make them stop (according to Newton's second law), there is no force on the glasses. In the absence of a force, the inertia of the glasses makes them continue moving in a straight line at a constant speed until they hit the windshield, just as Newton's first law says.

▼ Example 3.6

Scenario	Susie wants to go hunting with her dad. Her father decides to let her shoot his .308 calibre rifle at a target out in the country. She places the rifle against her shoulder and fires at the target. The rifle fires the bullet and Susie feels the painful kickback of the gun on her shoulder.
Explanation	The firing of the bullet and the kickback of the gun are explained by Newton's third law of motion. The gun pushes the bullet (action force), and the

Thinking About Newton's Laws of Motion

Learning to apply Newton's laws of motion to physical situations takes thought and practice. Students often get the laws confused or choose one of the laws to describe a certain situation when another law fits better. Your study process for Newton's laws of motion begins with memorizing the three laws verbatim. When answering an application question, always cite (i.e., quote) the law or part of a law which best describes a situation. Below are some tips and examples to aid you in selecting the best law for a particular situation and in crafting a well-formed answer to a typical quiz question.

First Law

1. This law applies to situations where no acceleration is present—in other words, when objects are moving at a constant speed or sitting still.
2. This law is about what happens *without* a net force, so apply it when there is no net force present.
3. Another common way to apply this law is when something is moving at a constant speed and then a collision or some other event occurs. If a piece of the moving object keeps moving, it does so because of its inertia and because there is no outside force on the piece to slow it down and stop it.

Typical questions with appropriate answers:

Q: A space ship is traveling in space at 20,000 m/s when it runs out of fuel. As far as the captain can tell, there are no stars or other planets in front of him or any where else nearby. What will happen?

A: There are no forces from gravity or fuel to accelerate the ship. Thus, according to Newton's first law, the ship will continue at traveling at 20,000 m/s in a straight line forever.

Q: A kid on a sled slides down a snow-covered hill but hits a rough spot of grass at the bottom. His sled stops suddenly but the kid keeps right on going. How is this explained?

A: There is no force present on the kid to stop him so his inertia keeps him going even when his sled stops due to the friction force of the grass (second law). The law of inertia (Newton's first) states that unless a net force is present, a moving object continues to move at a constant speed.

Second Law

1. Apply this law whenever something is speeding up or slowing down. When faced with an application question, ask yourself if the object in question is speeding up or slowing down. If so, then $a = F/m$ applies.

2. Sometimes an object stops for an instant when it changes direction while accelerating, such as when a ball is thrown straight up. The important point is that the ball is slowing before it stops and then speeding up after it stops. Thus, the ball is accelerating the entire time and Newton's second law applies. Only if an object remains at rest for some period of time does the first law apply.

Q: A woman accelerates her car to get on the freeway. How do Newton's laws describe this situation?

A: The car is accelerating according to Newton's second law, $a = F/m$. The force provided by the engine and the mass of the car determine how rapidly the car accelerates.

Q: A boy is jumping up and down on a trampoline. He flies up in the air and is at the very top of his flight at the instant before he begins falling back down. How do Newton's laws apply to this moment?

A: Anything close to earth is accelerated by gravity according to Newton's second law, $a = F/m$. The gravitational attraction of the earth produces a net force on the boy which accelerates him and causes him to fall.

Third Law

1. This law is often misapplied. To apply it correctly requires a phrase such as "A pushes on B and B pushes equally and oppositely on A." You must always refer to the *same two* things, each pushing equally and oppositely on the other.
2. Don't refer to this law just because there is a force mentioned in a question. Instead, look for language in the situation that sounds like two specific objects, each pushing equally and oppositely on the other.
3. Remember that "action" and "reaction" do not refer to events or processes. They refer to forces, that is, *pushes* or *pulls*. Plain and simple.

Q: Carl gets angry and slams his fist into the car windshield, breaking the windshield. Unfortunately, upon striking the windshield, Carl's knuckles are broken, too. Use Newton's laws to describe what happens.

A: Newton's third law states that for every action force there is an equal and opposite reaction force. When Carl's fist hits the windshield hard enough to break it, the windshield hits Carl's fist just as hard. Both the windshield and Carl's fist are broken by these equal and opposite forces.

Q: Jimmy jumps to the left off the stern of his canoe into the lake. Jimmy's leap causes the canoe to shoot off to the right. Describe this situation using Newton's laws.

A: Newton's third law states that for every action force there is an equal and opposite reaction force. Jimmy pushes the canoe to the right and the canoe pushes Jimmy to the left.

bullet pushes the gun equally and oppositely (reaction force). The force of the gun on the bullet is very high, and causes the bullet to accelerate out of the gun barrel. The force of the bullet on the gun is equal, and in the opposite direction. This high kickback force causes the gun to hit Susie's shoulder hard, indicating a second action-reaction pair: the gun hits Susie's shoulder and Susie's shoulder hits the gun.

▲

3.2.6 How a Rocket Works

Rockets are a great way to see Newton's third law in operation. In outer space, there is no friction and no air, so normal methods of accelerating or stopping a vehicle on earth cannot be used. On earth, cars accelerate by using the friction between the tires and the road to speed the car up or to slow it down. Propeller-driven airplanes and boats push against the air or water. These vehicles all use Newton's third law. The tires push against the road, the road pushes against the tires in the opposite direction. The boat propeller pushes against the water, and the water pushes against the propeller, equally and in the opposite direction.

But in space, there is nothing there to push against. No road, no water, no air. Rockets speed up or slow down by using Newton's third law of motion just like every other vehicle does, but a rocket has to do it without anything to push against. The only way to do this is to *push something out of the rocket*.

A rocket engine works by throwing the mass of its own burnt fuel (the products of combustion) out the rocket engine with a massive amount of force. In the parlance of rocketry, this force is called *thrust*. The combustion of the fuel is really only needed to get the fuel to fly out of the rocket with a great deal of force, producing a large thrust. In third law terms, we say, "The rocket engine pushes the combusted fuel gases." And the rocket pushes those gases very *hard*! This is the action force in the third law. The reaction force is, "The gases push the rocket engine." This push is *just as hard*. Since the rocket engine is, of course, connected to the rest of the space ship, the force of the fuel pushing on the rocket engine accelerates the entire space ship according to the second law. The action-reaction pair is illustrated in Figure 3.4.

To slow down, the space ship again has nothing to push against and so again must use a rocket engine. However, in this case the ship

Reaction force: Combustion gases push up on the rocket

Action force: Rocket pushes down on the combustion gases

Figure 3.4. Action-reaction pairs associated with the solid-rocket boosters of the Space Shuttle during liftoff.

uses *retro-rockets*, which are small rocket engines that point toward the front of the space ship. When the "retro-rocket pushes the gases" from the front of the space ship, the "gases push the retro-rocket" from the front, against the motion of the ship, thus slowing the ship down. As we saw in the example computations earlier, the retro-rockets cause a negative acceleration, and a corresponding negative force. That is, the force is on the front of the ship, against its motion, slowing it down.

We will revisit Newton's third law of motion when we address the topic of momentum in Chapter 5. But while we are here talking about rockets, this is a good place to mention that Newton's second and third laws of motion, taken together, are essentially equivalent to the *law of conservation of momentum* applied to the case of objects accelerating in one dimension. If you understand this statement, you will understand a lot about objects in motion. Stay tuned for more in Chapter 5!

Do You Know ...   *Does gravity cause elliptical orbits?*

Isaac Newton was 44 when his *Principia Mathematica* was first published. In it, he proved that if the planets are attracted to the sun by a force inversely proportional to the square of the distance, elliptical orbits result. Johannes Kepler, of course, was the first to describe the orbits as elliptical in his first law of planetary motion (1609), and Kepler also first suspected that the planets were attracted to the sun by a force inversely proportional to the square of the distance. Newton not only figured out the details of the law of universal gravitation, but derived a mathematical proof showing that the force of gravity leads to elliptical orbits.

One of the most brilliant physicists and teachers of the twentieth century was Richard Feynman, who won the Nobel Prize in Physics in 1965. Feynman was curious about Newton's proof and tried to follow it but finally admitted that he could not figure it out. So Feynman developed his own proof, based on the geometry of the ellipse and Newton's first two laws of motion. He delivered his proof "just for the fun of it" to one of his undergraduate physics classes at Caltech in 1964. (If you love physics and math, you might enjoy reading Feynman's lecture some day. It is presented, along with a detailed history of this subject, in the book *Feynman's Lost Lecture* by David Goodstein and Judith Goodstein.)

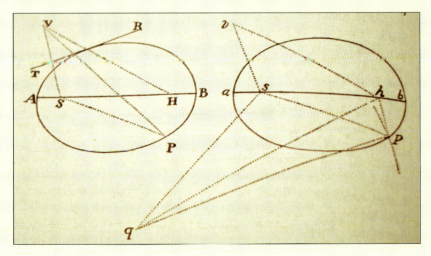

Chapter 3 Exercises

Newton's Second Law Practice Problems

Convert everything to MKS units before you start.

1. Consider a car with a mass of 1,880 kg. How much force does it take to give this car an acceleration of 1.50 m/s²?

2. A softball has a mass of 188.4 g. Determine its weight.

3. A certain baseball player can hit the ball with a force of 250.0 N. If the ball has a mass of 144,000 mg, what is its acceleration?

4. A rocket has an acceleration of 2.3 m/s² at launch. If the engines are putting out 230,000 N of thrust (force), what is the rocket's mass at launch?

5. A car accelerates at a rate of 0.0022 mi/hr². If it has a mass of 2.2 Mg, what is the force produced by the engine? (Be careful with the unit prefixes here.)

6. A woman weighs 125.1 lb. Determine her mass. (When dealing with forces in pounds, always convert them to newtons first, and then use the weight equation to get the mass.)

7. The same woman (same mass) weighs 20.9 lb on the moon. Determine the acceleration of gravity on the moon, g_m.

8. A gun accelerates a bullet from rest to 125.0 m/s in 22.00 ms. The exploding gunpowder provides a force of 142.0 N. What is the mass of the bullet? State your answer in grams.

9. How much force is required to accelerate a 4.5 kg shot-put from rest to 8.00 mi/hr in 500 ms?

10. A space ship is cruising at 2,500.0 km/hr with the engines off. The engines fire for 8.000 s with a thrust of 45,450 N, bringing the ship up to its new speed of 2,750 km/hr. What is the mass of the ship?

11. A child's toy car has a mass of 166 g and uses a force of 0.0450 N to accelerate the car. If the car begins from rest, how fast is it going after 2.1 s?

12. A proton has a mass of 1.673×10^{-18} μg. It is accelerated from rest to 0.05000% of the speed of light in 455 ns. Determine the force required to make the proton do this. Express your answer in GN.

13. A large ship has a mass of 6.548 Gg and is traveling at 8.35 mi/hr. The ship reverses its propellers and comes to a stop in 0.288 min. Determine the force the propellers must produce to do this.

14. A mouse is creeping along at 3.5 cm/s when a cat attacks it. The mouse accelerates from its initial speed to a new speed of 18.5 cm/s in 220 ms. What is the acceleration of the mouse, in m/s²?

15. A missile with a mass of 45,500 kg is accelerated from rest to 55 m/s in 6.4 seconds.
    a. What is the acceleration of the missile?
    b. What is the thrust of the missile's rockets while it is being launched?

16. How large a horizontal force must be exerted on an 8.5 g bullet to give it an acceleration of 18,500 m/s²?

Answers

1. 2,820 N
2. 1.85 N
3. 1,740 m/s²
4. 1.0×10^5 kg
5. 6.0×10^{-4} N
6. 56.8 kg
7. 1.64 m/s²
8. 24.99 g
9. 30 N
10. 5,230 kg
11. 0.57 m/s
12. 5.52×10^{-25} GN
13. 1.41×10^6 N
14. 0.68 m/s²
15. a) 8.6 m/s²; b) 3.9×10^5 N
16. 160 N

Newton's Laws of Motion Study Questions

1. A space ship is deep in outer space firing its rockets and accelerating. Then the rockets are switched off. What does the space ship do immediately after the rockets cease firing?

2. A space ship is near a planet and heading toward it when it switches off its rockets. What happens? According to Isaac Newton why does this happen? According to Albert Einstein why does it happen?

3. According to Newton's second law of motion, what is necessary to change the velocity of a moving object?

4. Distinguish between matter and mass.

5. Distinguish between mass and inertia.

6. Give four new examples applying the law of Inertia (Newton's first law) to actual situations.

7. Why is the first law called the law of inertia? What is inertia and how does it relate to this law?

8. Answer each of the following questions by arguing from Newton's laws of motion.
   a. If you start a pendulum swinging it eventually comes to a stop. Why? (Be careful with your answer. Sure, gravity is acting on the pendulum to make it swing. But is gravity what makes the pendulum slow down and stop?)
   b. Reconsidering the previous question, suggest a way to set up a pendulum so that it would almost never slow down, and would swing for a very long time. (Sorry, perfect perpetual motion is not possible, but think of a design that

comes as close as possible.)

   c.   You are an astronaut and are traveling in a space ship from earth to Mars. Once you get up to speed and leave earth's gravitational field you have a long journey ahead (about a year, actually). During this part of the journey, do your rocket engines need to be on full, medium, low, or off? Why?

   d.   As an astronaut, when do you ever need to have your engines on? Do you even need engines?

   e.   If Newton's first law is correct, why do we have to keep the engine on while we are driving down the highway at a constant speed?

9. Give two examples applying Newton's third law to actual situations in which objects remain at rest. In each case identify the action force, the reaction force, and the results of each of these forces.

10. Give two examples applying Newton's third law to actual situations in which the two objects do not remain at rest, but one or both of them accelerate instead. In each case identify the action force, the reaction force, and the results of each of these forces.

11. You and another space traveler are stranded in space at a dead stand still because your space ship had a fuel leak while at rest and you lost all your fuel. Your companion notices that your ship is pointed exactly toward the space station where you need to go, and suggests that you quickly heave your massive copy of *Les Misérables* (which you haven't finished reading) out the back window of the space ship. Outraged, you suggest that your companion take a flying leap out the back window himself. Would either of these actions really help? If so, which helps more? Answer these questions by arguing from Newton's laws of motion.

12. For the Soul of Motion Experiment, identify the explanatory variable and the response variable. Also identify two realistic possibilities for lurking variables.

Do You Know ...   *Where is Isaac Newton's tomb?*

Sir Isaac Newton's beautiful tomb is in Westminster Abbey, in London. Be sure to go see it if you go to England. Visitors are not normally allowed to take photographs inside the Abbey. But I asked for special permission to take the photo below. When the Abbey stewards learned that I wanted the photo for a science textbook, they graciously allowed me a moment to take this picture!

CHAPTER 4
Energy

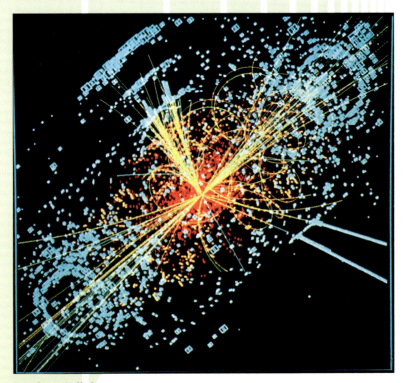

Large Hadron Collider

In the Large Hadron Collider (LHC) at CERN in Switzerland (see box on page 12), protons are accelerated and collided at extremely high energies. The purpose of these collisions is to help scientists discover more about the fundamental structure of matter. Theory predicts the existence of a particle called the Higgs Boson. The image above is a computer simulation of a Higgs detection event inside the CMS detector at the LHC. The CMS website states, "The lines represent the possible paths of particles produced by the proton-proton collision in the detector while the energy these particles deposit is shown in blue."

As you know, the speed of light is 300,000,000 m/s. To generate the energy needed to observe the Higgs Boson, protons are accelerated to a speed that is only 3 m/s slower than the speed of light! At this speed the protons only require 90 µs (0.000090 s) to travel 17 miles around the main underground tunnel of the LHC. This huge kinetic energy is way beyond the energy produced by any other particle accelerator yet constructed.

OBJECTIVES

Memorize and learn how to use these equations:

$$E_G = mgh \qquad E_K = \tfrac{1}{2}mv^2 \qquad v = \sqrt{\frac{2E_K}{m}} \qquad W = Fd$$

After studying this chapter and completing the exercises, students will be able to do each of the following tasks, using supporting terms and principles as necessary:

1. State the law of conservation of energy.
2. Describe how energy can be changed from one form to another, including:
   a. different forms of mechanical energy (kinetic, gravitational potential, elastic potential)
   b. chemical potential energy
   c. electrical energy
   d. elastic potential energy
   e. thermal energy
   f. electromagnetic radiation
   g. nuclear energy
   h. acoustic energy
3. Briefly define each of the types of energy listed above.
4. Describe two processes by which energy can be transferred from one object to another (work and heat), and the conditions that must be present for the energy transfer to occur.
5. Describe in detail how energy from the sun is converted through various forms to end up as energy in our bodies, as energy used to run appliances in our homes, or as energy used to power machines in industry.
6. Explain why the efficiency of any energy conversion process is less than 100%.
7. Calculate kinetic energy, gravitational potential energy, work, heights, velocities, and masses from given information using correct units of measure.
8. Define friction.
9. Using the pendulum as a case in point, explain the behavior of ideal and actual systems in terms of mechanical energy.
10. Explain how friction affects the total energy present in a mechanical system.

4.1   What is Energy?

4.1.1 Defining Energy

Defining energy is tricky. Dictionaries usually say, "the capacity to do mechanical work," which is not particularly helpful. Actually, there is no definition for energy that gets at what it actually *is*, so I will not try to define it. We are just going to accept that energy exists in the universe, it was put there by God when he made the universe, and it exists in many different forms. It is fairly obvious that a bullet traveling at 2,000 ft/s has more energy than a bullet at rest. This is why the high speed bullet can kill but the bullet at rest cannot. This study is mainly about tracking energy as it changes from one form to another, and calculating the quantities of three particular forms of energy.

4.1.2 The Law of Conservation of Energy

The *law of conservation of energy* is as follows:

> Energy can be neither created nor destroyed, only changed in form.

Energy can be in many different forms in different types of substances, such as in the molecules of gasoline, in the waves of a beam of light, in heat radiating through space, in moving objects, in compressed springs, or in objects lifted vertically on earth. As different physical processes occur—such as digesting food, throwing a ball, operating a machine, heating due to friction, or accelerating a race car—energy in one form is being converted into some other form. Energy might be in one form in one place, such as in the chemical potential energy in the muscles of your arm, and be converted through a process like throwing a ball to become energy in another form in another place, like kinetic energy in the ball.

4.1.3 Mass-Energy Equivalence

In 1905, Albert Einstein published his now-famous equation, $E = mc^2$. The E and m in this equation stand for energy and mass; c stands for the speed of light. With this equation, Einstein theorized that mass and energy are really just different forms of the same thing. That is, all mass has associated with it an equivalent amount of energy (given by $E = mc^2$), and vice versa. This theory of mass-energy equivalence is now considered to be a fundamental property of the universe.

The reason I mention mass-energy equivalence here is that since mass is a form of energy, matter must be taken into consideration for a completely accurate statement of the law of conservation of energy. In nuclear reactions, such as take place in the sun (fusion) or in nuclear power plants (fission), quantities of matter are converted completely into energy. Einstein's equation $E = mc^2$ also gives the amount of energy that appears when a quantity of matter in converted to energy in one of these nuclear processes. Thus, to be completely accurate, we need to state that the law of conservation of energy includes all mass as well, as one of the forms energy can take. Let's restate the conservation law with this in mind: "*mass-energy can be neither created nor destroyed, only changed in form.*"

Most of the problems we encounter in physics and chemistry don't involve nuclear reactions (thankfully). This means that for most purposes, we can consider the common forms of energy listed below without worrying about the complicated issue of mass-energy equivalence.

4.2   Energy Transformations

4.2.1 Forms of Energy

Here are some common forms energy can take:

Gravitational Potential Energy   This is the energy an object possesses because it has been lifted up in a gravitational field. If such an object is released and allowed to fall, the gravitational potential energy converts into kinetic energy. The term *potential* in the name of this form of energy indicates that the energy is stored and converts into another form of energy when released. There are other forms of energy listed below that use this term for the same reason.

Kinetic Energy This is the energy an object possesses because it is in motion. The faster an object is moving, the more kinetic energy the object has.

Electromagnetic Radiation This is the energy in electromagnetic waves traveling through space, or through media such as air or glass. This type of energy includes all forms of light, as well as radio waves, microwaves, and a number of other kinds of radiation. We will study electromagnetic waves in some detail in a later chapter.

Chemical Potential Energy This energy is in the chemical bonds of molecules. In the case of substances that burn, the chemical potential energy in the molecules is released in large quantities as heat and light when the substance is burned, making these substances useful as fuel.

Electrical Energy This is energy flowing in electrical conductors, such as from a power station to your house to power your appliances.

Thermal Energy This is the energy a substance possesses due to being heated. We will examine thermal energy more closely in the next chapter.

Elastic Potential Energy This is the energy contained in any object that has been stretched (like a bungee cord or a hunter's bow) or compressed (like a compressed spring).

Nuclear Energy This is energy released from the nuclear processes of fission (when the nuclei of atoms are split apart) or fusion (when atomic nuclei are fused together). As I mentioned previously, these processes convert mass into energy.

Acoustic (Sound) Energy This is the energy carried in sound waves, such as from a person's voice, the speakers in a sound system, or the noise of an explosion. Since sound waves are carried by moving air molecules, this is really a special form of kinetic energy.

4.2.2 Energy Transfer

Two more important energy-related terms are those associated with the process of energy being transferred from one place, substance, or object to another. These two terms are:

Work Work is a mechanical process by which energy is transferred from one object to another. Objects do not possess work like they do other forms of energy. Instead, one object "does work" on another object by applying a force to it and moving it a certain distance. When one object does work on another, energy is transferred from the first object to the second. We will study work in more detail later in the chapter.

Heat The term *heat* is used as a general description of energy in transit, flowing by various means from a hot substance to a cooler substance when a difference in temperature is present. We will study heat and the three ways it flows in more detail in Chapter 7. As with work, substances do not possess heat. What substances do possess is kinetic energy in their moving atoms, and we refer to this energy as the *internal energy* of the substance.

Let's look at a common example of energy changing from one form into others. We all know what happens when a person lights a firecracker. (It explodes!) What forms of energy

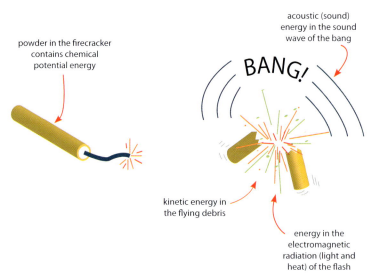

Figure 4.1. Chemical potential energy in a firecracker is converted into other forms of energy when the firecracker explodes.

are present during the explosion, and where did all this energy come from? As shown in Figure 4.1, the energy released in the explosion is the chemical potential energy in the molecular bonds of the chemicals inside the firecracker. When these chemicals burn, they release a lot of energy. And as you already know, an exploding firecracker gives off a flash of light and heat (both are forms of electromagnetic radiation), a loud bang, and the fragments of the firecracker are blown all over the place. Thus, the chemical potential energy in the powder inside the firecracker is converted into several different kinds of energy during the explosion.

Now consider how the law of conservation of energy applies to this explosion. All the energy present in the chemicals before the explosion is still present in various forms after the firecracker explodes. This is what "conservation" of energy means. We could represent the conservation of energy in a sort of equation like this:

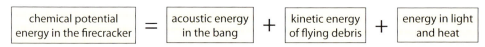

We will not do any calculations this complex in this course. But later in this chapter, you will learn how to use the principle of conservation of energy to solve problems involving three of the forms of energy we have seen so far.

4.2.3 The "Energy Trail"

Much of the energy we depend on here on earth comes to us from the sun. As we track the forms this energy takes in its journey from the sun to, say, the energy in our bodies, we might call this the "energy trail." (This way we can have fun yelling *Yee-Haw!* while we are studying this. Ask your teacher for a demonstration.) We will follow the energy trail beginning with the sun, through different processes of conversion, and arriving at different places where energy is commonly used.

Do You Know ...   *What is dark energy?*

When Einstein first developed his equations for the general theory of relativity (published in 1915), the equations implied that the universe must be either expanding or contracting. At the time, the prevailing view was that the universe was doing neither, so Einstein put a fudge factor in the equations to keep the universe static.

Just a few years later, astronomer Edwin Hubble made observations that enabled Belgian Georges Lemaître, a physicist and Catholic priest, to conclude that the universe was expanding, just like Einstein's original equations implied! So Einstein took the fudge factor out of the equations and said that putting it in was "the biggest blunder of his life."

In the 1990s, astronomers discovered that the expansion of the universe is *speeding up*—the expansion of the universe is *accelerating*. This seems impossible, because the gravitational attraction of the galaxies pulling on each other should be slowing the expansion rate down. In a classic illustration of how the Cycle of Scientific Enterprise works, no known theory was able to account for the cause of the acceleration, so scientists had to get busy theorizing about this mystery.

At present, the most accepted hypothesis for the cause of the acceleration is the presence of an unknown form of energy called dark energy that permeates all of space. Calculations indicate that on the basis of mass-energy equivalence, 68% of the energy in the universe is dark energy, 27% of the energy is in the form of dark matter (see the box on page 8), and only 5% of the energy is in the form of ordinary matter.

The sun's energy is produced by fusion reactions as the nuclei of hydrogen atoms "fuse" or stick together to form helium. This is so hard to do that we have not yet succeeded in doing it here on earth in a controlled way. However, we have succeeded in doing it in an uncontrolled way. Fusion is the same nuclear reaction as the main reaction in a thermonuclear bomb. A thermonuclear explosion is an uncontrolled nuclear fusion reaction.

When referring to this energy being produced in the sun, we will simply call it nuclear energy. The energy leaves the sun as electromagnetic radiation, a different form of energy, and travels through space to us. When this energy lands on earth most of it warms the ground, oceans, and atmosphere. This is very important for stabilizing the earth's climate and making earth habitable, but unless we collect the energy in a solar collector of some kind we are not able to use this energy directly.

However, some of the electromagnetic radiation streaming from the sun is captured by plants and converted into chemical potential energy in the molecules in the plants through the process of photosynthesis. These plants eventually become the foods we eat or the fuels we burn. Current theory holds that in ancient eras in the earth's history, many vast forests were buried and the plant matter was converted underground into what we now call "fossil fuels" (petroleum, coal, and natural gas). Some fuels come from living plants too, such as firewood from trees and automotive alcohol (ethanol) from corn. The energy in the molecules of these fuels is chemical potential energy that is converted into heat energy when the fuels are burned.

Your task is to describe the energy conversions each step of the way from the sun all the way to your breakfast cereal or your computer. Tables 4.1 through 4.4 illustrate a few examples of following the energy from the sun to different places it can end up here on earth. When asked to outline one of these pathways in the "energy trail," always list two things for each step of the way: (1) Where the energy is, and (2) what form the energy is in.

Where is the energy?	The Sun	Electro-magnetic waves in space	Plants on earth	Breakfast cereal	Muscles in the human body	Stretched bow	Flying arrow
What form is the energy in?	Nuclear energy	Electro-magnetic radiation	Chemical potential energy	Chemical potential energy	Chemical potential energy	Elastic potential energy	Kinetic energy

Table 4.1. Energy transformations from the sun to a flying arrow, assuming the archer was on a vegetarian diet.

Where is the energy?	The Sun	Electro-magnetic waves in space	Plants on earth	Chicken feed	Muscles in the bodies of chickens	Muscles in the human body	Moving kid on skate-board
What form is the energy in?	Nuclear energy	Electro-magnetic radiation	Chemical potential energy	Chemical potential energy	Chemical potential energy	Chemical potential energy	Kinetic energy

Table 4.2. Energy transformations from the sun to a kid on a skateboard, assuming the kid was eating chicken.

Where is the energy?	The Sun	Electro-magnetic waves in space	Plants on earth	Fossil fuel (crude oil, coal, natu-ral gas)	Spin-ning gas turbine generator at power station	Wires from the power station to your house	Heat from the coils in the toaster
What form is the energy in?	Nuclear energy	Electro-magnetic radiation	Chemical potential energy	Chemical potential energy	Kinetic energy	Electrical energy	Electro-magnetic radiation

Table 4.3. Energy transformations from the sun to the heat from a toaster in your house.

Where is the energy?	The Sun	Electro-magnetic waves in space	Plants on earth	Fossil fuel (coal)	Heat from burning coal	Steam in the boiler	Moving train
What form is the energy in?	Nuclear energy	Electro-magnetic radiation	Chemical potential energy	Chemical potential energy	Heat	Thermal energy	Kinetic energy

Table 4.4. Energy transformations from the sun to a moving steam locomotive.

4.2.4 The Effect of Friction on a Mechanical System

You probably already have a feel for what people mean by the term *friction*. Friction is a force present any time one object or material comes in contact with another object or material. The effect of friction is to oppose any relative motion between the two objects in contact. The cause of friction is rather complicated, but down at the atomic level friction has to do with the electrical attractions and repulsions between the charged particles in the atoms of the objects.

Friction makes it harder for one object to slide on top of another, which is good if you are talking about the friction between the tires of a car and the pavement. If there were no friction, cars could not start or stop or steer. (You may experience this physically if you are ever in the undesirable position of trying to drive a car on ice.) Friction is also very nice to have around any time one is attempting to grab something or clamp something. Without friction, things would just slip through our fingers. But friction is undesirable when the goal is to design the parts of a machine so they will slide smoothly against one another without wear or damage to the machine. And, of course, there is friction when an object moves through the air. This friction is usually called air resistance or *drag*.

In this course we are not considering friction in the calculations we do. However, in all real mechanical systems friction plays a significant role. Friction is caused when parts of the system rub against each other or when parts of the system move through a fluid such as air or water. Just as when you rub your hands together on a cold day, friction always results in heating. When the parts of a mechanical system such as a machine get warm from friction, heat flows from the warm parts into the cooler surrounding environment. (We will look more at how this happens in Chapter 7.) This heat energy flowing out of the system is energy that used to be in the system.

When heat energy flows out of a system due to friction, the law of conservation of energy still applies: no energy is created or destroyed. However, the energy remaining in the system is reduced by the amount of energy that flows out of the system due to heating from friction. A scientist or engineer may refer to energy "lost" due to friction. This does not mean the energy is destroyed or ceases to exist, only that it flows out of the system as heat and is no longer available as energy in the system. The net effect, of course, is that things slow down as energy gradually leaves the system as heat due to friction.

4.2.5 Energy "Losses" and Efficiency

For all the different forms of energy we have considered, there are many different kinds of processes that might be involved in converting energy from one form to another. Combustion is a process that converts chemical potential energy into heat. Photosynthesis converts electromagnetic energy from the sun into chemical potential energy in the cells of plants. The Industrial Revolution began when humans began learning how to design machines and systems to convert energy from various forms found in nature into forms that can be harnessed to do useful work for us.

Let's consider some process like this, such as an engine in a car converting the chemical potential energy in the gasoline into kinetic energy in the moving car. One of the facts of life on earth is that it is theoretically impossible for a conversion process to capture all the energy involved and convert it to a form that can do useful work. Whether we want it to or not, some of the energy always converts to heat, which radiates out into the environment. The laws of thermodynamics state that this is always the case.

This situation is represented in Figure 4.2. It is common to speak of the energy converted into heat as "lost." Keeping the law of conservation of energy in mind, it should be clear that what we mean by this is not that the energy ceases to exist, only that the energy escapes into the environment where it is no longer available to us in a usable form. It is lost from the system, not from the universe.

The *efficiency* of an energy conversion process is the ratio of the usable energy coming out of the process to the energy that goes into the process:

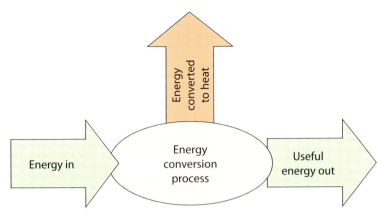

Figure 4.2. In any energy conversion process, some energy is converted to heat that is not available to do useful work.

$$Efficiency = \frac{E_{out \; (useable)}}{E_{in}}$$

Since some energy is always lost as heat, the efficiencies of our machines are always less than 100%. As physical examples, the efficiency of typical automobile engines is only around 15%, which means that 85% of the energy in the fuel is not used to propel the car. (That's a lot of lost energy.) Solar cells convert electromagnetic radiation from the sun into electricity. At present, the highest efficiency realized with these technologies is around 25%. The overall efficiency of the new electric cars is around 20–25%. This figure may seem low, but there are a lot of losses in generating the electrical power at a power station and transporting the power to the homes where people charge up their cars.

4.3   Calculations With Energy

4.3.1 Gravitational Potential Energy and Kinetic Energy

Two important forms energy can take in mechanical systems are gravitational potential energy, E_G, and kinetic energy, E_K. The gravitational potential energy an object possesses depends on how high up it is and the kinetic energy of an object depends on how fast it is moving. Both also depend on the object's mass. Gravitational potential energy is calculated as

$$E_G = mgh$$

where E_G is energy in joules (J), m is the mass (kg), $g = 9.80$ m/s^2, and h is the height (m).

Notice that if you know how much gravitational potential energy an object has and its mass, you can solve this equation for h to find out how high the object is above the ground. Simply divide both sides of the equation by mg and you have

$$h = \frac{E_G}{mg}$$

Notice also that the gravitational potential energy of an object is directly proportional to its height. If the height of an object increases by 50%, the gravitational potential energy of the object also increases by 50%. When calculating gravitational potential energy, the energy you calculate always depends on the location you choose to use as your zero reference for the height. This zero reference might be sea level, or the ground, or the floor of your classroom, or a table top. It doesn't matter, because the E_G an object has is always relative to where $h = 0$ is. Usually, the most logical and convenient location for $h = 0$ is clear from the context.

The equation for gravitational potential energy gives us another example of a derived MKS unit, the joule. Multiplying the units together for the terms on the right side of the E_G equation, we can see that a joule is made up of primary units as follows:

$$1 \text{ J } = 1 \text{ kg} \cdot \frac{m}{s^2} \cdot m = 1 \frac{kg \cdot m^2}{s^2}$$

You might compare this to the units listed at the bottom of page 22.

▼ Example 4.1

A golf ball has a mass of 45.9 g. While climbing a tree near a golf course, little Janie finds a golf ball stuck in a branch 9.5 ft above the ground. What is the gravitational potential energy of the golf ball at that height?

Start by writing the givens and doing the unit conversions, keeping one extra significant digit in your intermediate calculations.

$$m = 45.9 \text{ g} \cdot \frac{1 \text{ kg}}{1000 \text{ g}} = 0.0459 \text{ kg}$$

$$h = 9.5 \text{ ft} \cdot \frac{0.3048 \text{ m}}{ft} = 2.90 \text{ m}$$

$$E_G = ?$$

Now write the equation and complete the problem.

$$E_G = mgh = 0.0459 \text{ kg} \cdot 9.80 \frac{m}{s^2} \cdot 2.90 \text{ m} = 1.30 \text{ J}$$

These calculations all contain one extra significant digit. The given height only has two significant digits, so now we round our final result to two digits.

$$E_G = 1.3 \text{ J}$$

▲

▼ Example 4.2

An ant carries a grain of sugar up the side of a building to its nest on the roof. The mass of the grain of sugar is 0.0356 μg. After it has been carried to the roof, the E_G in the grain of sugar is 1.91 nJ. How high is the ant nest?

Write the givens and do the unit conversions.

$$m = 0.0356 \ \mu g \cdot \frac{1 \ g}{10^6 \ \mu g} \cdot \frac{1 \ kg}{1000 \ g} = 3.56 \times 10^{-11} \ kg$$

$$E_G = 1.91 \ nJ \cdot \frac{1 \ J}{10^9 \ nJ} = 1.91 \times 10^{-9} \ J$$

$$h = ?$$

Now write the equation, solve for h, and compute the result.

$$E_G = mgh$$

$$h = \frac{E_G}{mg} = \frac{1.91 \times 10^{-9} \ J}{3.56 \times 10^{-11} \ kg \cdot 9.80 \ \frac{m}{s^2}} = 5.47 \ m$$

Every value in this computation has three significant digits, as does this result, so the problem is complete.

Now we will look at another important form of energy, kinetic energy. Kinetic energy is a very important concept in physics because relates to many other concepts. Kinetic energy is calculated as

$$E_K = \tfrac{1}{2} mv^2$$

where E_K is the kinetic energy in joules (J), m is the mass (kg), and v is the velocity (m/s). The units for kinetic energy are joules, just as with all other forms of energy. Kinetic energy is proportional to the mass of an object and to the square of the object's velocity.

Notice that if you know how much kinetic energy an object has and its mass, you can solve this equation for v to find out how fast the object is moving. Since the algebra to do this may be unfamiliar to students in this course, you may want to just go ahead and memorize the equation for velocity as a function of kinetic energy. This equation is

$$v = \sqrt{\frac{2E_K}{m}}$$

▼ Example 4.3

An electron with a mass of 9.11×10^{-28} g is traveling at 1.066% of the speed of light. Determine the amount of kinetic energy the electron has and state your result in nJ.

Start by writing the givens and doing the unit conversions. To obtain the electron's velocity, we must multiply the speed of light (from Appendix A) by 0.01066.

$$m = 9.11 \times 10^{-28} \text{ g} \cdot \frac{1 \text{ kg}}{1000 \text{ g}} = 9.11 \times 10^{-31} \text{ kg}$$

$$v = 0.01066 \cdot 3.00 \times 10^8 \; \frac{\text{m}}{\text{s}} = 3.198 \times 10^6 \; \frac{\text{m}}{\text{s}}$$

$$E_K = ?$$

Now compute the kinetic energy.

$$E_K = \tfrac{1}{2} mv^2 = 0.5 \cdot 9.11 \times 10^{-31} \text{ kg} \cdot \left(3.198 \times 10^6 \; \frac{\text{m}}{\text{s}} \right)^2 = 4.658 \times 10^{-18} \text{ J}$$

The problem statement requires the result to be in units of nanojoules (nJ), so perform this conversion.

$$4.658 \times 10^{-18} \text{ J} \cdot \frac{10^9 \text{ nJ}}{\text{J}} = 4.658 \times 10^{-9} \text{ nJ}$$

Both the mass and the speed of light values have three significant digits, so rounding this result to three significant digits gives

$$E_K = 4.66 \times 10^{-9} \text{ nJ}$$

 Example 4.4

A kid fires a plastic dart from a dart gun. The mass of the dart is 21.15 g and its kinetic energy is 0.3688 J when it comes out of the dart gun. Determine the velocity of the dart.

Write the givens and do the unit conversions.

$$m = 21.15 \text{ g} \cdot \frac{1 \text{ kg}}{1000 \text{ g}} = 0.02115 \text{ kg}$$

$$E_K = 0.3688 \text{ J}$$

$$v = ?$$

Now complete the problem using the memorized velocity equation.

$$v = \sqrt{\frac{2E_K}{m}} = \sqrt{\frac{2 \cdot 0.3688 \text{ J}}{0.02115 \text{ kg}}} = 5.905 \; \frac{\text{m}}{\text{s}}$$

Both the mass and the kinetic energy values have four significant digits, so this result is rounded to four significant digits.

▲

4.3.2 Work

The way an object acquires kinetic energy or gravitational potential energy is that another object or person or machine does *work* on it. Work is the way mechanical energy is transferred from one machine or object to another. Work is a form of energy, but objects don't possess work. Work is the process by which energy is transferred from one mechanical system to another. Work is defined as the energy it takes to push an object with a certain (constant) force over a certain distance. Work is calculated as

$$W = Fd$$

where W is the work done on the object in joules (J), F is the force on the object (N), and d is the distance the object moves (m).

Notice from this equation that since work is energy, the units here come out to be

$$1\,J = 1\,N \cdot m$$

Let's take a moment to convince ourselves that the units here are the same as the units described above right after the E_G equation. The work equation says that joules are equal to newtons times meters. A newton is a force, and we know from Newton's second law of motion that force equals mass times acceleration, or $F = ma$. If we multiply all these units together we have

$$1\,J = 1\,N \cdot m = 1\,kg \cdot \frac{m}{s^2} \cdot m = 1\,\frac{kg \cdot m^2}{s^2}$$

These units are indeed the same as the units we worked out for gravitational potential energy a few pages back.

The concept of work is the basic principle governing how energy is transferred from one device to another in a *mechanical system*. For example, as depicted in Figure 4.3, if an electric motor is used to lift a piece of equipment, the motor must reel in a certain length, L, of steel cable, and it must pull on the cable with a certain force, F, while doing so. The pulling force times the length of cable is the amount of work done by the motor. And where does this work energy supplied by the motor go? Assuming 100% efficiency in the lifting motor and cables (and electric motors have very high efficiencies, so this is not a bad approximation), the energy all goes into the gravitational potential energy acquired by the piece of equipment being lifted. In actuality, since the efficiency of all systems is less than 100%, some of the energy leaves the system as heat. In the end, the gravitational potential energy of the lifted piece of equipment does not quite represent all of the energy the electric motor has to supply.

We see here that work and conservation of energy are very closely related. As another example, if a man pushes a kid on a bicycle over a short distance to get the

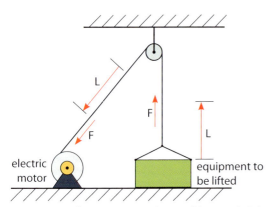

Figure 4.3. An electric motor raising an object to a height L by means of force F pulling on the cable.

kid going, the man delivers energy to the kid on the bicycle equal to the pushing force times the distance pushed. Ignoring friction for now, that work energy from the man is now in the kinetic energy of the kid on the bicycle.

There are two more important details to note about work. First, the equation for work, $W = Fd$, requires that the force applied to an object and the distance the object travels must lie in the same direction. As depicted in Figure 4.4, if a person lifts a bucket of water, then work is done on the bucket of water. The force is applied vertically and the bucket moves vertically, so the work done to lift a bucket of water is the force required to lift it, its weight, times the distance it is lifted. But a person carrying a bucket of water down the road is *not doing any work on the bucket.* This is because the force on the bucket to hold it up is vertical, but the distance the bucket is moving is horizontal. These two forces do not point in the same direction. In fact, they are at right angles to one another and no work is done on the bucket of water.

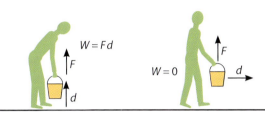

Figure 4.4. In raising the bucket (left) work is done on the bucket. In moving the bucket horizontally no work is done on the bucket.

The second detail is foreshadowed in the previous paragraph. People often say that the force required to lift a bucket is just a little larger than the weight, but this is not correct. If the upward force is at all greater than the weight, then we have a net upward force on the bucket. Recall from Chapter 3 that Newton's second law of motion says that a net force does not just raise the bucket, it *accelerates* the bucket, giving it kinetic energy. But for a bucket to move up with a *constant* speed requires no net force at all, according to the first law of motion. So after a little bump of force to get the bucket moving (which does briefly require a larger force and some energy), the bucket can be lifted to any height with a force equal to the bucket's weight. In physics problems of this kind, we normally just neglect the little bump of force necessary to get the bucket started and assume that the force required to lift the bucket is equal to the weight of the bucket.

So, a handy problem solving tip to keep in mind is this: *The force required to lift an object is equal to its weight.* Recall that you can always calculate the weight of an object from its mass as

$$F_w = mg$$

Do You Know ...   What is alpha radiation?

Nuclear radiation is emitted from radioactive substances during the process of nuclear decay. One form of radiation, called *alpha radiation*, consists of alpha particles streaming out of the radioactive substance. These incredibly small and fast moving alpha particles each consist of two protons and two neutrons. They receive their kinetic energy from radioactive atoms as matter in the atom converts into kinetic energy during the process of nuclear decay.

In Chapter 6, we encounter the story of Ernest Rutherford, who used the alpha particles from a compound called radium bromide to explore the structure of the atom. The alpha particles emitted from radium bromide are travelling at 15,000,000 m/s! This is 9,300 miles per second, a speed that is 5% of the speed of light!

▼ Example 4.5

An elevator in a skyscraper has a mass of 904.9 kg. Inside the elevator are three people whose masses are 67.8 kg, 55.9 kg, and 75.1 kg. Determine how much work the elevator motor does in lifting this elevator and the people inside it from the ground floor up to the 47th floor, 564 ft above the ground floor. Assume there is no friction and state the result in kJ.

Write the givens and do the unit conversions.

$$m = 904.9 \text{ kg} + 67.8 \text{ kg} + 59.9 \text{ kg} + 75.1 \text{ kg} = 1103.7 \text{ kg}$$

$$d = 564 \text{ ft} \cdot \frac{0.3048 \text{ m}}{\text{ft}} = 171.9 \text{ m}$$

$$W = ?$$

As I wrote just above, the force required to lift an object is equal to its weight. So next we need to compute the weight of the elevator and the people.

$$F_w = mg = 1103.7 \text{ kg} \cdot 9.80 \frac{\text{m}}{\text{s}^2} = 10{,}820 \text{ N}$$

My calculator has a lot more digits in it than this, but I see that several of the pieces of given information have three significant digits, and I only need one extra digit for intermediate calculations, so I round to four digits.

Now complete the problem.

$$W = Fd = F_w d = 10{,}820 \text{ N} \cdot 171.9 \text{ m} = 1{,}860{,}000 \text{ J}$$

This result is rounded to three significant digits. As a last step, we convert this value to kilojoules, as required by the problem statement.

$$W = 1{,}860{,}000 \text{ J} \cdot \frac{1 \text{ kJ}}{1000 \text{ J}} = 1860 \text{ kJ}$$

4.3.3 Applying Conservation of Energy

When an object is thrown or fired straight up from the ground, it leaves the ground with a certain velocity, and thus a certain amount of E_K. As it goes up, what happens to this E_K? It is converted to E_G, of course, as the object goes higher and higher and goes slower and slower. At the top of its flight, all the energy the object has at the ground in E_K has been converted into E_G. We can use the law of conservation of energy, along with the equations for E_G and E_K, to determine how high the object goes.

The same thing works in reverse. An object at a certain height has E_G. If the object is then released, as it falls the E_G is gradually and continuously converted into E_K. Just before it hits the ground, all the E_G it has at the top has been converted into E_K. We can use the law of conservation of energy, along with the equations for E_G and E_K, to find out how fast the object is going just before it strikes the ground.

In all the problems we do in this course involving conservation of energy, we are ignoring friction. In reality, friction is always present in any so-called mechanical system, such as moving objects or machines. In Section 4.2.4, we considered the effect friction has on mechanical systems, but in our computations we ignore it. Many physical systems can be approximated pretty well even if friction is ignored. In the conservation of energy experiment you will perform (the Hot Wheels Experiment), friction is low enough that the experimental velocity you measure should agree fairly well with the prediction.

Let us now look at a simple example of the conservation of energy in action. Figure 4.5 illustrates the application of conservation of energy to a person lifting a bucket and letting it drop. When a person lifts a bucket vertically, the person does work on the bucket. To compute this work, the force to lift the bucket is the weight of the bucket and the distance involved is the height it is lifted, so the work done on the bucket by the person is

$$W = F_w h$$

Since the weight, F_w, is equal to mg, this equation can be written

$$W = F_w h = mgh$$

Energy is transferred from the person (the chemical potential energy in the person's muscles) to the bucket, and the bucket now has gravitational potential energy equal to

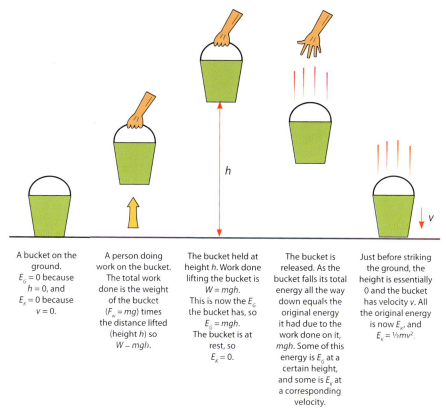

| A bucket on the ground. $E_G = 0$ because $h = 0$, and $E_K = 0$ because $v = 0$. | A person doing work on the bucket. The total work done is the weight of the bucket $(F_w = mg)$ times the distance lifted (height h) so $W - mgh$. | The bucket held at height h. Work done lifting the bucket is $W = mgh$. This is now the E_G the bucket has, so $E_G = mgh$. The bucket is at rest, so $E_K = 0$. | The bucket is released. As the bucket falls its total energy all the way down equals the original energy it had due to the work done on it, mgh. Some of this energy is E_G at a certain height, and some is E_K at a corresponding velocity. | Just before striking the ground, the height is essentially 0 and the bucket has velocity v. All the original energy is now E_K, and $E_K = \frac{1}{2}mv^2$. |

Figure 4.5. Conservation of energy applied to a lifted and falling bucket.

$$E_G = mgh$$

Right here we can see the conservation of energy at work. The work done by the person to lift the bucket is *mgh*. Where did that energy go? It went into the E_G the bucket has at the top, which is *mgh*, the same amount of energy. If the person releases the bucket, then as the bucket falls the gravitational potential energy begins to convert to kinetic energy. At any point as the bucket is falling, energy is conserved, which means that the total energy the bucket has is still the same as the energy it had at the top, but some of the energy is in kinetic energy and some of it is in gravitational potential energy. At the instant before the bucket hits the ground, there is no more gravitational potential energy because the height then is zero, so all the energy originally given to the bucket by the work done on it is in the kinetic energy of the bucket.

4.3.4 Conservation of Energy Problems

Now we look at a couple of example problems with objects falling down or flying up. In these kinds of problems, we use the basic principle of conservation of energy to find out how high an object goes or how fast an object is going just before it lands.

A helpful problem solving technique for these kinds of problems is to draw a little diagram for yourself to indicate whether the E_G is converting to E_K ($E_G \rightarrow E_K$), or vice versa ($E_K \rightarrow E_G$). This helps you keep track of what you are doing so you don't become confused. I will demonstrate this in the example problems we do below.

One more thing before we do those examples. To help you continue to remember how to calculate an object's mass from its weight, I like to design these problems by giving you the weight instead of the mass that you need. So just first do a separate little problem to obtain the object's mass. Then proceed with the energy calculations. The examples that follow make all these things clear.

▼ Example 4.6

A certain bucket of paint weighs 8.55 lb and is carried up a ladder until it is 4.750 ft above the ground. Sadly, the bucket then falls off the ladder. How fast is the bucket of paint moving just before it hits the ground and makes a colossal mess?

To start, use the weight equation to obtain the mass of the bucket. As always, we will first convert the given weight into MKS units.

$$F_w = 8.55 \text{ lb} \cdot \frac{4.45 \text{ N}}{\text{lb}} = 38.05 \text{ N}$$

$$m = ?$$

$$F_w = mg$$

$$m = \frac{F_w}{g} = \frac{38.05 \text{ N}}{9.80 \frac{\text{m}}{\text{s}^2}} = 3.883 \text{ kg}$$

Notice that I did those calculations with four significant digits. This is because the value for *g* has three significant digits and my intermediate results always have an extra digit before I round off at the end.

Just as a reminder, the beauty of working in the MKS unit system is that when we use MKS units in a calculation, the result always has MKS units. So when I divide newtons (N) by meters per second squared (m/s²), I don't have to worry about puzzling out any unit issues. I know this calculation gives me a mass, and the MKS unit for mass is the kilogram (kg).

Now that we have the mass, let's write down everything and begin the energy calculation.

$m = 3.883$ kg

$h = 4.750$ ft $\cdot \dfrac{0.3048 \text{ m}}{\text{ft}} = 1.448$ m

$v = ?$

Now here is our energy diagram for this problem:

$E_G \rightarrow E_K$

This tells me that in this problem, all the E_G we have to begin with converts into E_K as the bucket falls. So I need to calculate the E_G first and then use this amount of energy as the E_K for calculating the velocity.

$E_G = mgh = 3.883$ kg $\cdot 9.80 \; \dfrac{\text{m}}{\text{s}^2} \cdot 1.448$ m $= 55.10$ J

Since this gravitational potential energy converts to kinetic energy we now have

$E_K = 55.10$ J

Finally, we use this value in the velocity equation to obtain our final result.

$v = \sqrt{\dfrac{2E_K}{m}} = \sqrt{\dfrac{2 \cdot 55.10 \text{ J}}{3.883 \text{ kg}}} = 5.33 \; \dfrac{\text{m}}{\text{s}}$

This result is rounded to three significant figures as required.

▼ Example 4.7

A baseball batter hits a baseball straight up with a velocity of 180 ft/s. A regulation baseball has a mass of 144.3 g. Ignoring air friction, how high does the baseball go before it comes to a stop?

I will work this out using the same method as in the previous problem. First, write down the givens and do the unit conversions.

$$m = 144.3 \text{ g} \cdot \frac{1 \text{ kg}}{1000 \text{ g}} = 0.1443 \text{ kg}$$

$$v = 180 \ \frac{\text{ft}}{\text{s}} \cdot \frac{0.3048 \text{ m}}{\text{ft}} = 54.9 \ \frac{\text{m}}{\text{s}}$$

$$h = ?$$

Now draw the energy diagram that indicates what is happening in this problem. The kinetic energy the ball has as it leaves the bat converts to gravitational potential energy as it rises, so

$$E_K \to E_G$$

Since we are starting with kinetic energy this time, compute that first.

$$E_K = \tfrac{1}{2} m v^2 = 0.5 \cdot 0.1443 \text{ kg} \cdot \left(54.9 \ \frac{\text{m}}{\text{s}}\right)^2 = 217 \text{ J}$$

Since this energy is converting to gravitational potential energy, at the top of the ball's flight we have

$$E_G = 217 \text{ J}$$

Now use this value to solve for the height.

$$E_G = mgh$$

$$h = \frac{E_G}{mg} = \frac{217 \text{ J}}{0.1443 \text{ kg} \cdot 9.80 \ \frac{\text{m}}{\text{s}^2}} = 153 \text{ m}$$

Finally, recall that the given velocity only has two significant digits, so we must round this result to two digits, which gives

$$h = 150 \text{ m}$$

4.3.5 Energy in the Pendulum

A swinging pendulum provides us with one final example of the conservation of energy in action. To begin, note that because of friction between the swinging pendulum and the air and friction in the pivot at the top, any actual pendulum loses energy to the environment as heat. This is why any actual free-swinging pendulum always comes to a stop.

But let's imagine a perfect pendulum, one that loses no energy due to friction. We call this an *ideal pendulum*. In an ideal pendulum, no energy leaves the "system" (the swinging pendulum) as heat and the pendulum just keeps on swinging without slowing down. (Actually, it's a bit more complicated because of the rotation of the earth, so even in a vacuum with a magnetic bearing at the pivot the pendulum still slows down, so don't start getting visions of a perpetual motion machine! But our imaginary ideal pendulum is also free from such influences.)

From what you know about the forms of energy and energy conservation, you can probably already see how energy transformation will work in this ideal pendulum. As shown in Figure 4.6, we let the height of the pendulum when it is at rest (that is, not swinging) be our reference for height measurements. This means when the pendulum is straight down its height is zero and its gravitational potential energy is also zero. When the pendulum swings up to its highest point, it momentarily comes to rest. At this moment, its velocity is zero and so its kinetic energy is zero. Put these facts together and let the pendulum start swinging.

Because of the conservation of energy, the pendulum always has the same total amount of energy no matter where it is. When someone lifts the pendulum to get it started, the person does work on the pendulum equal to the force it takes to lift it times the height (just as we saw with the bucket example). Since the pendulum is ideal, no energy

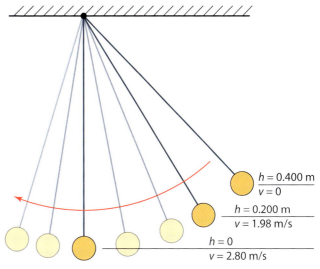

Figure 4.6. Conservation of energy in a swinging pendulum.

leaves the system as it swings. This means that no matter where the pendulum is, the total energy in the system is always the same and is equal to the amount of energy put into the system in the first place by the work done on it to get it up to where it is released. As the pendulum swings down, E_G converts into E_K, and as the pendulum swings up, E_K converts back into E_G. At all times, the total energy the pendulum possesses always equals the sum of the E_G and the E_K, and this sum always adds up to the same value no matter where the swinging pendulum is.

Just to run through a quick calculation, let's say the mass at the end of the pendulum is 2.00 kg and we lift it up 0.400 m above its lowest point to release it. The total E_G at this starting point is mgh, which gives $E_G = 7.84$ J. The kinetic energy here is zero, so now we know that the pendulum has a total of 7.84 J of energy no matter where it is.

How fast is it going when it is halfway down, 0.200 m high? Well, the E_G at that position is 3.92 J, so the E_K is 7.84 J – 3.92 J = 3.92 J. Using this kinetic energy value in the velocity equation gives a velocity of 1.98 m/s. At the bottom, all the energy (7.84 J) is kinetic energy, so the velocity is 2.80 m/s. As you see, if you know how high the pendulum is at any point, you can determine how fast it is moving, and vice versa.

Chapter 4 Exercises

Energy Study Questions

1. Write out the stages in the "energy trail" for the following sequences. For each stage, list where the energy is and what form it is in.
    a. From the sun to a ball thrown straight up at the top of its flight.
    b. From the sun to a galloping horse.
    c. From the sun to water stored in a water tower, pumped up there by the city's electric pumps.
    d. From the sun to a moving motorcycle.
    e. From the sun to an electric blow dryer for drying hair.
    f. From the sun to a diesel truck parked at the top of a steep hill.

2. Write down the law of conservation of energy from memory. Then write a paragraph explaining the law in your own words. Include some examples in your explanation.

3. From an energy standpoint, what is an "ideal system"? (Think of the ideal pendulum discussed at the end of the chapter.)

4. If you run out of gas, your car soon comes to a stop. What happens to all the kinetic energy the car has before running out of gas? Where does it go?

Classroom Energy Computation Examples

These examples are written here so that as your teacher works examples in class you can focus on the solutions rather than on worrying about getting the problems written down.

1. Water is pumped into a high water tower. If the total mass of the water is 1.00×10^5 kg and the tower is 240 feet high, what is the gravitational potential energy (E_G) of the water in the tower?

2. A bullet of mass 25 g is fired at a velocity of 556 ft/s. How much kinetic energy (E_K) does the bullet have?

3. A man lifts a bucket of sand 75 cm above the ground. If the bucket of sand has a mass of 12,500 g, how much work does the man do on the bucket?

4. Referring again to the previous problem, after the bucket of sand is lifted, how much E_G does the sand have?

5. If the man releases the bucket of sand and lets it drop, what is its velocity the instant before it strikes the ground?

6. A boy carries a water balloon up to the top of a ladder to let it drop. The mass of the water balloon is 255.8 g and the ladder is 10.4 ft tall.
    a. How much E_G does the balloon have at the top of the ladder?
    b. If the balloon is released, how fast is it going just before it splats on the ground?

Answers

1. 72,000,000 J
2. 360 J
3. 92 J
4. 92 J
5. 3.8 m/s
6. a) 7.95 J; b) 7.88 m/s

Energy Calculations Set 1

1. A load of building materials is hoisted to the top of a building under construction. If the total mass of the material is 1.31×10^3 kg and the building is 177.44 feet high, what is the gravitational potential energy (E_G) of the material at the top of the building?

2. A car of mass 2,345 kg is traveling at a speed of 31 mph. How much kinetic energy (E_k) does the car have?

3. A woman lifts a bucket of water 61.7 cm above the ground. If the bucket of water has a mass of 17.5 kg, how much work does the woman do?

4. How much E_G does the woman's bucket have after being lifted?

5. If the woman releases the bucket of water and lets it drop, what is its velocity the instant before it strikes the ground?

6. A kid shoots an arrow straight up. The arrow has a mass of 122 g and leaves the bow with a velocity of 13.75 m/s. How high does it go above the point where it is shot from the bow?

7. A girl drops a stone into the water from a bridge. The stone has a mass of 325 g and the bridge is 36.1 m above the water. How fast is the stone moving just before it hits the water?

8. A worker slides a carton across the floor. The force of friction between the carton and the floor is 735 N. If the worker pushes the carton 26 m, how much work does he do?

Answers

1. 694,000 J
2. 230,000 J
3. 106 J
4. (This answer is top secret!)
5. 3.48 m/s
6. 9.65 m
7. 26.6 m/s
8. 19,000 J

Energy Questions Set 2

1. A carpenter hauls twenty 80.0-pound bundles of shingles up onto a roof, which is 8.5 m above the ground.

a. Compute the total mass of the shingles using the conversion factor for pounds to newtons found in Appendix A, and then the weight equation to get the mass.
b. Compute the work the carpenter has to do to get the shingles up onto the roof. State your answer in joules (J).
c. Using the equation for gravitational potential energy (E_G), compute the E_G of the shingles while they are on the roof.
d. If the entire stack of shingles slides off the roof, how much kinetic energy does the stack have at the following times:
   i. At the moment it first begins to slide.
   ii. Just before it hits the ground.
e. Compute how fast the stack is falling just before it hits the ground.

2. A car weighing 3,193 lb rests at the top of a hill, then begins to roll down the hill. (The engine is off.) Assume it rolls to the bottom of the hill with negligible friction.
   a. If the hill is 16 m high compared to the flat road at the bottom, compute the E_G of the car while it is at the top of the hill.
   b. After the car rolls to the bottom of the hill, where is the energy that is in the E_G of the car while it is at the top of the hill?
   c. Compute the E_K of the car when it reaches the bottom of the hill.
   d. Compute the velocity of the car when it reaches the bottom of the hill.
   e. Explain how conservation of energy relates to this problem.
   f. Explain specifically how friction would change the results of the problem in a more realistic example.

Answers

1. a) 727 kg; b) 6.0×10^4 J; c) 6.0×10^4 J; d) 0 J, 6.0×10^4 J; e) 13 m/s
2. a) 230,000 J; c) 230,000 J; d) 18 m/s

Energy Questions Set 3

1. Consider a large, ideal (that is, frictionless) pendulum with a steel ball weighing 27.05 lb on the end. This ball is lifted to a height of 185 cm and released. Since the pendulum is frictionless, the ball swings back and forth forever.

   a. How much work is done to lift the ball?
   b. What is the E_G of the ball after it is lifted, but before it starts falling?
   c. At the bottom of the ball's pathway, what is its E_G?
   d. At the bottom of the ball's pathway, what is its E_K?
   e. How fast is the ball going at the bottom?
   f. Use friction and energy considerations to explain the difference between this pendulum and an actual pendulum.

2. A group of city water pumps pushes water up into a water tower 197 feet high. The water tower holds 6.016×10^6 kg of water. Determine the amount of mechanical work the pumps must do to fill the water tower and state your answer in GJ.

3. Imagine a new frictionless roller coaster that uses magnetic levitation so that the

cars float above the rails without actually touching them. Imagine also that the aerodynamic design of the cars is so brilliant that there is essentially no air friction. The car has a mass of 5,122 kg. From the top of a 25.0 m-hill the car rolls down to a valley where the track is right on the ground. Assuming the roller coaster begins at rest at the top of the hill, determine how fast it is traveling when it reaches the bottom of the valley.

4. A boy weighing 104.6 lb runs up the steps two at a time. There are 13 steps, each one 16.5 cm high.
   a. How much work does he do to get to the top of the steps?
   b. If he steps over the hand rail and drops back down to the ground, how fast is he moving just before he lands?

5. An object with mass $m = 351$ g is sliding on a frictionless surface at 500.00 cm/s when it begins going up a ramp. What is the height, h, when it stops?

Answers

1. a) 223 J; b) 223 J; c) 0 J; d) 223 J; e) 6.02 m/s
2. 3.54 GJ
3. 22.1 m/s
4. a) 998 J; b) 6.48 m/s
5. 1.28 m

Do You Know ...   *Why are there pendulums in clocks?*

Galileo first discovered that the period of a swinging pendulum depends only on its length (a fact you confirmed in the Pendulum Experiment). Because of this, pendulums are used to regulate the motion of the mechanical systems in clocks. The weight on the end of the pendulum is supported by a nut on a threaded rod, and the vertical position of the nut is adjusted to give the pendulum the precise length needed for the clock to run at the correct speed.

The pendulum in a grandfather clock is kept in motion by the gravitational potential energy in the weights hanging inside the cabinet. As the weights slowly descend, their gravitational potential energy is transferred to the energy in the swinging pendulum. The gravitational potential energy in the weights is replenished when a person does work on them, raising them back to their highest position to begin descending again. Because of friction, the pendulum would eventually stop swinging without receiving energy from the continuous action of the descending weights.

As the pendulum swings, it continuously converts its energy from kinetic energy to gravitational potential energy and back again.

CHAPTER 5
Momentum

CERN

CERN is a major research facility in Switzerland, well known for cutting-edge experiments in atomic physics. Researchers smash high-speed beams of protons into each other, causing the protons to disintegrate into other particles. The trick is to use sophisticated detectors, like the one shown above, to measure the velocity and charge of the particles produced by the collision. Scientists use conservation of momentum and other physical principles to analyze the particles produced by the collision. If the detector measures the velocities of several known and unknown particles, the velocities and known masses can be used in a conservation of momentum calculation to determine an unknown particle's mass.

From CERN's website, as quoted on wikipedia:

The photo above shows the silicon strip tracker of the CMS detector. Shown here are three concentric cylinders, each comprised of many silicon strip detectors (the bronze-colored rectangular devices, similar to the CCDs used in digital cameras). These surround the region where the protons collide.

How fast are the protons moving when they collide? Turn back to page 84 to find out.

OBJECTIVES

Memorize and learn how to use this equation:

$$p = mv$$

After studying this chapter and completing the exercises, students will be able to do each of the following tasks, using supporting terms and principles as necessary:

1. Use the momentum equation to calculate mass, velocity, and momentum using correct units of measure.
2. Define *interaction*.
3. State the law of conservation of momentum.
4. Distinguish between elastic and inelastic collisions.
5. Use the principle of conservation of momentum to solve problems involving linear interactions between objects with equal or unequal masses in elastic collisions, with one object initially at rest.
6. Describe how the law of conservation of momentum relates to Newton's second and third laws of motion.

5.1   Defining Momentum

One of the basic principles of physics, and a fairly simple one to understand, is the principle of *momentum*. The study of momentum is important in physics because momentum relates very closely to Newton's laws of motion. In the study of momentum, we also encounter another of the important conservation laws in physics. There are several conservation laws and in understanding them you actually understand a lot about the way nature works. We will get to the conservation law soon, but first we will begin with a simple definition.

The momentum of an object is the product of its mass and its velocity. In physics, we use the variable p to represent momentum. (You probably wish the variable was M, but it isn't. Oh well.) Thus, the equation for momentum is

$$p = mv$$

where p is the object's momentum (kg·m/s), m is the mass of an object (kg), and v is the velocity of the object (m/s). There are no special units of measure for momentum. The units are just the product of the mass and velocity units.

This equation for momentum is obviously pretty simple, so let's go ahead and look at a few example calculations.

▼ Example 5.1

A proton at the Large Hadron Collider at CERN in Switzerland is traveling at 7,600 miles per second. The mass of a proton is 1.673×10^{-21} g. Determine the momentum of this proton.

Begin as always by writing down the givens and converting the units to MKS units. We observe that the given velocity has two significant digits, and so we will be careful, as usual, to retain an extra digit in our intermediate calculations.

$$v = 7600 \ \frac{\text{mi}}{\text{s}} \cdot \frac{1609 \ \text{m}}{1 \ \text{mi}} = 12,200,000 \ \frac{\text{m}}{\text{s}}$$

$$m = 1.673 \times 10^{-24} \ \text{g} \cdot \frac{1 \ \text{kg}}{1000 \ \text{g}} = 1.673 \times 10^{-27} \ \text{kg}$$

$$p = ?$$

Now write down the equation, insert the values, and compute the result.

$$p = mv = 1.673 \times 10^{-27} \ \text{kg} \cdot 12,200,000 \ \frac{\text{m}}{\text{s}} = 2.0 \times 10^{-20} \ \frac{\text{kg} \cdot \text{m}}{\text{s}}$$

This result has been rounded to two digits, as required.

▼ Example 5.2

A man and his motorcycle have a combined weight of 957 lb. If the rider is running his bike at 65.5 mph, determine the combined momentum of the bike and the man on it.

This problem is worked in two parts, just like other problems we have seen in the last couple of chapters. We need the mass to determine the momentum, but the mass is not given in the problem. However, the weight is given and we know how to calculate the mass from the weight, so we do that first. Begin, as always, by writing down the given information and converting all the units to MKS units. From the given information we see that our result must have three significant digits, so all the intermediate calculations must have four digits.

$$F_w = 957 \ \text{lb} \cdot \frac{4.45 \ \text{N}}{1 \ \text{lb}} = 4259 \ \text{N}$$

$$v = 65.5 \ \frac{\text{mi}}{\text{hr}} \cdot \frac{1609 \ \text{m}}{1 \ \text{mi}} \cdot \frac{1 \ \text{hr}}{3600 \ \text{s}} = 29.27 \ \frac{\text{m}}{\text{s}}$$

$$p = ?$$

Next, we write down the weight equation and use it to determine the mass.

$$F_w = mg$$

$$m = \frac{F_w}{g} = \frac{4259 \ \text{N}}{9.80 \ \frac{\text{m}}{\text{s}^2}} = 434.6 \ \text{kg}$$

Now we are ready to calculate the momentum.

$$p = mv = 434.6 \ \text{kg} \cdot 29.27 \ \frac{\text{m}}{\text{s}} = 12,700 \ \frac{\text{kg} \cdot \text{m}}{\text{s}}$$

Since each of the given values has three significant digits and all of our unit conversion factors have at least three digits, our result must have three digits. The result is rounded accordingly.

▼ Example 5.3

A hockey puck has a mass of 165 g. In 2011, a world record for hockey puck speed was set by a man who gave the hockey puck a momentum of 8.134 kg·m/s. With that momentum, how fast is that puck moving? State your result in miles per hour.

We first list the givens and convert the units.

$$p = 8.134 \ \frac{\text{kg} \cdot \text{m}}{\text{s}}$$

$$m = 165 \ \text{g} \cdot \frac{1 \ \text{kg}}{1000 \ \text{g}} = 0.165 \ \text{kg}$$

$$v = ?$$

Now write the momentum equation and solve for v.

$$p = mv$$

$$v = \frac{p}{m}$$

Now insert the values and calculate the result.

$$v = \frac{p}{m} = \frac{8.134 \ \frac{\text{kg} \cdot \text{m}}{\text{s}}}{0.165 \ \text{kg}} = 49.30 \ \frac{\text{m}}{\text{s}}$$

Finally, perform the final unit conversion required by the problem.

$$v = 49.30 \ \frac{\text{m}}{\text{s}} \cdot \frac{1 \ \text{mi}}{1609 \ \text{m}} \cdot \frac{3600 \ \text{s}}{1 \ \text{hr}} = 110.3 \ \frac{\text{mi}}{\text{hr}}$$

This result has 4 digits of precision, but we need to round it to three because the given mass has three significant digits. But we can't just drop the last digit and write 110 mph, because this value only has two digits. The only way to represent our result with three digits is to put the result in scientific notation. Doing this, our final result is

$$v = 1.10 \times 10^2 \ \frac{\text{mi}}{\text{hr}}$$

5.2   Conservation of Momentum

5.2.1 The Law of Conservation of Momentum

The *law of conservation of momentum* is another of the famous conservation laws in physics. The first conservation law we encountered was the law of conservation of energy in the previous chapter.

The law of conservation of momentum is about what happens during a collision between two objects. The law applies to a collision between cars, a collision between billiard balls, or a collision between two atoms. It doesn't matter what the objects are or how it is they collide; the law of conservation of momentum always applies.

In fact, the objects don't even have to touch for the law of conservation of momentum to apply. Two protons fired directly at one another in a physics experiment actually never "touch." They just get so close together that their repulsive electric charge makes them bounce apart. Because of this, we often use the term *interaction* instead of collision. Examples of interactions are two cars colliding, two charged particles repelling one another, two magnets pushing apart, and the sun and a planet pulling on each other with gravitational attraction.

The law of conservation of momentum is:

> In any interaction, the total momentum of the objects involved is conserved.

In physics, when we say that a quantity is "conserved" during a certain event, we simply mean that the total amount of this quantity before the event is equal to the total amount of the same quantity after the event. In the case of momentum, we are talking about the total of all the momentum values of all the objects involved in the interaction. Now, to apply the law of conservation of momentum to specific types of problems is not very difficult. But before we get to problem solving, it will be helpful to consider two different types of interactions between two objects.

5.2.2 Elastic and Inelastic Collisions

In terms of energy and momentum, there are two basic types of interactions. In an *elastic collision*, objects bounce perfectly apart without any damage and the total kinetic energy of the moving objects is conserved, just as the momentum is conserved. In other words, no kinetic energy is converted to heat or friction as the objects collide.

In an *inelastic collision*, the kinetic energy of the moving objects is not conserved. This is because in inelastic collisions, some of the initial kinetic energy is converted to other forms of energy during the collision. Some of the original kinetic energy is converted to heat due to friction as the objects collide, and some of the original energy may also be used to bend or break parts of the objects, as with most car collisions. Bending and breaking requires work to be done, and this work energy comes from the initial kinetic energy in the objects. When some of the initial kinetic energy is converted into these different forms, the kinetic energy of the objects after the collision is less than it is before the interaction.

In the real world, there are no perfectly elastic collisions. All interactions involve energy changing forms. However, many interactions may be modeled very accurately as elastic collisions. Collisions between two pool balls, between two marbles, and between two steel ball bearings are all highly elastic and are typically modeled this way. The same thing applies to interactions between subatomic particles, such as two protons colliding in a particle accelerator.

5.2.3 Problem Solving Assumptions

To make our problem solving with conservation of momentum manageable, we are going to establish some simplifying assumptions. The first is that we will only deal with elastic collisions. As I mentioned in the previous section, there are real interactions so close to being elastic that we can study them as if they are perfectly elastic.

Our second assumption is that the objects in the collisions we study always move in a straight line with each other. A third assumption that applies to problems in this text is that prior to the collision, one of the objects is at rest but free to move when hit. This assumption simplifies the math quite a bit. In summary, there are three working assumptions that apply to all the problems we address in this text:

Our working assumptions for conservation of momentum problems:

- The collision is elastic.
- All the motion before and after the collision is in a single straight line.
- Before the collision, one of the objects is at rest but free to move.

A major issue for us to consider next is whether the two objects involved in the collision have the same mass. Of course, in the case of billiard balls the two objects do have the same mass. When the two objects have the same mass and one of them is at rest, the outcome of an elastic collision is totally predictable. The initially moving object always stops dead and the other object begins moving at the same speed as the original object and in the same direction. This is illustrated in Figure 5.1 with two billiard balls.

The figure shows a yellow ball with mass m at rest and a white ball with mass m heading toward it with velocity v. Since the yellow ball is at rest, its velocity is zero and it has no momentum. The white ball is moving with velocity v so the white ball has momentum equal to mv. The total momentum before the collision is simply the momentum of the white ball (mv) because it is the only ball moving.

After the collision, the yellow ball is, of course, moving off to the right. But since the masses are equal, the white ball comes to a complete stop. This means that for momentum to be conserved, the yellow ball has to move off after the collision with the same velocity as the white ball had originally.

This case seems simple enough. But in general, we need to consider what happens when two objects with different masses interact. Now consider that you have had many experiences watching things bump into each other. In addi-

Figure 5.1. The interaction of two billiard balls (equal masses) colliding.

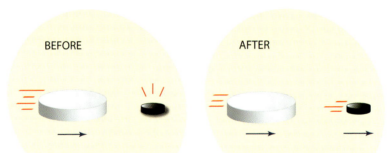

When a large mass collides with a small mass at rest, both masses move off in the original direction of motion.

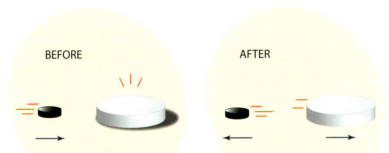

When a small mass collides with a large mass at rest, the small mass bounces back, while the large mass moves off in the original direction of motion.

Figure 5.2. The interaction of unequal masses colliding: a large mass striking a small mass (top), and a small mass striking a large mass (bottom).

tion to billiard balls, you may have seen bumper cars colliding, or toy cars, or marbles. You have seen things collide all your life. From all these experiences, you know that when a moving object collides with a stationary object there are three possibilities for what the objects do after the collision.

The first possibility is the one we just considered. When the objects have equal masses, the moving object stops dead and the stationary object begins moving. The second possibility is that the first moving object continues moving in the same direction but at a slower speed than before, so that both objects move off after the collision in the same direction. This is exactly what happens when the first moving object has a greater mass than the stationary object. Because the moving object is more massive, it keeps moving in the same direction rather than bouncing back. The third possibility is that the first moving object bounces backward, going back toward where it originally came from, while the stationary object moves off in the original direction. This is what happens when the moving object is less massive than the stationary one. The two possibilities for collisions with unequal masses are illustrated in Figure 5.2. For this illustration, I chose two disk-shaped objects.

5.2.4 The Directionality of Momentum

The third of the three scenarios we have just considered raises an important question. In the third scenario, after the collision occurs the two objects are moving in opposite directions. Will motion in two different directions make a difference in our mathematical approach to solving problems using the law of conservation of momentum? Yes, indeed! Momentum is one of the types of quantities in physics that have a *direction* associated with them. That is, they *point* somewhere. Quantities like this are called *vectors*. We have already encountered several other variables that are vectors, including velocity, acceleration, and force.

Hopefully, it seems intuitive to you that the direction of these quantities is important. A positive or negative velocity is the difference between driving north and driving south. A positive or negative acceleration is the difference between speeding up and slowing down. A positive or negative force is the difference between pushing an object to the right or pushing it to the left.

In fact, we have seen a problem or two in previous chapters when a force or acceleration came out to be a *negative* quantity. We didn't worry about this too much when it came up. But to study conservation of momentum, we must pay attention to the direction objects are moving.

We model vector quantities mathematically by using positive and negative signs. All we do is identify the two possible directions, such as left and right, and call one of them positive and one of them negative. It doesn't matter which is which, but typically we identify motion to the right as positive and motion to the left as negative. If we do this, all the math works. I illustrate this in the next section when we use the law of conservation of momentum to solve problems.

While we are on the subject, you will be interested to know there are other variables in physics that have no direction associated with them, such as mass, energy, and temperature. Quantities like these are called *scalars*. The temperature at a certain place in the room you are in right now has no direction associated with it, that is, it doesn't point anywhere. Mass is another variable that has no direction. To be sure, the force of gravity acting on a mass *does* have a direction (down), which is why force is a vector. But mass has no direction and is a scalar quantity.

5.2.5 Solving Problems with Conservation of Momentum

We are now ready to use the law of conservation of momentum to solve problems! Let's begin by stating the law of conservation of momentum in the form of an equation. In our work, we are assuming that before the collision only one of the objects is in motion. Let us call this object 1. The other object is at rest, and so let us call it object 2. After the collision, we always have either only one object moving (which is object 2 in the case when the masses are equal) or both objects moving.

The law of conservation of momentum says that the total momentum of all the objects before the collision must equal the total momentum of all the objects after the collision. Assuming the most general case, that both objects are moving after the collision, this statement is written mathematically this way:

$p_{1i} = p_{1f} + p_{2f}$

In this equation, the subscripts *i* and *f* mean *initial* and *final*, just as with the velocities in the acceleration equation. The subscripts 1 and 2 refer to *object 1* and *object 2*. Before the collision, the *initial* (before) part of the problem, the only object moving is object 1, so its momentum, on the left side of the equation, is p_{1i}. The momenta (plural for momentum) of the two objects after the collision, the *final* (after) part of the problem, are added together on the right side of the equation. The equation simply sets the momentum before the collision equal to the total momentum after the collision, as the law of conservation of momentum says.

Now, we considered three different collision cases above, so let's look at solution strategies for each of these three cases.

Case 1: Equal Masses

If the masses are equal, we know the initially moving object stops at the collision, so $p_{1f} = 0$, and the conservation of momentum equation becomes

$$p_{1i} = p_{2f}$$

Putting in the masses (both equal to *m*) and velocities for the two momentum terms, we have

$$mv_{1i} = mv_{2f}$$

But wait! We can divide both sides by *m* and get

$$v_{1i} = v_{2f}$$

Well, well. We just proved what I told you about equal mass collisions back when we began. If one object is at rest, then after the collision when object 1 stops, object 2 takes off with the same velocity object 1 had originally.

Case 2: Larger Mass Initially Moving

In this case, our procedure depends on what the problem asks us to solve for. In general, we follow these steps:

- Write the given information, as usual.
- Write the conservation of momentum equation from page 115.
- Solve the conservation of momentum equation for the momentum term that relates directly to the unknown quantity you seek to determine (a mass or velocity).
- Calculate the other momentum values from the given information.
- Add or subtract these values to determine the momentum that relates to your unknown.
- Use the momentum equation to solve for the unknown mass or velocity.

Case 3: Smaller Mass Initially Moving

We follow the same steps as for the previous case. The only difference is that since the small mass bounces back, its momentum is negative in the conservation of momentum equation. Thus, the conservation of momentum equation is

$$p_{1i} = \left(-p_{1f}\right) + p_{2f}$$

These three cases are illustrated in the following examples. Let's get to it!

▼ Example 5.4

Joe and Bill are riding bumper cars at the State Fair. The two cars are identical, and Joe and Bill weigh the same. Bill brings his car to a stop and while he is stopped Joe rams him going 0.95 meters per second. Assuming this collision is perfectly elastic, what does each car do after the collision?

No calculations are even necessary to solve this problem. Since the weights of the two cars are equal, the masses are equal as well. This means that Joe's car stops completely and Bill's car moves off at the same speed Joe was going, or 0.95 m/s.

▼ Example 5.5

A science teacher has an air track that operates like an air hockey table. Aluminum gliders glide in a straight line on the track almost without friction, and the spring bumpers on the gliders mean the collisions between the gliders are almost perfectly elastic. The teacher places a glider with a mass of 225.0 g on the track at rest. The teacher then takes another glider with a mass of 450.0 g and sends it down the track at a velocity of 36 cm/s toward the smaller glider. After the collision, the larger mass has a velocity of 12 cm/s. What is the velocity of the smaller mass after the collision?

Write down the givens and convert everything to MKS units.

$$v_{1i} = 36\ \frac{cm}{s} \cdot \frac{1\ m}{100\ cm} = 0.36\ \frac{m}{s}$$

$$v_{1f} = 12\ \frac{cm}{s} \cdot \frac{1\ m}{100\ cm} = 0.12\ \frac{m}{s}$$

$$m_1 = 450.0\ g \cdot \frac{1\ kg}{1000\ g} = 0.4500\ kg$$

$$m_2 = 225.0\ g \cdot \frac{1\ kg}{1000\ g} = 0.2250\ kg$$

$$v_{2f} = ?$$

The unknown velocity we must find, v_{2f}, relates directly to the momentum p_{2f}. So write the general conservation of momentum equation, and solve it for p_{2f}. Since the larger mass is the one moving initially, both masses have positive momentum after the collision.

$$p_{1i} = p_{1f} + p_{2f}$$
$$p_{2f} = p_{1i} - p_{1f}$$

Now calculate the values of the two momentum terms on the right side and use them to compute p_{2f}.

$$p_{1i} = m_1 v_{1i} = 0.4500 \text{ kg} \cdot 0.36 \frac{\text{m}}{\text{s}} = 0.162 \frac{\text{kg} \cdot \text{m}}{\text{s}}$$

$$p_{1f} = m_1 v_{1f} = 0.4500 \text{ kg} \cdot 0.12 \frac{\text{m}}{\text{s}} = 0.0540 \frac{\text{kg} \cdot \text{m}}{\text{s}}$$

$$p_{2f} = 0.162 \frac{\text{kg} \cdot \text{m}}{\text{s}} - 0.0540 \frac{\text{kg} \cdot \text{m}}{\text{s}} = 0.108 \frac{\text{kg} \cdot \text{m}}{\text{s}}$$

Now that we have the momentum for the second mass after the collision, we use it to determine its velocity. Write the simple momentum equation for the second mass and solve it for v.

$$p_{2f} = m_2 v_{2f}$$

$$v_{2f} = \frac{p_{2f}}{m_2} = \frac{0.108 \frac{\text{kg} \cdot \text{m}}{\text{s}}}{0.2250 \text{ kg}} = 0.48 \frac{\text{m}}{\text{s}}$$

This value has been rounded to two significant digits, as the given information requires.

For the last example, we will look at an actual collision between atoms. Mathematically, this will be one of the most difficult examples in the book. This is no problem; the most difficult example has to appear at some point, doesn't it? Well, here it is! The fun thing about it is that it is a real-life example from real history. Don't let the math intimidate you! You can do it! We will walk carefully through this example together. As we go along, just remember our steps for solving problems such as this:

• Write the given information, as usual.

• Write the appropriate conservation of momentum equation, based on whether the large mass is initially moving or at rest.

• Solve the conservation of momentum equation for the momentum term that relates directly to the unknown quantity you seek to determine.

• Calculate the other momentum values from the given information.

• Add or subtract these values to determine the momentum that relates to your unknown.

• Use the momentum equation to solve for the unknown mass or velocity.

Now to the example.

▼ Example 5.6

In 1909, Ernest Rutherford discovered the atomic nucleus with his famous "gold foil experiment." In this experiment, small "alpha particles" were fired at much larger gold atoms and some of them bounced straight back, a perfect example of the third type of collision we studied. (FYI, an alpha particle contains two protons and two neutrons, but we don't need to know this to solve this problem.) Let's calculate the rebound velocity for an alpha particle after a collision with a single atom of gold, assuming the gold atom is free to move. An alpha particle has a mass of 6.646×10^{-27} kg and the much larger gold atom has a mass of 3.271×10^{-25} kg. The gold atom is at rest when Rutherford fires the alpha particle at it with a velocity of (incredible as it may sound) 15,000,000 m/s. After the collision, the gold atom moves off at a whopping 597,000 m/s. What is the rebound velocity of the alpha particle after the collision?

We begin by writing down the given information, all of which is already in MKS units.

$$v_{1i} = 15,000,000 \ \frac{\text{m}}{\text{s}}$$

$$m_1 = 6.646 \times 10^{-27} \ \text{kg}$$

$$m_2 = 3.271 \times 10^{-25} \ \text{kg}$$

$$v_{2f} = 597,000 \ \frac{\text{m}}{\text{s}}$$

$$v_{1f} = ?$$

Right away we notice that the initial velocity of the alpha particle, v_{1i}, has only two significant digits, so I make a mental note that my intermediate calculations need to have three significant digits.

Next, we write down the conservation of momentum equation for this type of collision. This is an example of a small mass colliding with a larger mass at rest. We know that in this case, the small mass bounces back and thus its final momentum is negative. We will go ahead and write in this negative sign, which makes the conservation of momentum equation to be

$$p_{1i} = \left(-p_{1f}\right) + p_{2f}$$
$$p_{1i} = p_{2f} - p_{1f}$$

The velocity we are solving for, v_{1f}, relates to p_{1f}, so we solve our conservation of momentum equation for p_{1f}. I will do this by first adding p_{1f} to both sides, and then subtracting p_{1i} from both sides. I will show all of the algebra:

$$p_{1i} + p_{1f} = p_{2f} - p_{1f} + p_{1f}$$
$$p_{1i} + p_{1f} = p_{2f}$$
$$p_{1i} + p_{1f} - p_{1i} = p_{2f} - p_{1i}$$
$$p_{1f} = p_{2f} - p_{1i}$$

Now we proceed as in the previous example. Calculate the two momenta on the right side of the above equation, put them into the equation, and solve for p_{1f}.

$$p_{2f} = m_2 v_{2f} = 3.271 \times 10^{-25} \text{ kg} \cdot 597{,}000 \ \frac{\text{m}}{\text{s}} = 1.95 \times 10^{-19} \ \frac{\text{kg} \cdot \text{m}}{\text{s}}$$

$$p_{1i} = m_1 v_{1i} = 6.646 \times 10^{-27} \text{ kg} \cdot 15{,}000{,}000 \ \frac{\text{m}}{\text{s}} = 9.97 \times 10^{-20} \ \frac{\text{kg} \cdot \text{m}}{\text{s}}$$

$$p_{1f} = 1.95 \times 10^{-19} \ \frac{\text{kg} \cdot \text{m}}{\text{s}} - 9.97 \times 10^{-20} \ \frac{\text{kg} \cdot \text{m}}{\text{s}} = 9.53 \times 10^{-20} \ \frac{\text{kg} \cdot \text{m}}{\text{s}}$$

Finally, use that momentum value with the momentum equation to determine v_{1f}.

$$p_{1f} = m_1 v_{1f}$$

$$v_{1f} = \frac{p_{1f}}{m_1} = \frac{9.53 \times 10^{-20} \ \frac{\text{kg} \cdot \text{m}}{\text{s}}}{6.646 \times 10^{-27} \text{ kg}} = 14{,}300{,}000 \ \frac{\text{m}}{\text{s}}$$

The initial velocity of the alpha particle has only two significant digits, so now we round our answer to two digits. Doing so gives

$$v_{1f} = 14{,}000{,}000 \ \frac{\text{m}}{\text{s}}$$

That alpha particle comes at the gold atom moving 15 million meters per second and when it rebounds it is going 14 million meters per second! That is fascinating.

5.3   Momentum and Newton's Laws of Motion

Back on the first page of this chapter, I mentioned that momentum is closely related to Newton's laws of motion. To conclude this chapter, we need to explore this important connection.

In Chapter 2, we learned that acceleration is a measure of how fast a velocity is changing. The equation for acceleration is

$$a = \frac{v_f - v_i}{t}$$

Using the symbol Δ (the Greek letter delta) to mean change, this equation for acceleration can be written more simply as

$$a = \frac{\Delta v}{t}$$

In this equation, Δv just means the *change in velocity* (which is $v_f - v_i$) and so $\Delta v/t$, which is the acceleration, just means *how fast* the velocity is changing, or the velocity's *rate of change*. Let's now insert this expression for acceleration into Newton's second law of motion:

Do You Know ...

What is angular momentum?

Conservation of momentum also applies to rotating objects, where it is referred to as *angular momentum*. With angular momentum, the direction of the momentum lies along the *axis of rotation*. (This may sound weird, but this is the way the math works!) If you try to change the direction of the axis while an object is spinning, something else has to change with it so that the total angular momentum of the system remains the same.

Conservation of angular momentum is the principle behind the behavior of spinning tops and gyroscopes, like the one in the photograph. Conservation of angular momentum is also the reason a bicycle balances on two wheels and why leaning a bicycle while it is moving causes it to turn, even without turning the handle bars.

$$a = \frac{F}{m}$$

$$\frac{\Delta v}{t} = \frac{F}{m}$$

Finally, let's move the mass m to the other side of this equation (multiply both sides by m) and see what we've got.

$$\frac{m\Delta v}{t} = F$$

This equation says that the mass of an object multiplied by how fast its velocity is changing is equal to the net force acting on it. In this chapter, you have learned that the momentum of an object is equal to its mass times its velocity. Well, if mass times velocity is momentum, then *mass times rate of change of velocity is the rate of change of momentum*. As an equation, this is written

$$\frac{\Delta p}{t} = F$$

This equation is actually the mathematical form of Newton's statement in his second law of motion. Newton's actual words were (translated from Latin), "The change of motion is proportional to the motive force impressed." Newton defined "motion" as "velocity and quantity of matter conjointly," meaning *mv*, what we now call momentum. This means that the rate of change of momentum is what Newton means by his phrase "change of motion." Thus, we come to this profound restatement of Newton's second law: *the rate of change of an object's momentum is equal to the net force acting on it.*

Now let's consider Newton's third law of motion, and apply this important discovery about the effect of a net force. Newton's third law says that a force applied in one direction is always accompanied by an equal force applied in the opposite direction. But above we saw that a net force is equal to the rate of change in an object's momentum. Well, if action force is equal and opposite to reaction force, then the rate of change of momentum in one direction is equal to the rate of change of momentum in the opposite direction. This is just another way of stating the law of conservation of momentum!

Chapter 5 Exercises

Basic Momentum Computations

1. A car with a mass of 1,350 kg is cruising at 35 mi/hr. How much momentum does it have?

2. A small particle has a mass of 8.71×10^{-4} grams. If it is fired from a test apparatus at 725.5 m/s, how much momentum does it have?

3. A baseball with mass 144.3 g is thrown at 99.55 mph. How much momentum does it have?

4. An M114 Howitzer fires an artillery shell with a muzzle velocity of 1,847 ft/s. If the momentum of the shell is 25,565 kg·m/s, determine the mass (kg) and weight (lb) of the shell.

5. An electron has a mass of 9.11×10^{-28} g. If the electron is accelerated to 33.75% of the speed of light, what is its momentum when it is fully up to speed?

6. A proton with a mass of 1.673×10^{-24} g has a momentum of 4.5516×10^{-23} kg·m/s. Determine the velocity of this proton. State your answer in inches per second.

7. A freight train weighing 47,055 tons is traveling at 55.75 mph. Determine its momentum. (One ton is defined as exactly 2,000 pounds.)

8. The earth's mass is 5.98×10^{24} kg. The length of the earth's orbit is 584 million miles. Determine the earth's momentum. (That is, its linear momentum. It also has angular momentum due to its rotation, but that is not our topic.) (Hint: You must compute the earth's velocity using the length of its orbit and the time it takes to complete one orbit. Have fun!)

9. A small bug is speeding across the floor at its top speed of 13.25 in/s. The momentum of the bug is 5.9805×10^{-4} kg·m/s. Determine the mass of the bug and state your answer in grams.

10. A human cannonball is fired from his cannon with a momentum of 2,177.9 kg·m/s. If the man weighs 188.5 lb, determine his velocity as he leaves the cannon.

Answers

1. 21,000 kg·m/s
2. 6.32×10^{-4} kg·m/s
3. 6.420 kg·m/s
4. 45.41 kg, 1.00×10^2 lb
5. 9.22×10^{-23} kg·m/s
6. 1.071×10^6 in/s
7. 1,060,000,000 kg·m/s
8. 1.78×10^{29} kg·m/s
9. 1.777 g
10. 25.4 m/s

Conservation of Momentum Computations

1. A proton is fired at another proton at rest and hits it dead on. All protons have a mass of 1.673×10^{-27} kg. If the first proton is traveling at 8,672 km/s before the collision, what does each proton do immediately after the collision?

2. A large car is parked in a slightly sloping driveway when its brakes fail. The car rolls down the driveway and bumps into a smaller car parked at the bottom of the driveway. The large car is moving at 2.00 mph just before the collision occurs and 1.00 mph just after the collision. The large car has a mass of 3,450 kg and the small car's mass is 1,150 kg. Assuming this collision is elastic (not a bad assumption at very slow speeds), determine the velocity of the small car immediately after the collision. State your answer in mph.

3. For the previous problem, calculate the total kinetic energy of the cars before the collision and after the collision and verify they are the same. (This demonstrates that kinetic energy is conserved and that the collision is elastic.)

4. A small block of wood slides and collides with a larger block of wood that has fives times the mass. After the collision, the small block rebounds at 0.7500 m/s, while the large block moves off at 0.3750 m/s in the direction of the original motion. If the mass of the small block is 375.00 g, what is the initial velocity of the small block before the collision occurs?

5. Two kids are playing around with a couple of hockey pucks. While one hockey puck is at rest, the kids slam it at 9.75 m/s with the other puck. Each hockey puck has a mass of 165 g. What does each hockey puck do immediately after they collide?

6. Two electric model trains are sitting on a long stretch of track. One train has a mass of 5.50 kg and rolls into the second train which is at rest. The second train is much smaller, with a mass of 550.0 g. After the bump, the large and small trains have velocities of 3.2727 cm/s and 7.2727 cm/s, respectively. What is the original velocity of the large train before the bump?

7. For the previous problem, calculate the total kinetic energy before the collision and after the collision and verify they are the same.

8. A young baseball pitcher is practicing by throwing baseballs at a larger solid plastic ball that has exactly ten times the mass. The pitcher throws the 144.3-gram

baseball with a velocity of 65.00 mph and, after a direct, hit the baseball bounces straight back at 53.1818 mph. What velocity does the large plastic ball have just after the baseball hits it? State your result in mph.

Answers

1. $v_{1f} = 0$; $v_{2f} = 8{,}672$ km/s
2. $v_{2f} = 3.00$ mph
3. $E_{Ki} = E_{Kf} = 1380$ J
4. $v_{1i} = 1.125$ m/s
5. $v_{1f} = 0$; $v_{2f} = 9.75$ m/s
6. $v_{1i} = 4.00$ cm/s
7. $E_{Ki} = E_{Kf} = 0.00440$ J
8. $v_{2f} = 11.82$ mph

Momentum Study Questions

1. Give two examples of the law of conservation of momentum in action.

2. Distinguish between elastic and inelastic collisions.

3. What evidence is there that car crashes are usually not elastic collisions?

4. Two protons are fired at each other at different speeds and interact head-on. One proton bounces straight back the way it came. The possibilities for what the second proton does after the interaction are:

   a. It continues moving the way it was going, but at a different speed.
   b. It stops dead.
   c. It bounces back the way it came just like the first one does.

   Which of these possibilities can happen according to the law of conservation of momentum? Explain your answer carefully.

5. Distinguish between momentum and inertia.

6. A "Newton's cradle" consists of a frame with several hard steel balls hanging in it. The balls can all swing like pendulums, but when they are at rest they are all at the same height and just touch each other side by side. Explain why it is that regardless of how many balls are raised up and released from one side, when they reach the bottom and smack into the balls that were not raised the same number of balls goes swinging up on the other side.

7. Hard steel balls are used in a Newton's cradle because they make almost perfectly elastic collisions, so they keep bouncing for a long time. Imagine a Newton's cradle made of, say, hacky sacks hanging in place of the usual steel balls. A collision between hacky sacks is certainly not elastic. What do you think happens in a Newton's cradle made of five hanging hacky sacks if two hacky sacks are raised up on one side and released? Explain your answer.

8. Explain in a few sentences how Newton's second law of motion relates to the momentum of an object.

9. Explain in a few sentences how Newton's third law of motion and the law of con-

servation of momentum are related.

Do You Know ...

What is a hydraulic jump?

When fast-moving water is flowing in an open channel, the water can reach a position where it suddenly widens out, slows down, and the level of the surface water *rises*. The rising of the water surface is called a *hydraulic jump*. You can see a hydraulic jump in the ring in a sink around where the fast-moving water suddenly slows and rises in level. Another beautiful example is the sudden rise in the water level at the Upper Spokane Falls, shown in the center photograph.

These are both examples of *stationary jumps*. An example of a moving hydraulic jump is in the wonderful *tidal bore* photographed at Upper Cook Inlet, Alaska. This tidal bore is a moving wall of water traveling upstream against the tidal current in the inlet.

Other examples of hydraulic jumps may be seen in the water flowing from a dam through a spillway and in white water rapids where people in a raft will experience a sudden rise in the level of the water surface.

The height of a tidal jump can be calculated using the principle of conservation of momentum.

CHAPTER 6
Atoms, Matter and Substances

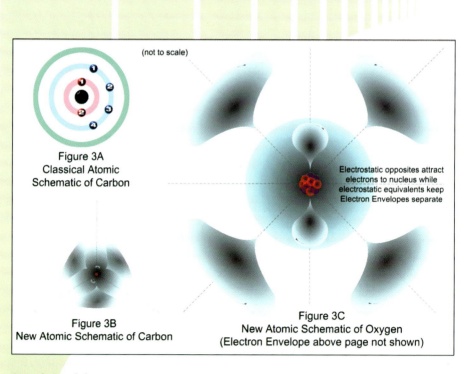

(not to scale)

Figure 3A
Classical Atomic
Schematic of Carbon

Electrostatic opposites attract
electrons to nucleus while
electrostatic equivalents keep
Electron Envelopes separate

Figure 3B
New Atomic Schematic of Carbon

Figure 3C
New Atomic Schematic of Oxygen
(Electron Envelope above page not shown)

Atomic Models

In this chapter, you will read about Ernest Rutherford's 1909 gold foil experiment, in which Rutherford discovered the atomic nucleus. But efforts at modeling the atom did not stop back in the early twentieth century. The image above was developed in 2008, indicating that new atomic models are still being developed. The large image on the right depicts an oxygen atom, with its nucleus in the center. The light blue regions are the "orbitals" where the eight electrons in an oxygen atom are understood to reside. I think the tiny black marks in these orbitals are the artist's attempt to depict the electrons gadding about in there.

OBJECTIVES

Memorize and learn how to use this equation:

$$\rho = \frac{m}{V}$$

After studying this chapter and completing the exercises, students will be able to do each of the following tasks, using supporting terms and principles as necessary:

1. Define and describe *atom* and *molecule*.
2. State the five points of John Dalton's atomic model and the year it was published.
3. Write a brief description of each of the following three major experiments:

    *   J.J. Thomson's cathode ray tube experiment
    *   Robert Millikan's oil drop experiment
    *   Ernest Rutherford's gold foil experiment

    For each written description, include the following:

    *   A brief description of the experimental setup (one or two sentences)
    *   The name of the scientist and the year the experiment was performed
    *   The major discoveries or new atomic model that resulted

4. Write brief descriptions of the Bohr and quantum models of the atom.
5. Use the density equation to calculate the density, volume, or mass of a substance.
6. Calculate the volume of right rectangular solids and use the volume in density calculations.
7. Draw the "family tree" of substances, including in it the following terms: substance, pure substance, alloy, mixture, compound, heterogeneous mixture, homogeneous mixture, colloid, element, and solution.
8. Name two examples for each substance listed in the family tree.
9. Define each of the terms in the substances family tree.
10. Distinguish between compounds and mixtures.
11. Describe the two basic types of structures atoms can form when bonding together to form compounds.
12. Recognize and state the chemical symbols for 24 common element names and vice versa (see Table 6.2).
13. Use the concept of internal energy to define the three common phases of matter—solids, liquids, and gases—and distinguish between them at the molecular level.
14. Explain how evaporation and sublimation occur.
15. Define *heat of fusion* and *heat of vaporization*.

6.1   Atoms and Molecules

Let's begin with a summary of the basic facts about atoms and molecules. Much of this information you probably already know, but I will include some new details.

All matter is made of atoms, which are the smallest basic units ordinary matter is composed of. An *atom* of a given *element* is the smallest unit of matter that possesses all the properties of that element.

Atoms are almost entirely empty space. Each atom has an incredibly tiny nucleus in the center which contains all the atom's *protons* and *neutrons*. Since the protons and the neutrons are in the nucleus, they are collectively called *nucleons*. The masses of protons and neutrons are very nearly the same, although the neutron mass is about 0.1% greater. Each proton and neutron has nearly 2,000 times the mass of an electron, so the nucleus of an atom contains practically all of the atom's mass. Outside the nucleus is a weird sort of cloud surrounding the nucleus containing the atom's *electrons*. Sometimes people have modeled electrons as orbiting the nucleus like planets orbiting the sun, but we have known for a century that this is not at all an accurate description of what is going on with electrons. It is actually very hard to say what is *really* going on, but we are not really going to get into that in this course.

I wrote above that atoms are almost entirely empty space because the nucleus is incredibly small compared to the overall size of the atom with its electron cloud. It's so easy for us to glibly pass over that remark without pausing to consider what it means. So here's an analogy to help you visualize this. Assume that we imagine an atom to be the size of the huge Roman Coliseum, shown in Figure 6.1. Those tiny figures near the center are people. Imagine that one of those people has a flower on his lapel, held there by a straight pin. Suppose he removes the pin in order to straighten his flower. For a moment he holds the pin up and looks at the pinhead. Then he looks up all around at the distant edges of the massive Coliseum. If our atom were the size of the Coliseum, the nucleus of that atom would be the size of that pinhead! If you have ever been inside a football stadium you can try to imagine this for yourself. So what fills all this vast space inside the atom? Nothing, not even air! (Air is also made of atoms.) The inside of an atom is empty space.

Moving along with our discussion of atoms, you probably also know that neutrons have no electric charge. Protons and electrons each contain exactly the same amount of charge, but the charge on protons is positive and the charge on electrons is negative. If an atom has no net electric charge, it contains equal numbers of protons and electrons.

Figure 6.1. Engraving of the Roman Coliseum by Giovanni Piranesi.

Atoms are about 1/5,000th the size of the wavelengths of visible light, which means light does not reflect off atoms and there is no way to see them. The same is true of *molecules*. Molecules are tiny clusters of atoms chemically bonded together. When atoms of different elements are bonded together in a molecule, they form a *compound*, which we will discuss more later in this chapter. But sometimes atoms of the same element bond together in molecules. Examples of this are oxygen, nitrogen, and hydrogen molecules, which are each composed of a pair of atoms. These three gases are examples of *diatomic gases*, so called because their atoms bond together in pairs to form molecules of gas.

6.2   The History of Atomic Models

The story of atomic theory starts back with the ancient Greeks. As we look at how the contemporary model of the atom developed, we will hit on some of the great milestones in the history of chemistry and physics along the way.

6.2.1 Ancient Greece

In the fifth century BC, the Greek philosopher Democritus (Figure 6.2) proposed that everything was made of tiny, indivisible particles. Our word atom comes from the Greek word *atomos*, meaning "indivisible." Democritus' idea was that the properties of substances were due to characteristics of the atoms they are made from. So atoms of metals were supposedly hard and strong, atoms of water were assumed to be wet and slippery, and so on. At this same time, there were various views about what the basic elements were. One of the most common views was that there were four elements—earth, air, water, and fire—and that everything was composed of these.

Figure 6.2. Greek philosopher Democritus (c. 460–370 BC).

Not much real chemistry went on for a very long time. During the medieval period, of course, there were the alchemists who sought to transform lead and other materials into gold. But this cannot be done by the methods available to them, so their efforts were not successful.

6.2.2 John Dalton's Atomic Model

But in the 17th century, things started changing as scientists became interested in experimental research. The goal of the scientists described here was to figure out what the fundamental constituents of matter are. This meant figuring out how atoms are put together, what the basic elements are, and understanding what is going on when various chemical reactions take place. The nature of earth, air, fire, and water were under intense scrutiny over the next 200 years.

In 1803, English scientist John Dalton (Figure 6.3) produced the first scientific model of the atom. *Dalton's*

Figure 6.3. English scientist John Dalton (1766–1844).

129

The Five Principles in Dalton's Atomic Model

1. All substances are composed of tiny, indivisible particles called atoms.
2. All atoms of the same element are identical.
3. Atoms of different elements have different weights.
4. Atoms combine in whole number ratios to form compounds.
5. Atoms are neither created nor destroyed in chemical reactions.

atomic model was based on five main points, which you need to know. These are listed in the accompanying box.

The impressive thing about Dalton's atomic theory is that even today the last three of these points are regarded as correct and the first two are at least partially correct. On the first point, it is still scientifically factual that all substances are made of atoms, but we now know that atoms are not indivisible. This should be obvious, since you know that atoms themselves are composed of protons, neutrons, and electrons. The second point is correct in every respect but one. Except for the number of neutrons in the nucleus, every atom of the same element is identical. However, we now know that atoms of the same element can vary in the number of neutrons they have in the nucleus. These varieties of nuclei are called *isotopes*.

6.2.3 New Discoveries

Moving on toward the later part of the 19th century, many individual elements had been discovered by this time. Another huge milestone was the publication of the first Periodic Table of the Elements by Russian chemist Dmitri Mendeleev in 1869 (Figure 6.4). Mendeleev had discovered that there are patterns in the properties of the elements that he could use to order the elements into rows and columns. When he arranged the

Figure 6.4. Russian scientist Dmitri Mendeleev (1834–1907).

elements this way, he noted that there appeared to be gaps between some of the elements. This led him to predict several elements which had not yet been discovered and he used the

"Ekasilicon" Prediction		Germanium Discovered	
Date Predicted	1871	Date Discovered	1886
Atomic Mass	72	Atomic Mass	72.6
Density	5.5 g/cm³	Density	5.32 g/cm³
Color	dark gray	Color	light gray

Table 6.1. Predicted and actual characteristics of germanium.

patterns in his new table to predict the properties of these unknown elements. One of these was the element germanium, which Mendeleev called *ekasilicon*. Germanium is right below silicon in the Periodic Table of the Elements.

Mendeleev predicted the color, density, and atomic weight of this unknown element. His predictions turned out to be quite accurate once the element was discovered 15 years later. Table 6.1 shows a comparison between Mendeleev's ekasilicon prediction and the actual properties of germanium. Figure 6.17 later in this chapter shows the modern periodic table.

Our attention now shifts from the chemists to the physicists as we take a look at the most famous developments that led to our contemporary understanding of how atoms are structured.

English scientist J.J. Thomson (Figure 6.5) worked at the Cavendish Laboratory in Cambridge, England. In 1897, he conducted a series of landmark experiments that revealed the existence of electrons. Because of his work, he won the Nobel Prize in Physics in 1906 and was knighted in 1908. Thomson's ingenious setup is shown in Figure 6.6.

Thomson placed electrodes from a high-voltage electrical source inside a very elegantly made, sealed-glass vacuum tube. This apparatus generates a so-called *cathode ray* from the negative electrode (1), called the *cathode*, to the positive one, called the *anode* (2). A cathode ray is simply a beam of electrons, but this was not known at the time. (In the era prior to flat-screen displays, television sets and computer monitors used cathode rays to hit the screen from behind, creating the picture on the screen.) The anode inside Thomson's vacuum tube had a hole in it for some of the electrons to escape through, which created a beam of "cathode rays" heading toward the other end of the tube (5).

Figure 6.5. English physicist J.J. Thomson (1856–1940).

Thomson placed the electrodes of another voltage source inside the tube (3), above and below the cathode ray and discovered that the beam of electrons deflected when this voltage was turned on. He also placed magnetic coils on the sides of the tube (4) and discovered that the electrons also deflected as they passed through the magnetic field produced by the

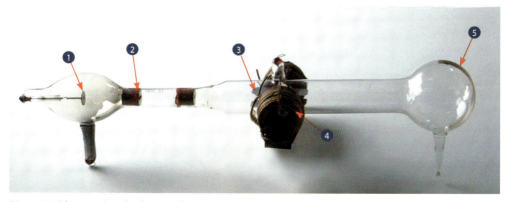

Figure 6.6. Thomson's cathode-ray tube.

131

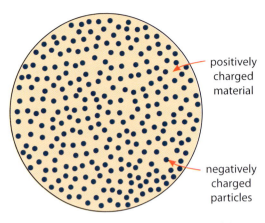

Figure 6.7. Thomson's "plum pudding" model of the atom.

positively charged material

negatively charged particles

coils. The deflection of the beam toward the positive electrode led Thomson to theorize that the beam was composed of negatively charged particles, which he called "corpuscles." (The name *electron* was first used a few years later by a different scientist.) By trying out many different arrangements of cathode ray tubes, Thomson confirmed that the ray was negatively charged. Then using the scale on the end of the tube to measure the deflection angle (5), he was able to determine the charge-to-mass ratio of the individual electrons he had discovered, which is 1.8×10^{11}. (The units of this value, as we would now say, are coulombs/kilogram, or C/kg.) When you think about all the work involved in the hand-blowing of every glass tube, the skill of getting the metal parts inside the glass vacuum tube, and the whole idea of this experiment, you have to admit that this is a *very* cool experiment. *What a great scientist!*

Thomson went on to theorize that electrons came from inside atoms. He developed a new atomic model that envisioned atoms as tiny clouds of massless, positive charge sprinkled with thousands of the negatively charged electrons, as depicted in Figure 6.7. This model is usually called the *plum pudding* model, so let's call it that. But since American students rarely know what plum pudding is, you might think of it as the "watermelon model." The red meat of the watermelon is like the overall cloud of positive charge, and the seeds are like the negatively charged electrons scattered around inside it.

Figure 6.8. American scientist Robert Millikan (1868–1953).

In 1911, American scientist Robert Millikan (Figure 6.8) devised a brilliant experiment that allowed him to determine the charge on individual electrons, 1.6×10^{-19} coulombs (C). Using the value of the charge/mass ratio determined by Thomson, Millikan then calculated the mass of the electron, 9.1×10^{-31} kg. Millikan's apparatus is sketched in Figure 6.9.

Millikan's famous experiment is called the *oil drop experiment*. Inside a metal drum about the size of a large bucket, Millikan placed a pair of horizontal metal plates connected to an adjustable high-voltage source. The upper plate had a hole in the center and was connected to the positive voltage, the lower plate to the negative. He used an atomizer spray pump (like cartoons have on bottles of perfume) to spray in a fine mist of watchmaker's oil above the positive plate. Some of these droplets fell through the hole in the upper plate and moved into the region between the plates. Connected through the side of the drum between the two plates was a telescope eyepiece and lamp Millikan used to see the oil droplets between the plates.

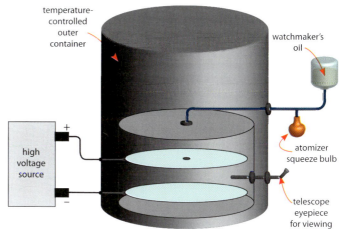

The process of squirting in the oil droplets with the atomizer sprayer caused some of the droplets to acquire a charge of static electricity. This means the droplets had excess electrons on them and were thus negatively charged. They picked up these extra electrons by friction as the droplets squirted through the rubber sprayer tube. As Millikan

Figure 6.9. Millikan's oil drop experiment.

looked at an oil droplet through the eyepiece and adjusted the voltage between the plates, he could make the charged oil droplet hover when the voltage was just right. Millikan took into account the weight of the droplets and the viscosity of the air as the droplets fell and determined that every droplet carried a charge that was a multiple of 1.6×10^{-19} C. From this, he deduced that this must be the charge on a single electron, which it is. Millikan won the Nobel Prize in Physics in 1923 for this work.

The last famous experiment in this basic history of atomic models was initiated in 1909 by one of J.J. Thomson's students at Cambridge, New Zealander Lord Ernest Rutherford (Figure 6.10). Rutherford was already famous when this experiment occurred, having just won the Nobel Prize in Chemistry the previous year. Rutherford's *gold foil experiment* resulted in the discovery of the atomic nucleus. To understand this experiment, you need to know that an *alpha particle* (or α-particle, using the Greek letter alpha) is a particle

composed of two protons and two neutrons. Alpha particles are naturally emitted by some radioactive materials in a process called *nuclear decay*.

Rutherford's experimental setup is sketched in Figure 6.11. Rutherford created a beam of α-particles by placing some radioactive material (radium bromide) inside a lead box with a hole in one end. The α-particles from the decaying radium atoms streamed out of the hole at very high speed. (We looked at these particles in the previous chapter and saw that their velocity is 15,000,000 m/s.) Rutherford aimed the α-particles at a thin sheet of gold foil that was only a few hundred atoms thick. Surrounding the gold foil was a ring-shaped screen coated with a material that glowed momentarily when hit by an α-particle. Rutherford could then determine where the α-particles went after encountering the gold foil.

Figure 6.10. New Zealander and physicist Lord Ernest Rutherford (1791–1867).

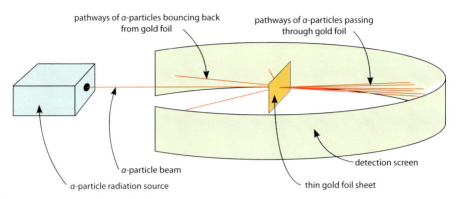

pathways of α-particles bouncing back
from gold foil

pathways of α-particles passing
through gold foil

α-particle beam

α-particle radiation source

detection screen

thin gold foil sheet

Figure 6.11. Rutherford's gold foil experiment.

Rutherford found that most of them went straight through the foil and struck the screen on the other side. However, occasionally an α-particle (one particle out of every several thousand) would deflect with a large angle. Sometimes these deflected particles would bounce almost straight back. This was astonishing and was not at all what Rutherford had expected.

Recall that Thomson had theorized that the positive charge in the atom was dispersed around throughout the atom. Because of this, Rutherford expected the massive and positively charged α-particles to blow right through the gold foil. Most of them did, but when some of them ricocheted backward the astonished Rutherford commented that it was like firing a huge artillery shell into a piece of tissue paper and having it bounce back and hit you! Rutherford's work led to his new proposal in 1911 for a model of the atom. Rutherford's model includes the key points in the next box, which you should know.

Key Principles in Rutherford's Atomic Model

1. The positive charge in atoms is concentrated in a tiny region in the center of the atom, which Rutherford called the nucleus.
2. Atoms are mostly empty space.
3. The electrons, which contain the atom's negative charge, are outside the nucleus.

And again, just what do we mean when we say that atoms are mostly empty space? Just think of that pinhead inside the Roman Coliseum! In 1917, Rutherford became the first to "split the atom." In this experiment he used α-particles again, this time striking nitrogen atoms. His work led to the discovery of the positively-charged particles in the atomic nucleus, which he named protons.

It took another twenty years before James Chadwick (Figure 6.12), another Englishman, discovered the neutron. Before World War I, Chadwick studied under Rutherford (at Cambridge, of course). Then the war began. Not only did the war interrupt the progress of the research in general, but Chadwick was a prisoner of war in Germany. Working back in England after the war, he discovered the neutron in 1932 and received the Nobel Prize in Physics in 1935 for his discovery. And, of course, ten years later he was knighted, too.

Chadwick's discovery of the neutron enabled physicists to fill in a lot of blanks in their understanding of the basic structure of atoms. But years before Chadwick made his discovery, Rutherford's atomic model was already being taken to another level through the work of Niels Bohr.

6.2.4 The Bohr and Quantum Models of the Atom

Danish physicist Niels Bohr (Figure 6.13) is one of the major figures in the development of the modern physics of the 20th century. For his work in the development of quantum physics, he received the Nobel Prize in Physics in 1922. In 1913, Bohr theorized that atoms are like little planetary solar systems with the electrons orbiting the nucleus the way planets orbit the sun. The negatively charged electrons are held in their orbits by electrical attraction to the positively charged nucleus. And a most significant aspect of the model is that each electron in an orbit possesses a specific amount of energy. If an electron absorbs additional energy somehow, it moves to an orbit that is for electrons with higher energy. If an electron emits energy somehow, it moves to an orbit that is for electrons with lower energy.

Figure 6.12. English physicist James Chadwick (1891–1974).

Bohr's "planetary model" is usually depicted as illustrated in Figure 6.14. The model allows a maximum of two electrons in the first energy level, eight electrons in each of the second and third energy levels, and higher numbers in higher levels. In this depiction of Bohr's model, orbits that are farther out from the nucleus are for electrons possessing higher energy.

Even though we now know that electrons do not orbit the nucleus like planets, Bohr was completely right about the electrons possessing specific energies. This concept is now well established in contemporary atomic theory. Of course, when you look at Figure 6.14 it is very difficult to think about the electron energy levels without thinking of them as *spatial*. But a discussion of the spatial arrangement of electrons in atoms is far beyond our objectives for this text. So I advise you to forget thinking about how the electrons are arranged spatially and simply think in terms of how much energy they have. Thinking of electrons as possessing specific energies brings us to our contemporary understanding of atoms, which I will now describe.

Our contemporary model or theory about atoms is called the *quantum model*. The quantum model was pieced together by a host of scientists over a period of decades in the early to mid-20th century. The quantum model includes Bohr's contribution that all electrons possess certain energies. But the big step forward in the quantum

Figure 6.13. Danish physicist Niels Bohr (1885–1962).

model was that energy itself was *quantized*. To say that energy is quantized is to say that there is a smallest unit of energy available in nature. This smallest chunk of energy is called a *quantum* of energy and all energies in the universe are multiples of this one. Energy is not a continuous quantity; it comes in lumps.

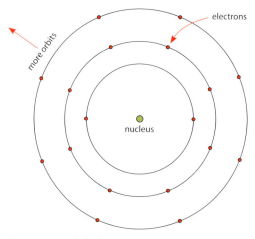

Figure 6.14. Bohr's planetary atomic model.

Our contemporary understanding that energy is quantized is analogous to our understanding that other things are quantized, too, as depicted in Figure 6.15. M&Ms and eggs are quantized; they come in lumps. You cannot buy 5.213 eggs because eggs only come in quanta of one egg. Water is also quantized. Water appears smooth, but is actually composed of huge numbers of individual water molecules. The smallest amount of water that can exist is one molecule of water, and all quantities of water are multiples of this smallest quantity, the single molecule. In the same way, we now understand energy to be quantized, too. There exists a smallest quantum of energy and all energies that exist are multiples of the single quantum of energy (which is unbelievably small). So, in the quantum model of the atom, every electron has a

M&Ms are quantized.

Eggs are quantized.

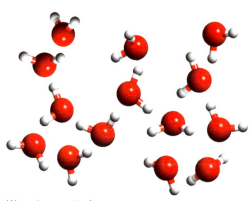

Water is quantized.

Figure 6.15. Things that are quantized: they come in quanta.

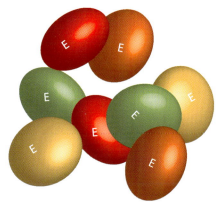

Energy is quantized. (But it doesn't look like this! Energy is not an object.)

certain amount of energy, which is a specific multiple of the smallest quantum of energy. Electrons cannot possess *any* amount of energy. The energy they possess must be a specific multiple of the smallest unit of energy that can exist, the quantum.

6.3   Volume and Density

6.3.1 Calculations with Volume

In the next section, we address the topic of density. To compute densities, you must be comfortable with the various units for volume. There are quite a few of these, so here we will look into volume units a bit.

You are already familiar with the common unit for volume in the U.S. Customary System, the gallon. A standard plastic jug of milk from a grocery store contains one gallon of milk. In the MKS system, the standard volume unit is the cubic meter (m^3). Two more common metric volume units are the cubic centimeter (cm^3) and the milliliter (mL). Another common unit used in scientific work is the liter (L). The liter is not an official SI unit, but it is used all the time anyway. Images depicting these different quantities are shown in Figures 6.16a through 6.16f.

As you can see from Figure 6.16a, a cubic meter (m^3) is quite a large volume, enough to hold several people! In 6.16b, one-liter (1 L) and one-gallon (1 gal) containers are shown side-by-side to help you visualize the fact that there are 3.786 liters in one gallon. Figure 6.16c shows two different objects, each occupying a volume of one liter (1 L). The orange one has dimensions of 10 cm × 10 cm × 10 cm to show that 1 L is equal to 1,000 cm^3. There are 1,000 liters in one cubic meter, as suggested by Figure 6.16d. Figure 6.16e shows a small cube with a volume of one cubic centimeter (cm^3) and a 10 × 10 array of 100 of these cubes, which obviously has a volume of 100 cm^3. As you can see by comparing Figure 6.16a and 6.16e, a volume of 100 cm^3 is clearly *not* equal to 1 m^3, a common mistake made by students learning to work with unit conversions.

A milliliter (mL) is one one-thousandth of a liter (0.001 L). It is also the case that one milliliter is equal to one cubic centimeter. You may also find it interesting to know that it takes about 40 drops of liquid to make one milliliter (1 mL), as suggested by Figure 6.16f. This means that a single drop has a volume of about 0.025 mL. By convention, the liter is generally used only in reference to liquid volumes. The cubic meter and cubic centimeter are commonly used for everything.

Now that we have reviewed the basic volume units, it would be good for you to turn again to Section 2.2.2 on Converting Units of Measure and re-read the paragraphs relating to volume units (Examples 2.1 and 2.4). The exercises at the end of this chapter will give you a substantial amount of practice calculating volumes and converting volume units. Play close attention to those exercises. Mastering volume units and unit conversions is critical to success in science!

6.3.2 Density

Earlier in the chapter, I mentioned Dmitri Mendeleev's predictions about unknown elements. His predictions included estimates for the densities of the elements. *Density* is an example of a what is called a *physical property*. All substances have a certain density, as well as other physical properties such as color, boiling point, hardness, and electrical conductivity.

a. The white frame encloses a volume of one cubic meter.

b. One-gallon and one-liter containers.

c. Each of these objects has a volume of one liter.

d. To fill the 1-m³ cube would require 1,000 1-L cubes.

e. One cubic centimeter (left) and 100 cubic centimeters (right).

f. It takes about 40 drops to equal 1 mL, or 1 cm³.

Figure 6.16. Various volume units.

Density is a measure of how much matter is packed into a given volume for different substances. No doubt you are already familiar with the concept of density. You know that if you hold equally sized balloons in each hand, one filled with water and one filled with air, the water balloon weighs more because water is denser than air. You know that for equal weights of sand and Styrofoam packing peanuts, the volume of the packing material is much larger because the packing material is much less dense. And you probably also know that objects less dense than water float, while objects denser than water sink.

The equation for density is

$$\rho = \frac{m}{V}$$

where the Greek letter ρ (spelled rho and pronounced like row, which rhymes with snow) is the density in kg/m³, m is the mass in kg, and V is the volume in m³. These are the variables and units in our familiar MKS unit system. However, since laboratory work typically involves only small quantities of substances, it is quite common for densities to be expressed in g/cm³. In the examples that follow, I will illustrate the use of both these units. If all you are doing is using the density equation, then either of these units is fine. However, if you are using the density with other equations in a multi-step calculation, such as calculating the weight of an object, then you certainly need to stick with MKS units as always.

Finally, note that the density of water at room temperature is

$$\rho_{water} = 998 \ \frac{\text{kg}}{\text{m}^3} = 0.998 \ \frac{\text{g}}{\text{cm}^3}$$

This value is useful to know because water comes up in many different applications. So, the density of water is the third and final physical constant listed on the memory list for *Introductory Physics* (see Appendix A). Now is the time to commit this value to memory.

▼ Example 6.1

The density of germanium is 5.323 g/cm³. A small sample of germanium has a mass of 17.615 g. Determine the volume of this sample.

Begin by writing the givens.

$$\rho = 5.323 \ \frac{\text{g}}{\text{cm}^3}$$
$$m = 17.615 \ \text{g}$$
$$V = ?$$

Now write the equation and solve for the volume. Then insert the values and compute the result.

$$\rho = \frac{m}{V}$$
$$V = \frac{m}{\rho} = \frac{17.615 \ \text{g}}{5.323 \ \frac{\text{g}}{\text{cm}^3}} = 3.309 \ \text{cm}^3$$

This value has four significant figures, as it should.

▼ Example 6.2

Determine the density of a block of plastic that weighs 18.25 N and has dimensions 4.0 in × 2.5 in × 9.50 in. State your result in g/cm³.

We begin by writing down the given information and converting the length units to centimeters.

$$F_w = 18.25 \text{ N}$$

$$l = 4.0 \text{ in} \cdot \frac{2.54 \text{ cm}}{\text{in}} = 10.16 \text{ cm}$$

$$w = 2.5 \text{ in} \cdot \frac{2.54 \text{ cm}}{\text{in}} = 6.35 \text{ cm}$$

$$h = 9.0 \text{ in} \cdot \frac{2.54 \text{ cm}}{\text{in}} = 22.86 \text{ cm}$$

$$\rho = ?$$

As you can see from the density equation, to calculate the density we need the mass. We obtain the mass from the given weight. We also need the volume, which we calculate from the given dimensions. Since the weight calculation is always performed with MKS units, the resulting mass will be in kilograms. We convert the mass to grams, since the problem statement requires our answer to be in g/cm³.

To calculate the mass, we write

$$F_w = mg$$

$$m = \frac{F_w}{g} = \frac{18.25 \text{ N}}{9.80 \ \frac{\text{m}}{\text{s}^2}} = 1.86 \text{ kg} \cdot \frac{1000 \text{ g}}{\text{kg}} = 1860 \text{ g}$$

Then we calculate the volume

$$V = l \cdot w \cdot h = 10.16 \text{ cm} \cdot 6.35 \text{ cm} \cdot 22.86 \text{ cm} = 1475 \text{ cm}^3$$

Finally, the density is

$$\rho = \frac{m}{V} = \frac{1860 \text{ g}}{1475 \text{ cm}^3} = 1.26 \ \frac{\text{g}}{\text{cm}^3}$$

Rounding to two significant figures, we have

$$\rho = 1.3 \ \frac{\text{g}}{\text{cm}^3}$$

6.4   Types of Substances

6.4.1 Major Types of Substances

A *substance* is anything that contains matter. There are many different types of substances, but they fall into two broad categories — *pure substances* and *mixtures*. I am going to describe pure substances first, but it will be helpful to know that a *mixture* is made any time different substances are combined together without a chemical reaction occurring. If a chemical reaction does occur when different substances are combined, a pure substance called a *compound* is formed. Thus, these are the two major types of substances that are composed of combinations other substances—compounds and mixtures. The major distinction between mixtures and compounds is that if no chemical reaction occurs when substances are combined, the resulting substance is a mixture. If a chemical reaction does occur when substances are combined, a compound (or perhaps more than one compound) is formed. We will come back to compounds shortly.

There are two kinds of *pure substances*, one of which is compounds. The other kind of pure substance is a group of substances called *elements*. We discuss elements first.

6.4.2 Elements

In previous science classes, you may have seen or studied the Periodic Table of the Elements, which lists all the known elements. This famous table, shown in Figure 6.17, on the next page, plays a major role in the study of physics and chemistry. As you read below about elements, enjoy the great photographs of a few of the elements in Figures 6.18 through 6.22.

The factor that defines each element in the periodic table is the number of protons the element has in the nucleus of each of its atoms. This number, called the *atomic number,* is the number used for ordering elements in the periodic table. For example, the atomic number of carbon is 6. This means that an atom of carbon has six protons. All carbon atoms have six protons: if an atom does not have six protons it is not a carbon atom, and if an atom does have six protons it is a carbon atom. An element is therefore a type of atom, classified according to the number of protons the atom has. Elemental carbon is any lump of atoms that contain only six protons apiece. Oxygen, atomic number 8, is another example of an element. Pure oxygen is a gas (ordinarily) that contains only atoms with eight protons each. Other examples of elements you have heard of are iron, gold, silver, neon, copper, nitrogen, lead, and many others.

For every element there is a *chemical symbol* which is used in the periodic table and in the chemical formulas for compounds (which we will discuss next). For some elements, a single upper case letter is used, such as N for nitrogen and C for carbon. For other elements, an upper-case letter is followed by one lower-case letter, such as Na for sodium and Mg for magnesium. (The three-letter symbols at the right side of Period 7 beginning with U are just placeholders until official names and two-letter symbols are selected by the appropriate governing officials.)

Some of the chemical symbols are based on the Latin names, such as Ag for silver, which

Figure 6.18. Synthetically-grown crystals of gold (Au), atomic number 79.

141

Figure 6.17. The Periodic Table of the Elements.

	1	2	3	4	5	6	7	8	9	10	11	12	13	14	15	16	17	18
1	1 H Hydrogen 1.0079																	2 He Helium 4.0026
2	3 Li Lithium 6.941	4 Be Beryllium 9.0122											5 B Boron 10.811	6 C Carbon 12.011	7 N Nitrogen 14.0067	8 O Oxygen 15.9994	9 F Fluorine 18.9984	10 Ne Neon 20.1797
3	11 Na Sodium 22.9898	12 Mg Magnesium 24.3050											13 Al Aluminum 26.9815	14 Si Silicon 28.0855	15 P Phosphorus 30.9738	16 S Sulfur 32.066	17 Cl Chlorine 35.4527	18 Ar Argon 39.948
4	19 K Potassium 39.098	20 Ca Calcium 40.078	21 Sc Scandium 44.9559	22 Ti Titanium 47.88	23 V Vanadium 50.9415	24 Cr Chromium 51.9961	25 Mn Manganese 54.9380	26 Fe Iron 55.847	27 Co Cobalt 58.9332	28 Ni Nickel 58.6934	29 Cu Copper 63.546	30 Zn Zinc 65.39	31 Ga Gallium 69.723	32 Ge Germanium 72.61	33 As Arsenic 74.9216	34 Se Selenium 78.96	35 Br Bromine 79.904	36 Kr Krypton 83.80
5	37 Rb Rubidium 85.468	38 Sr Strontium 87.62	39 Y Yttrium 88.9059	40 Zr Zirconium 91.224	41 Nb Niobium 92.9064	42 Mo Molybdenum 95.94	43 Tc Technetium 98.9072	44 Ru Ruthenium 101.07	45 Rh Rhodium 102.9055	46 Pd Palladium 106.42	47 Ag Silver 107.8682	48 Cd Cadmium 112.411	49 In Indium 114.82	50 Sn Tin 118.710	51 Sb Antimony 121.76	52 Te Tellurium 127.60	53 I Iodine 126.9045	54 Xe Xenon 131.29
6	55 Cs Cesium 132.905	56 Ba Barium 137.327	71 Lu Lutetium 174.967	72 Hf Hafnium 178.49	73 Ta Tantalum 180.9479	74 W Tungsten 183.85	75 Re Rhenium 186.207	76 Os Osmium 190.2	77 Ir Iridium 192.22	78 Pt Platinum 195.08	79 Au Gold 196.9665	80 Hg Mercury 200.59	81 Tl Thallium 204.3833	82 Pb Lead 207.2	83 Bi Bismuth 208.9804	84 Po Polonium 208.9824	85 At Astatine 209.9871	86 Rn Radon 222.0176
7	87 Fr Francium 223.0197	88 Ra Radium 226.0254	103 Lr Lawrencium 262.11	104 Rf Rutherfordium 261.11	105 Db Dubnium 262.114	106 Sg Seaborgium 263.118	107 Bh Bohrium 262.12	108 Hs Hassium (265)	109 Mt Meitnerium (266)	110 Ds Darmstadtium (281)	111 Rg Roentgenium (281)	112 Cn Copernicium (285)	113 Uut Ununtrium (284)	114 Fl Flerovium (289)	115 Uup Ununpentium (288)	116 Lv Livermorium (293)	117 Uus Ununseptium (294)	118 Uuo Ununoctium (294)

■ liquid at room temperature

■ radioactive

57 La Lanthanum 138.9055	58 Ce Cerium 140.115	59 Pr Praseodymium 140.9077	60 Nd Neodymium 144.24	61 Pm Promethium 144.9127	62 Sm Samarium 150.36	63 Eu Europium 151.965	64 Gd Gadolinium 157.25	65 Tb Terbium 158.9253	66 Dy Dysprosium 162.50	67 Ho Holmium 164.9303	68 Er Erbium 167.26	69 Tm Thulium 168.9342	70 Yb Ytterbium 173.04
89 Ac Actinium 227.0278	90 Th Thorium 232.0381	91 Pa Protactinium 231.0359	92 U Uranium 238.0289	93 Np Neptunium 237.0482	94 Pu Plutonium 244.0642	95 Am Americium 243.0614	96 Cm Curium 247.0703	97 Bk Berkelium 247.0703	98 Cf Californium 251.0796	99 Es Einsteinium 252.083	100 Fm Fermium 257.0951	101 Md Mendelevium 258.10	102 No Nobelium 259.1009

stands for *argentum*, and Pb for lead, which stands for *plumbum*. There is no need for you to memorize the periodic table. That would truly be torture for most us. However, there are many elements that one encounters so frequently in scientific study that you should memorize the chemical symbols for them. I have listed 24 elements in Table 6.2 that you should know for this course. For each one, you should be able to give the symbol from the name and vice versa. Spelling matters, so be sure to notice the spellings in elements like sulfur and fluorine.

Figure 6.19. Synthetic crystals of silver (Ag), atomic number 47.

6.4.3 Compounds

Now that you know what elements are, we can describe *compounds* in greater detail. Compounds and elements are the two types of pure substances. As the name implies, a compound is formed when two or more elements are chemically bonded together, which is always the result of a *chemical reaction*. A chemical reaction is any process in which connecting bonds between atoms are formed or broken. It takes a chemical reaction to bond atoms together, and it takes a different chemical reaction to break them apart.

A compound may consist of only two elements, as in the case of water, which is composed of hydrogen and oxygen, and carbon dioxide, which is composed of carbon and oxygen. All compounds can be represented by a *chemical formula*, which specifies the elements that are in the compound, and in what ratios. For example, the chemical formulas for water and carbon dioxide are written as H_2O and CO_2, respectively. Other compounds are composed of several elements, such as acetic acid, denoted by the formula CH_3COOH (or $C_2H_4O_2$), which consists of carbon, hydrogen, and oxygen. Another example is ammonium phosphate, which is denoted by the formula $(NH_4)_3PO_4$ and is composed of nitrogen, hydrogen, phosphorus, and oxygen.

When atoms bond together to form a compound, the atoms in the compound can be arranged in two different basic types of structures. In many cases, the atoms join together in small, individual, identical groups called *molecules*. Water is composed of molecules, each molecule consisting of two hydrogen atoms and one oxygen atom, as depicted in Figure 6.23. You should notice the characteristic elbow shape of a water molecule. This shape is responsible for many of water's unusual properties. (For example, see the box on page 154.) Other examples of compounds composed of molecules are ammonia (NH_3), sulfur dioxide (SO_2), and methane (CH_4).

Figure 6.20. A synthetic crystal of bismuth (Bi), atomic number 83. The surface is a thin, iridescent oxide coating.

Do You Know ...

What structures can carbon atoms make?

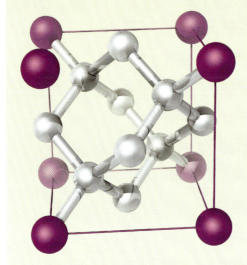

Carbon atoms can combine together in many different amazing crystal and molecular shapes. Diamonds are made of carbon atoms arranged as shown in the image to the left. Each sphere is a carbon atom, and the purple ones mark the corners of a so-called *unit cell*, the repeating structure inside a diamond. A diamond crystal consists of huge numbers of these unit cells all stacked together with their corners in common.

Sixty carbon atoms can join together in the same pattern of hexagons and pentagons that marks the surface of a soccer ball. This molecule, C_{60}, is called *buckminsterfullerene*, named after Buckminster Fuller, the architect who invented the geodesic dome. (People also like to call these buckyballs.)

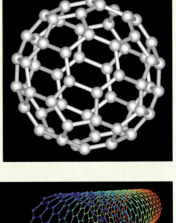

Recently, scientists have been learning how to make and apply tiny tubes of carbon atoms called *carbon nanotubes*. The carbon atoms can join together in three different lattice structures to form these tubes. Carbon nanotubes have the highest tensile strength of any known material. Their strength is analogous to a cable with the diameter of a toothpick supporting a weight of over 14,000 pounds! Nanotubes can even be cut and unrolled to make *graphene*, a flat sheet of one-atom thick carbon, and one of the strongest materials ever tested.

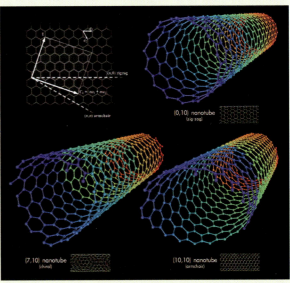

Name	Symbol	Name	Symbol	Name	Symbol
Aluminum	Al	Iron	Fe	Potassium	K
Calcium	Ca	Lead	Pb	Silicon	Si
Carbon	C	Magnesium	Mg	Silver	Ag
Chlorine	Cl	Mercury	Hg	Sodium	Na
Copper	Cu	Neon	Ne	Sulfur	S
Gold	Au	Nickel	Ni	Tin	Sn
Helium	He	Nitrogen	N	Uranium	U
Hydrogen	H	Oxygen	O	Zinc	Zn

Table 6.2. Names and chemical symbols for 24 common elements.

The other basic way atoms combine is in a continuous, geometric arrangement. These arrangements are called *crystals*, and the structure the atoms in the compound make when they join together is called a *crystal lattice*. Atoms form into a crystal lattice by first becoming *ions*—atoms which have gained or lost one or more electrons. When an atom gains or loses electrons, it also gains or loses negative charges since each electron carries a negative charge. Thus, ions are charged particles. Particles of opposite charge are attracted together and form into the crystal lattice. The structure of a crystal lattice depends on the charges of the ions involved.

The number of different arrangements atoms can make in a lattice is endless, and these arrangements are responsible for many of the unusual properties crystals possess. But what all lattices have in common is the regular arrangement of the atoms into repeating, geometrical patterns. A sketch of the very simple crystal structure for sodium chloride (NaCl, that is, table salt) is shown in Figure 6.24. The outlines on the crystal are there to help you visualize this 3-D depiction.

One last point to note about compounds is that the properties of a compound are completely different from the properties of any of the elements in the compound. Consider water, oxygen, and hydrogen. Oxygen is an invisible gas that you breathe in the air and that supports combustion. Hydrogen is an invisible, flammable gas. Water is composed of ox-

Figure 6.21. A vial of bromine (Br), atomic number 35, one of only two elements that are liquids at room temperature.

Figure 6.22. Rare-earth element thulium (Tm), atomic number 69, which is soft enough to cut with a knife.

145

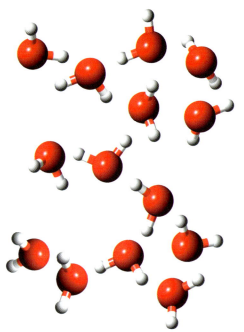

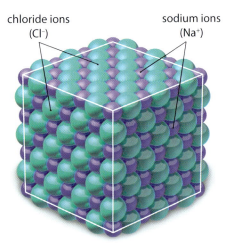

chloride ions
(Cl⁻)

sodium ions
(Na⁺)

Figure 6.24. Sodium and chlorine atoms ionize and join together in a crystal lattice to form sodium chloride.

Figure 6.23. Water molecules in liquid water. Oxygen (O) atoms are shown in red; hydrogen (H) atoms are shown in white.

ygen atoms bonded to hydrogen atoms, but one cannot breathe water and water neither combusts nor supports combustion. Or consider sodium chloride, which is the table salt you commonly apply to your food. We all have to have salt in our diets, and we find it tasty. But both sodium and chlorine, the two elements of which sodium chloride is composed, are deadly dangerous in their pure, elemental forms.

So far, we have been discussing one major category of substances, pure substances, which, again, consists of elements and compounds. The other major category is *mixtures*. Any time substances are mixed together without a chemical reaction occurring, a mixture is formed. Remember—if a chemical reaction occurs compounds are formed, not mixtures.

6.4.4 Heterogeneous Mixtures

If you toss vegetables in a salad, you've made a mixture. If you put sugar in your tea or milk in your coffee, you've made a mixture. If you mix up a batch of chocolate chip cookie dough, a bowl of party mix, or the batter for a vanilla cake, you've made a mixture. There are two basic kinds of mixtures, *heterogeneous mixtures* and *homogeneous mixtures* (pronounced het-ur-oh-JEEN-ee-us and home-oh-JEEN-ee-us).

In heterogeneous mixtures, you can see with your eyes or a microscope that at least two different substances are in the mixture. The foods shown in Figure 6.25 are examples of heterogeneous mixtures in which the different substances are plainly visible. You can easily see several different substances in these mixtures. Other examples of heterogeneous mixtures are pizza, salads, garden soil, and muddy water.

A common type of heterogeneous mixture is the *colloidal dispersion*, often simply referred to as a *colloid*. A colloid is a mixture in which there are two different phases of substances present. Substances typically exist in three different phases, solid, liquid, and gas (or vapor). If substances in two different phases are mixed together and they stay in their different phases, they make a colloid. A common example of a colloid is clouds or fog, both of which have water droplets (liquid) mixed in air (gas). Paint is another good example, which has tiny solid particles mixed into a liquid. Other common examples of colloids are marshmallows (gas bubbles mixed in a solid), mayonnaise (liquid drops dispersed in another liquid), and shaving cream (gas bubbles mixed in a liquid).

In some heterogeneous mixtures, the different substances can only be seen under a microscope. A sample of water from a still, clear stream may appear to be pure water. But examination under a microscope will show that the water is full of particulates and microorganisms.

6.4.5 Homogeneous Mixtures

In a homogeneous mixture, the substance is completely uniform, as with the glass of Gatorade and the stainless steel pan shown in Figure 6.26. Even under a microscope, they appear completely uniform, a yellow-green liquid and a silvery metal. However, the sports drink is a mixture, not a compound. If we could examine the drink at the

Figure 6.25. The potato soup and rice dish are heterogeneous mixtures.

molecular level, we would see water molecules, sugar molecules, and other particles all jumbled together. Other examples of homogeneous mixtures are salt water, coffee with sugar, vinegar, ammonia cleaning solution, and Kool-Aid.

Another name for homogeneous mixtures is *solutions*. Solutions appear everywhere in both everyday life and chemistry. A solution is made when one substance, called the *solute*, is *dissolved* into another substance, called the *solvent*. This process is illustrated in Figure 6.27. The solute can be a gas, like the CO_2 dissolved in a soft drink, a liquid, such as flavoring or syrup, or a solid, such as sugar or salt. (In fact, the solvent can be a solid too, but we have to melt it first to make a solution with it.) Water is by far the most common liquid solvent, and water solutions are found everywhere in nature, from oceans to lakes to the human body. When two liquids are mixed into a solution, the one that has the most volume is generally called the solvent, although it doesn't really matter. One could just as easily say that the two liquids are dissolving into each other.

There is a special class of *solid solutions* called *alloys*. An alloy is a solid solution of metals. Usually, to make an alloy the metals must be melted first so that they are liquids. But once the metals are melted, they can be thoroughly mixed together and allowed to cool to form the alloy. There are three particular alloys that we encounter all the time, so it is handy know what they are made of.

Figure 6.26. Sports drinks and stainless steel are homogeneous mixtures (solutions).

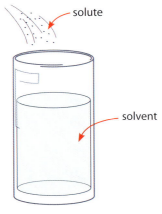

Figure 6.27. The two components of a solution.

Steel is an alloy made of iron with a small amount of carbon mixed in. There are many different steel alloys, including many different alloys called "stainless" steel because they don't rust (like the frying pan in Figure 6.26). Steel is one of the most popular building materials in the world, and some alloys of steel include other metals in addition to the iron and carbon.

Other well-known alloys are *brass*, which is an alloy of copper and zinc, and *bronze*, an alloy of copper and tin often used as a sculptural medium. Both brass and bronze contain copper, but it can be hard to remember which one is made with zinc and which one with tin. Here is my helpful way to remember which is which: "Bronze has a z, so it doesn't need the Z. Brass doesn't have a z, so it gets the Z." The Z here stands for *zinc*. Clever, huh?

As one more final example of alloys, most cars these days are available with "alloy wheels." These wheels are made mostly of aluminum, but there are small quantities of other metals or silicon blended in to give the wheels the desired mechanical properties. There are many different alloys of aluminum, all with slightly different mechanical or electrical properties. This is because aluminum is the material of choice for a great variety of applications due to its high strength relative to its low weight. (This is called the strength-to-weight ratio.) So quite a few different aluminum alloys have been developed for different purposes. (And by the way, in every country in the world except America, this element is called *aluminium*.)

The relationships between all these different types of substances are summarized (or will be soon) in Figure 6.28.

Do-It-Yourself Graphics!

Now that we have reviewed the different types of substances, make a chart, a sort of "family tree" of the substances. Your chart should include the ten terms listed in item 7 of the Objectives List for this chapter. Most physical science books show this chart in the text. I am not going to because you will remember it much better if you develop it yourself based on the information presented here in this section.

Start at the top with the term SUBSTANCES, and draw in the various substances in a tree underneath, showing which types are subsets of other types. You will have ten terms (nine boxes) in all in your family tree when you finish. While you are at it, write in two examples for each type of substance you list, and don't use any particular example more than once.

Figure 6.28. The family tree of substances (to be drawn by you).

6.5   Phases and Phase Changes

6.5.1 Phases of Matter

Our final topic in this chapter is *phases* (or *states*) of matter and phase changes. No doubt you have studied these before. Our only objective here is for you to be able to distinguish the four phases of matter from each other at the atomic or molecular level.

The four basic phases of substances are *solid*, *liquid*, *gas* or *vapor*, and *plasma*. The third phase may be called gas or vapor, depending on whether the substance normally exists in nature as a gas. If a substance normally exists in nature as a gas, we refer to a substance in this phase as a gas. Oxygen and nitrogen are examples of gases. (In fact, air is a homogenous mixture—that is, a solution—composed almost entirely of these two gases.) But if a substance is normally a liquid or a solid at room temperature and we boil some of it, we conventionally refer to the substance after it has been converted to the vapor phase as a vapor. The humidity in the air, for example, is water vapor.

The *internal energy* of a substance distinguishes the three phases from one another. We discuss internal energy more in the next chapter, but for now we can say that the internal energy of a substance is the sum of the kinetic energies of all the atoms or molecules in the substance. Every atom and every molecule is always in motion, and you know this means each atom or molecule has a certain amount of kinetic energy, equal to $\frac{1}{2}mv^2$. If we add up all these kinetic energies for a given quantity of a substance, we have the internal energy of the substance. As you will see, the internal energy is related strongly to the temperature of the substance. Heating a substance increases the internal energy and the temperature of the substance. Cooling a substance decreases the temperature and the internal energy of the substance.

Now let's relate the internal energy and temperature of a substance to its different phases. If a substance has a very low internal energy, this means its temperature is low and the substance is in the *solid* phase. (For a substance normally in the liquid phase we would say it is now frozen.) Since the atoms have a low internal energy, the atoms' kinetic energies are low and they are moving relatively slowly. This low energy state for the atoms means the atoms cling strongly to one another by the electrical attraction of their charged particles (protons and electrons), forming a solid, so the atoms or molecules are not free to move around. They are stuck in place. The atoms *do* have kinetic energy, so they are not at rest. But they do not have enough energy to break free of the electrical attractions holding them all together, so their kinetic energy forces them to vibrate in place. This is what is going on with the atoms in any solid. Figure 6.29 depicts the atoms in a crystal solid. Of course, molecules can form solids as well, but in many cases, the molecules in molecular solids are not organized in an orderly fashion the way the atoms are in crystals.

Now let's imagine that we heat this substance, raising the internal energy and the temperature of the substance. Every pure substance has a particular temperature at which the higher kinetic energy in the atoms overcomes the electrical attraction between the atoms, allowing the atoms to come apart. This is the *melting point* of the substance.

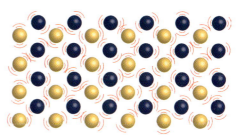

Figure 6.29. Atoms in a solid have low internal energy. They are bound together by electrical attraction and vibrate in place.

Figure 6.30 illustrates what is happening with the energy and temperature in a substance during the phase transition from solid to liquid (*melting*) or vice versa (*freezing*). At the left end of the curve, the substance is a solid. As the substance is heated, its temperature increases so the curve slopes up, indicating that increased internal energy in the substance correlates to increased temperature. When the temperature reaches the melting point, the particles have enough energy to stay apart at that same temperature if they are given an

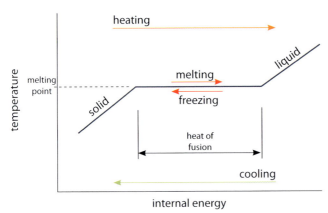

Figure 6.30. When a solid reaches the melting point, it melts completely only when an amount of energy equal to the heat of fusion for that substance is added. While this energy is being added, the temperature remains the same.

extra boost of energy to shake them loose from each other. This quantity of energy is called the *heat of fusion*.

At the melting point, as heat is added to the substance during melting, the energy allows one molecule after another to break free of the solid group. This is what melting is, and during melting the temperature of the substance stays the same. The steady temperature is indicated by the horizontal section of the curve in the figure. For water, the temperature where this happens is 0°C, and this is why ice water is always 0°C.

Adding more heat allows more water molecules to break free from the crystal lattice of the ice, but it does not raise the temperature. After the full heat energy equal to the heat of fusion has been added to the substance, the substance will be 100% liquid. Adding more heat after this causes the temperature to increase, as shown by the sloping part of the curve at the right end.

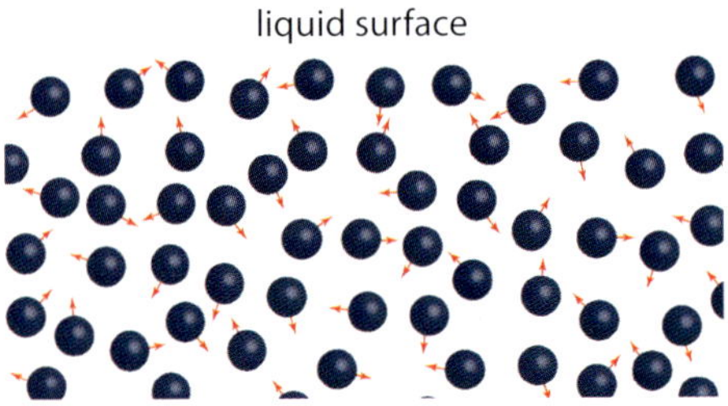

The diagram in Figure 6.30 is one type of *phase diagram*. (A more common type of phase diagram is shown in the box on page 153.) We can also read the diagram in reverse. When a liquid substance is cooled, the temperature drops until the freezing point is reached (the same temperature as the melting point). As more heat is removed, the liquid particles begin to cling strongly to one another because they do not have enough kinetic energy to stay apart any more, and the electrical attractions pull them together into the solid. After an amount of energy equal to the heat of fusion is removed, the entire substance is a solid. Further cooling after that drops the temperature of the solid below the freezing point.

Figure 6.31. Atoms or molecules in a liquid have higher internal energy. They are loosely bound together and are free to move around.

After the substance melts, it is a *liquid*. In the liquid phase, the atoms are still held together loosely, which is why liquids stay together in an open container or in puddles if the liquid is spilled. The atoms in a liquid move very fast, and there are a lot of them. The result is that in a liquid, the atoms are continuously colliding with each other at very high speeds, encountering an

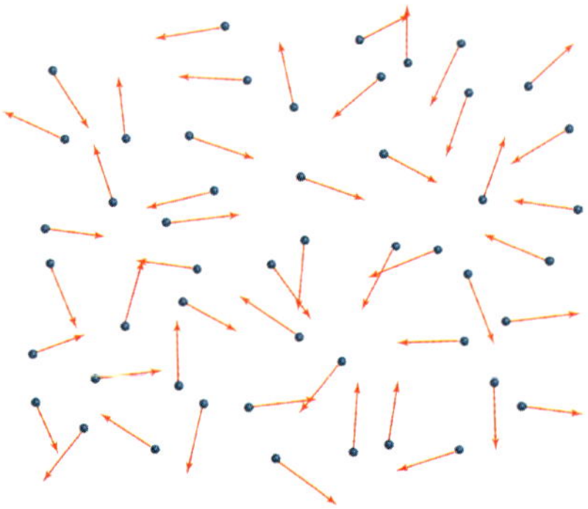

Figure 6.32. Atoms in a gas have high internal energy. They are not bound together at all. They have higher velocities and spread apart as far as their environment allows them to.

Do You Know ...

The variety of natural crystal structures is endless, and their beauty often defies description. The image to the right is a sample of chalcopyrite. Can you believe this? Do you think our creator loves color? The image at the lower right shows an example of crystal twinning, in this case with crystals of pyrite. I love the mathematical regularity shown in the way the three cubes lock together. The crystals in the image at the lower left formed inside a hollow stone called a *geode*. Groundwater can dissolve minerals out of rocks, leaving empty cavities in which beautiful new crystals can grow. These crystal deposits are secret—unknown to anyone—until someone cuts the geode open to show the lovely crystals inside.

Why are crystals so fascinating?

Creation is a wonderful and mysterious gift. And what a rich gift!

incredible number of collisions every second. Figure 6.31 depicts the atoms or molecules in a liquid.

If we heat the substance further, the temperature increases. Eventually, the temperature reaches the *boiling point*. At this temperature, the internal energy of the atoms or molecules is so high that they no longer stay together and the substance enters the *gas* phase. Instead of staying together, if the atoms are not contained they have so much energy they fly away at great speed. Because of their higher speeds, the atoms or molecules spread apart, still colliding with one another, but spreading out as much as they are able. If they are contained in a container of some kind, they still fly around at great speed, but instead of flying off into the atmosphere they collide furiously with each other and with the walls of the container. These collisions between the particles and the container walls are the cause of the gas pressure in

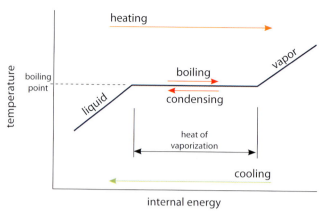

Figure 6.33. When a liquid reaches the boiling point, it vaporizes completely only when an amount of energy equal to the heat of vaporization for that substance is added. While this energy is being added, the temperature remains the same.

the container. We will consider this more closely in the next chapter, which includes a mind-blowing example showing just how fast these atoms move and how frequently they collide. Figure 6.32 depicts fast-moving atoms or molecules in a gas or vapor.

Figure 6.33 shows another phase diagram, this time illustrating the phase transition from liquid to vapor or gas (*boiling* or *vaporization*) or vice versa (*condensation*). This diagram is a lot like the previous one, and the logic behind it is the same. For the particles to completely break free from one another so they are not clinging to one another at all, an amount of energy called the *heat of vaporization* is required. During boiling, the temperature stays the same until the substance has completely vaporized. For water, the temperature at which this happens is 100°C.

All these phase change processes work the same way in reverse when cooling the substance. If we cool a vapor down to the *condensation temperature* (the same temperature as the boiling point), the atoms have low enough energies that they begin clinging to one another in a liquid. This is called *condensation*. Cool the substance further to the *freezing point* (the same temperature as the melting point) and the energies are low enough that the atoms lock together in a solid where they can only vibrate in place.

The heat of fusion and heat of vaporization depend on the electrical attractions that exist between the particles of a particular substance. This means these values are different for

	Heat of Fusion (kJ/kg)	Heat of Vaporization (kJ/kg)
water	334	2,260
iron	247	6,090
copper	209	4,730

Table 6.3. Example values for heat of fusion and heat of vaporization.

every substance. Some representative values are shown in Table 6.3. One kilojoule (1 kJ) is the amount of energy consumed by a 60-watt light bulb in 17 seconds. It takes a lot of heat to melt one kilogram of water—more than required for iron or copper. But when it comes to vaporization, the metals leave water in the dust for energy required.

The fourth phase of matter is the *plasma*. A plasma is an *ionized gas*. Recall from the discussion of compounds in Section 6.4.3 that a charged atom is called an *ion*. In the presence of heat or a strong electric field, atoms can be stripped of some of their electrons and become ionized. If this happens in a gas, a plasma is formed.

The electric fields present in thunderstorms can produce a plasma—the bright light of a lightning bolt is the light given off by ionized air molecules, a common plasma. Neon signs like the one shown in Figure 6.34 contain gases (such as neon) that become plasmas when the electrical supply is energized. The electricity connected to the glass tube of gas ionizes the gas, producing a plasma that emits light. Plasmas are gases, and their key features arc the same as those for gases. But note also that since plasmas are made of ions, they are electrically conductive. And one more fact with which you can impress your friends: the orange colors emitted by the flames in a fire are due to gases ionized by the heat of the fire—another example of a plasma.

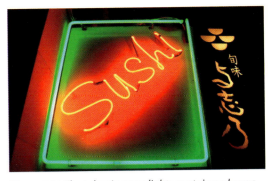

Figure 6.34 The tubes in neon lights contain a plasma that emits light.

6.5.2 Evaporation

Now that you know about the internal energy difference between the different phases, you may be wondering how substances can evaporate. *Evaporation* is the process of a substance slowly converting to a vapor without first being heated to the boiling point. If, as we have seen, substances change phase by being heated or cooled and changing the internal energy, how can a substance evaporate?

We have seen that the sum of the kinetic energies of the molecules in a substance is equal to the internal energy, and that the internal energy is related to the temperature. If we look more closely at the molecules in any substance—at any temperature—we see that the atoms or molecules in the substance are not all moving at the same speed. They are all moving at the same *average* speed and have the same average energy. But there is always a *distribution* of speeds, with many molecules moving at close to the average speed, some molecules moving slower than average, and some molecules moving faster than average. Because of this distribution of kinetic energies in the molecules, there are always some molecules with a much higher energy than average. These high-energy molecules sometimes have enough energy to break free from the other molecules in the liquid. When this happens, the molecule finds itself zooming around in the vapor state. This is evaporation—molecules leaving the others because of their higher than average energy, rather than because the entire substance has been heated to the boiling point.

Do You Know ...

The boiling and freezing points of a substance depend on the pressure and temperature. To the right is another type of phase diagram for water, showing the different phases like regions on a map. The phase transition from solid to liquid is shown at atmospheric pressure, 101.325 kilopascals (kPa), equal to 1 atmosphere (atm). This diagram indicates that at 0.6 kPa and 0.01°C, all three phases of water—solid, liquid, and vapor—can exist happily together. This set of conditions is called the *triple point*.

What is a triple point?

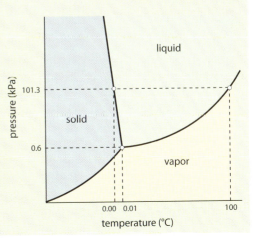

Do You Know ...   What causes the crystal structure of ice?

The sketch below depicts the crystal structure of ice—frozen water. The three-dimensional, six-sided crystal structure allows ice to form in many different shapes. The gray and blue lines show the attraction between hydrogen (white) and oxygen (red) atoms in the water molecules, a phenomenon called *hydrogen bonding*. The bonds for two hexagonal rings are shown in blue. Ice crystals are formed from different combinations of these hexagonal rings, which is why snowflakes exhibit hexagonal patterns in their structures.

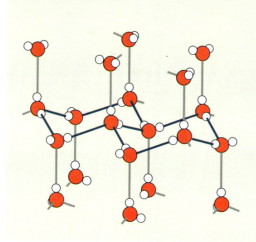

As water cools, these attractions due to hydrogen bonding cause the water molecules to arrange themselves into this structure. In doing so, they have to move farther apart from each other than they are when the liquid temperature is higher. As the molecules spread apart and move into position, the volume of the water increases so that water becomes less dense right before it freezes. Water is most dense at 4°C (1000.0 kg/m³), and begins getting less dense at temperatures below that. Thus, ice is less dense than water, and this is why ice floats.

Water boils at the relatively low temperature of 0°C. In a bowl of water at room temperature, 22°C, it is not difficult for a few H_2O molecules to have enough energy to place them up at the boiling temperature, allowing evaporation to occur. Water even does this when it is a solid (ice). This is why ice cubes in the freezer get smaller and smaller as they age. Water molecules on the surface of the ice cubes are evaporating! The boiling point of alcohol is even lower than that of water, so alcohol evaporates even more readily and more quickly than water. Other substances, such as oils, have much higher boiling points. Olive oil, for example, boils at around 300°C. It is nearly impossible for a molecule of olive oil in a bowl of oil at room temperature to have enough kinetic energy to put its temperature up almost 300 degrees above average, and so olive oil evaporates extremely slowly. The same reasoning applies to other oils as well.

6.5.3 Sublimation

The process of a solid converting directly to a gas without going through the liquid phase is called *sublimation*. Dry ice is frozen carbon dioxide (CO_2). It just so happens that at atmospheric pressure, the molecules of carbon dioxide go directly from the solid phase to the gas phase as they gain energy by

Figure 6.35. In water, sublimating dry ice produces a lovely fog. The fog is produced because the cold CO_2 condenses water vapor in the air above the bowl.

being warmed up. Other substances do this too, but dry ice is the only common substance that sublimates at atmospheric pressure.

Figure 6.35 shows a bowl of dry ice sublimating—gaseous CO_2 is being produced from the block of dry ice. The fog produced by dry ice is misleading and fools nearly everyone. The fog is not the CO_2; gaseous CO_2 is invisible. But the frozen CO_2 is so cold that water vapor (also invisible) near the bowl condenses into a fog of water droplets—tiny droplets of liquid water.

Chapter 6 Exercises

Atomic Model Study Questions

1. Practice writing brief descriptions of the atomic models developed by Democritus, Dalton, Thomson, Rutherford, and Bohr.

2. Practice writing brief descriptions of the famous experiments performed by Thomson, Millikan, and Rutherford. For each, relate the experiment to the major discoveries that resulted.

3. Describe the quantum model of the atom and explain how it differs from the Bohr model.

Volume, Mass, and Weight Exercises

In addition to volume, we are hitting on mass and weight here to make sure your skills are nice and sharp for the density exercises to follow. Remember to use horizontal bars for every unit fraction and to show every conversion factor with units in each of your calculations. Also remember that every given quantity represents a measurement of some kind and that this affects the number of significant digits in your result. Every conversion factor that is not exact also affects the number of significant digits in the result.

1. Convert 98.34 kg/m^3 into g/cm^3.

2. Convert 42 mL into gallons.

3. Given a weight of 18.5 lb, determine the mass.

4. Convert 3.6711×10^4 g/mL into kg/m^3.

5. Convert 1.957×10^4 in^3 into cm^3.

6. Convert 455 mL into m^3.

7. Given a mass of 46,000 kg, determine the weight in both newtons and pounds.

8. Convert 32.11 L into in^3.

9. Given a weight of 14.89 N, determine the mass (kg) and the weight in pounds.

10. Convert 36.00 cm^3 into m^3.

11. Convert 9.11 m^3 into cm^3.

12. Convert 4.11×10^5 m^3 into L.

13. Given a weight of 55,789 lb determine the mass.

14. Convert 5.022 g/cm³ into kg/m³.

15. Given a weight of 50,000 N, determine the mass (kg) and the weight in pounds.

16. Convert 1.75×10^{-6} m³ into cm³.

17. Convert 100.5 ft³ into m³.

18. Convert 37 m³ into in³.

19. Convert 750 cm³ into L.

20. Convert 5,755,000 gal into m³.

Answers

1. 0.09834 g/cm³	2. 0.011 gal	3. 8.40 kg	4. 36,711,000 kg/m³
5. 320,700 cm³	6. 0.000455 m³	7. 450,000 N, 1.0×10^5 lb	8. 1,959 in³
9. 1.52 kg, 3.35 lb	10. 3.6×10^{-5} m³	11. 9.11×10^6 cm³	12. 4.11×10^8 L
13. 25,300 kg	14. 5,022 kg/m³	15. 5,000 kg, 10,000 lb	16. 1.75 cm³
17. 2.846 m³	18. 2,300,000 in³	19. 0.75 L	20. 21,790 m³

Density Exercises

1. What is the density of carbon dioxide gas if 0.196 g of the gas occupies a volume of 100.1 mL?

2. Oil floats because its density is less than that of water. Determine the volume of 550 g of a particular oil with a density of 955 kg/m³. State your answer in mL.

3. A factory orders 15.7 kg of germanium. The density of germanium is 5.32 g/cm³. Calculate the volume of this material and state your answer both in m³ and cm³.

4. A block of wood 3.00 cm on each side weighs 5.336×10^{-2} lb. What is the density of this block?

5. A graduated cylinder contains 23.35 mL of water. An irregularly shaped stone is placed into the cylinder, raising the volume to 27.79 mL. If the mass of the stone is 32.1 g, what is the density of the stone?

6. Silver has a density of 10.5 g/cm³, and gold has a density of 19.3 g/cm³. Which has a greater mass, 5 cm³ of silver, or 5 cm³ of gold?

7. Five mL of ethanol has a mass of 3.9 g, while 7.5 liters of benzene has a mass of 6.6 kg. Which one is more dense?

8. Iron has a density of 7,830 kg/m³. An iron block is 2.1 cm × 3.5 cm at the base and has a mass of 94.5 g. How tall is the block?

9. A student measures out 22.5 mL of mercury and finds the mass to be 306 g. De-

termine the density of mercury and state your answer in kg/m³.

10. The density of water is greatest at 4.0°C. At this temperature, the density is 1,000.0 kg/m³. Determine the mass of 5.6 L of water at this temperature.

11. A large contemporary water tower can hold over 3 million gallons of water. How much does 3.0×10^6 gallons of water weigh? State your answer in pounds.

12. Given that the densities of lead and gold are 11.34 g/cm³ and 19.32 g/cm³ respectively, answer the following questions:
    a. Which weighs more, a cubic meter of gold or a cubic meter of lead?
    b. Which is larger, a cube of gold with a mass of 1 gram, or cube of lead with a mass of 1 gram?

13. A standard Olympic competition pool is 50.0 m long, 25.0 m wide, and at least 2.00 m deep. Determine the volume, in gallons, of the water this pool holds and then use the value for the density of water at room temperature to determine the weight of this water in tons. (A ton is defined as exactly 2,000 lb.)

Answers

1. 1.96×10^{-3} g/mL
2. 580 mL
3. 0.00295 m³, or 2,950 cm³
4. 0.897 g/cm³
5. 7.23 g/cm³
6. The answer to this one is top secret.
7. This one is, too.
8. 1.6 cm
9. 13,600 kg/m³
10. 5.6 kg
11. 24,700,000 lb
12. Another secret!
13. 6.60×10^5 gal; 2,750 tons

Substance Study Questions

1. Distinguish between mixtures and compounds.
2. Practice drawing the substances "family tree" (Figure 6.28) a few times on different days. Name two examples for each different type of substance in the diagram.
3. Describe the two basic types of structures atoms can make when they form compounds and give examples of each.

Phases of Matter Study Questions

1. Write an explanation for the fact that substances can exist in three different phases (solids, liquids, and gases). In your descriptions, use the concept of internal energy to explain how the three phases differ from one another.
2. Explain how evaporation and sublimation occur.

CHAPTER 7
Heat and Temperature

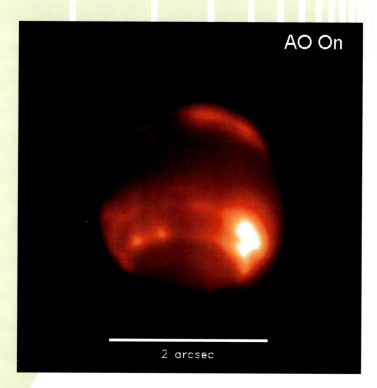

AO On

2 arcsec

Neptune

From NASA's website, posted 1999:

Neptune has never looked so clear in infrared light. Surprisingly, Neptune radiates about twice as much energy as it receives from the sun. A fascinating feature of the photograph above is that it was taken far from distant Neptune, through the Earth's normally blurry atmosphere. The great clarity of this recently released image was made possible by "rubber mirror" adaptive optics technology. Here, mirrors in the Palomar High Angular Resolution Observer (PHARO) instrument connected to the 200-inch Hale Telescope flex to remove the effects of turbulence in the Earth's atmosphere.

OBJECTIVES

Learn how to use these equations:

$$T_C = \frac{5}{9}(T_F - 32°) \qquad T_F = \frac{9}{5}T_C + 32° \qquad T_K = T_C + 273.2 \qquad T_C = T_K - 273.2$$

After studying this chapter and completing the exercises, students will be able to do each of the following tasks, using supporting terms and principles as necessary:

1. Define and distinguish between *heat, internal energy, thermal energy, thermal equilibrium, specific heat capacity,* and *thermal conductivity.*
2. State the freezing/melting and the boiling/condensing temperatures for water in °C, °F, and K.
3. Convert temperature values between °C, °F, and K using given formulas.
4. Describe and explain the three processes of heat transfer: conduction, convection, and radiation.
5. Describe how temperature relates to the internal energy of a substance and to the kinetic energy of its molecules.
6. Explain the kinetic theory of gases and use it to explain why the pressure of a gas inside a container is higher when the gas is hotter and lower when the gas is cooler.
7. Apply the concepts of specific heat capacity and thermal conductivity to explain how common materials such as metals, water, and thermal insulators behave.

7.1   Measuring Temperature

7.1.1 Temperature Scales

There are a lot of temperature scales around because scientists in different countries have been investigating materials at different temperatures for a long time. We will use only three different scales in this text: the Fahrenheit scale you have been using all your life (if you grew up in America), and the two main SI scales, the Celsius scale and the Kelvin scale.

The Celsius scale (formerly known to some as the "centigrade" scale) is based on the freezing and boiling temperatures of water (at atmospheric pressure). The scale was specifically designed so that water freezes at 0°C and boils at 100°C. On this scale, room temperature is around 20–25°C. For historical reasons, the Celsius and Fahrenheit scales use the strange term "degrees." This is odd when you think about it. We don't use a term like this with any other units of measure.[1] The Kelvin scale does not use this term. A temperature of 300 K is read as, "three hundred kelvins."

A degree on the Celsius scale is the same as a "degree" on the Kelvin scale; the degrees are the same size. But the Kelvin scale is an absolute scale, which means there are no negative temperatures on the Kelvin scale. The temperature of 0.0 K is referred to as "absolute zero." According to the laws of physics as we understand them, this temperature cannot be achieved. A region might be refrigerated down to a small fraction of 1.0 K, but nothing can ever reach 0 K. (And nothing can ever be less than 0 K, as I will explain shortly.)

---

1   Except for measuring angles, where the degree is the entire unit.

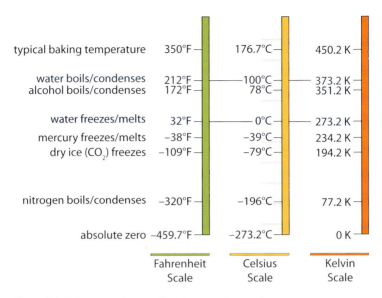

typical baking temperature	350°F	176.7°C	450.2 K
water boils/condenses	212°F	100°C	373.2 K
alcohol boils/condenses	172°F	78°C	351.2 K
water freezes/melts	32°F	0°C	273.2 K
mercury freezes/melts	−38°F	−39°C	234.2 K
dry ice (CO$_2$) freezes	−109°F	−79°C	194.2 K
nitrogen boils/condenses	−320°F	−196°C	77.2 K
absolute zero	−459.7°F	−273.2°C	0 K

Fahrenheit Scale   Celsius Scale   Kelvin Scale

Figure 7.1. Reference values on three temperature scales.

The temperatures for freezing/melting or boiling/condensing of water are used a lot in scientific work and you should know them. They are summarized, along with some other interesting temperature values, in Figure 7.1.

7.1.2 Temperature Unit Conversions

The only computations we are going to do in this chapter are conversions of temperature values from one scale to another. The equations involved can be confusing for students in this course, so I normally do not require students to memorize the conversion equations; I usually give all four of them to students on the quizzes. So performing these conversions should be easy if you are careful. To convert a temperature in degrees Fahrenheit (T_F) into degrees Celsius (T_C) use this equation:

$$T_C = \frac{5}{9}\left(T_F - 32°\right)$$

Using a bit of algebra, we can work this around to give us an equation that can be used to convert Celsius temperatures to Fahrenheit values:

$$T_F = \frac{9}{5}T_C + 32°$$

To convert a temperature in degrees Celsius into kelvins (T_K) use this equation:

$$T_K = T_C + 273.2$$

Again, some algebra gives us the equation the other way around.

$$T_C = T_K - 273.2°$$

One note about significant digits is in order for these calculations. Figuring out the significant digits with these two equations is more complicated than we want for our purposes here, so when doing temperature conversions I always use this special rule: *When performing temperature conversions, state every answer with one decimal place.*

▼ Example 7.1

The normal temperature for the human body is 98.6 °F. Express this value in degrees Celsius and kelvins.

Since the given value is in degrees Fahrenheit, write down the equation that converts values from °F to °C.

$$T_C = \frac{5}{9}\left(T_F - 32°\right)$$

Now calculate the Celsius value.

$$T_C = \frac{5}{9}\left(98.6° - 32°\right) = 37.0 \ °C$$

Now we are able to use the Celsius value to compute the Kelvin value.

$$T_K = T_C + 273.2 = 37.0 + 273.2 = 310.2 \ K$$

▲

▼ Example 7.2

The melting point of aluminum is 933.5 K. Express this temperature in degrees Celsius and degrees Fahrenheit.

Write down the equation that converts Kelvin values to Celsius values.

$$T_C = T_K - 273.2$$

From this we calculate the Celsius value as

$$T_C = T_K - 273.2 = 933.5 - 273.2 = 660.3 \ °C$$

Next, write down the equation for converting a Celsius value to a Fahrenheit value.

$$T_F = \frac{9}{5}T_C + 32°$$

With this equation calculate the Fahrenheit value.

$$T_F = \frac{9}{5}T_C + 32° = \frac{9}{5} \cdot 660.3 \ °C + 32° = 1,220.5 \ °F$$

7.2   Heat and Heat Transfer

Before we proceed further, a word is in order about the difference between atoms and molecules. As you learned in the previous chapter, in substances such as water and oxygen gas, the atoms are bonded together in clusters called molecules. In substances like these, the smallest particle of the substance is a molecule. So it makes no sense to refer to an atom of water. The smallest particle of water is a water molecule, represented by the formula H_2O. There are three atoms in this molecule (two hydrogen atoms and one oxygen atom). In other substances like gold and helium gas, the atoms don't form molecules, but remain alone. So here it is appropriate to speak of an atom of gold or an atom of helium gas. In the discussion that follows, I will use the terms atom and molecule interchangeably to mean the smallest particle of the substance.

7.2.1 How Atoms Possess Energy

There are two fundamental ways an individual atom can possess energy. The first is in the atom's kinetic energy, which is one of the major subjects in this chapter. The second way is in the energy states of the atom's electrons. The location of an electron in an atom depends on the electron's energy. Recall the brief discussion in the previous chapter of neon lights as examples of plasmas. The reason a plasma emits colored light is because electrons in the atoms in the plasma have been zapped up to high energy levels by the energy source (such as the high-voltage electricity powering a neon light). High-energy electrons in a plasma always eventually drop back down to lower energy states by emitting quanta of energy. These quanta are photons of light. We will leave further details on electron energy for your chemistry course.

7.2.2 Internal Energy and Thermal Energy

As you recall from Chapter 4, *heat* is energy in transit, flowing from one substance to a cooler substance. Substances do not possess heat. A substance can possess other forms of energy (various forms of kinetic or potential energy), but heat is the term for energy in the process of flowing from one substance to another because of a difference in temperature.

Recall from Chapter 6 that the *internal energy* of a substance is the sum of all the kinetic energies possessed by the atoms or molecules of a substance. Atoms or molecules are constantly in motion, vibrating or translating, or both. Atoms in solids cannot fly around, so they vibrate in place, but atoms in gases are free to translate, or zoom around (see the box for details about that zooming). Either way, since atoms are always moving they always possess kinetic energy. If you add up the kinetic energy of every particle in a certain substance, that is its internal energy. Although it is somewhat complicated to explain, the "temperature" of a substance is very closely related to its internal energy. The higher the temperature, the higher the internal energy and vice versa.

The internal energy of a substance does not include the kinetic or potential energy the substance possesses as a whole. For example, the atoms in a stone are vibrating and the energy of these vibrations is the stone's internal energy. However, if the stone is hurled into the air, the kinetic energy of the stone in flight is not considered to be part of the stone's internal energy.

In general, there are two ways the internal energy of a substance can change. The first is for the substance to be heated or cooled. Energy a substance possesses because it has been heated is called *thermal energy*. Obviously, any kind of substance can be heated, a process that adds thermal energy to a substance and thus increases its internal energy. The second

Do You Know ...   How fast are air molecules moving?

When we say that air molecules are zooming around all the time, how fast are they actually going? The answer is rather shocking. At room temperature an average air molecule is moving at about 1,100 mph, or 500 m/s! This speed is 1.5 times the speed of sound, also called Mach 1.5. Of course, this is just the average speed; about half the molecules are moving faster than that! This sounds bizarre.

But it gets weirder so hold on to your hat. We know there are *lots* of molecules in the air, so if they are moving this fast don't they bump into each other a lot? YES! In every cubic centimeter of air there are about 2.7×10^{19} molecules. Do you have *any idea* how many molecules this is? If there were this many people on earth there would be over 4,800 people *per square foot* everywhere on the planet, including the oceans!

With this many molecules all moving so fast, the average distance a molecule travels before colliding with another molecule is much smaller than the wavelength of light, and an average molecule collides 5 times every nanosecond (billionth of a second). Sprightly little fellows, eh? In sum, have a look at a little chunk of air in front of you about the size of a sugar cube. In this little space, there are 27,000,000,000,000,000,000 molecules and they each bump into another molecule 5,000,000,000 times per second moving at Mach 1.5 between every hit.

God's world is amazing—jaw-dropping in fact. And heck, this is just ordinary old air. How cool do you think the DNA molecules are inside your body? But you'll have to wait for biology to learn about that.

way the internal energy can be changed applies primarily to gases. Unlike solids and liquids, gases are *compressible*. When a gas is compressed, work is being done on the gas and this work energy goes into the gas and increases its internal energy, and thus its temperature, too.

7.2.3 Absolute Zero

Now that you know about internal energy, you can better understand what *absolute zero* is, the zero temperature on the Kelvin temperature scale. The temperature of a substance varies directly with its internal energy. A temperature of 0 kelvins (absolute zero) means no internal energy at all, or in other words, the atoms are standing still! Atoms can't move any slower than standing still, and thus, there is no temperature lower than absolute zero. So far as we know, there is no place in the universe where the temperature is absolute zero, although physicists in low-temperature research labs have succeeded in achieving temperatures of only a few millionths of a kelvin above absolute zero. (As you might imagine, very weird things happen at temperatures that low.)

7.2.4 Thermal Equilibrium

The laws of thermodynamics say that if a substance is at a different temperature than its environment heat flows from the warmer of the two into the cooler of the two. We are all familiar with this principle, even if we don't know the actual law. (FYI, it is the second law of thermodynamics that says heat flows this way.) *Thermal equilibrium* is the state in which an object is at the same temperature as its environment. When this happens, heat flow between them ceases.

7.3   Heat Transfer Processes

When a temperature difference exists, heat flow occurs. This means that energy flows from where the temperature is higher to where the temperature is lower. *How* the heat flows depends on whether the energy is flowing through a solid, through a fluid, or through a region that will permit electromagnetic radiation to exist, such as a vacuum. Below, we consider the three processes by which heat transfers from one substance or location to another.

7.3.1 Conduction

The first heat transfer process is *conduction*, illustrated in Figure 7.2. This process occurs in solids. In a solid, the atoms or molecules of the substance vibrate in place but are not free to flow around. Imagine a great number of atoms in a substance attached together in a flexible grid, as if they are connected to one another by springs. If one side of the substance is heated, the internal energy of the atoms there goes up, which means they begin vibrating more vigorously because their kinetic energy is increasing. These vibrations are transferred to the other atoms nearby because the atoms are all linked together, as atoms always are in a solid. The kinetic energy in these vibrations continues to spread to atoms farther and farther away from the location where the material is being heated. As atoms gain more kinetic energy, their temperature goes up. This is heat flow by conduction.

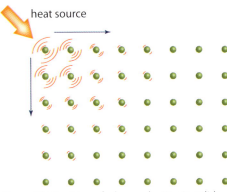

Figure 7.2. Heat transfer by conduction in solids. A heat source gives kinetic energy to nearby atoms, which begin vibrating vigorously. These vibrations are passed from one atom to the next within the solid substance. In this figure, the blue arrows indicate the direction of the spread of the vibrations, which is the spread of heat, in the substance.

A well-known example of conduction is the way heat transfers in the metal of a frying pan. Initially, the atoms in the pan are all at room temperature (and are thus at thermal equilibrium with the room). When the center of the pan is heated, the atoms there gain internal energy and begin vibrating more vigorously. These vibrations are passed atom to atom and the temperature increases more and more, farther and farther away from the location where the source of heat is. Eventually the vibrations get all the way out to the handle of the pan and the handle itself gets hot, even though it is nowhere near the heat source.

7.3.2 Convection

The second heat transfer process is *convection*, illustrated in Figure 7.3. This process occurs in *fluids*, that is, liquids and gases. The particles of a fluid are free to flow around and mix and mingle and collide with other particles. Particles with high internal energy are hot and move very rapidly (because the internal energy is the kinetic energy of the particles). These particles mix with cooler particles and when they do they collide with them and transfer some of their high energy to the lower energy particles. This is just like balls colliding on a pool table. When a fast ball hits a slow one, the fast one slows down and the slow one speeds up as kinetic energy is transferred from the ball with more energy to the one with less. Unlike pool balls, particles in liquids and gases never stop moving, so over time all the particles share the energy evenly. When this happens, they are all more or less at the same average temperature (thermal equilibrium).

A good example of convection at work is in the old radiators commonly used to heat homes in the days before central heating systems (and still in use in many older homes, particularly up north). Hot water is pumped through the inside of the radiator, making the iron radiator hot to the touch. When air molecules collide with the hot surface, they pick up kinetic energy from the rapidly vibrating atoms in the metal. These hot, fast moving molecules then gradually work their way through the room, colliding with the slower, cooler molecules and exchanging energy so the cooler molecules begin moving faster, which means they are hotter.

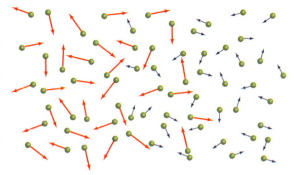

Figure 7.3. Heat transfer by convection in fluids. Hot, fast-moving molecules (red arrows) from some kind of hot source gradually work their way into a place full of cooler, slower moving molecules (blue arrows). As the molecules repeatedly collide, the hot molecules transfer kinetic energy to the cooler ones. The loss of kinetic energy cools the hot molecules and the gain of kinetic energy warms the cool ones.

7.3.3 Radiation

The third heat transfer process is *radiation*, illustrated in Figure 7.4. Radiation is the term for heat energy transferred by electromagnetic waves. As we saw in Section 4.2, electromagnetic waves travel through space and through the atmosphere to carry energy from the sun to the earth. Light, infrared radiation, ultraviolet radiation, microwaves, X-rays; these are all the same thing—electromagnetic radiation (or waves). The only thing that distinguishes these different terms for radiation is the *wavelength* of the waves. When we consider heat transfer, we are mainly talking about *infrared* (IR) radiation, which consists of invisible electromagnetic waves with wavelengths in the range of roughly 750 to 2,500 nanometers.

In Chapter 9, we will look at why this electromagnetic heat energy is called "infrared radiation" and we will discuss how infrared radiation relates to other electromagnetic radiation such as radio waves and light. But here it is important to point out that the hotter an object is, the more infrared energy it radiates. When you feel the heat radiating from a hot object such as a branding iron, it is the IR radiation you are feeling. When you hold your hands in front of a fire, it is the IR radiation from the coals you are

Figure 7.4. Heat transfer by radiation from hot objects. All hot objects radiate infrared electromagnetic waves, which we feel as heat. The amount of this radiation depends strongly on the temperature of the object—for hotter and hotter objects the amount of infrared radiation emitted goes up drastically. I drew the waves with red and orange lines to indicate the flow of heat, but infrared waves are, of course, invisible.

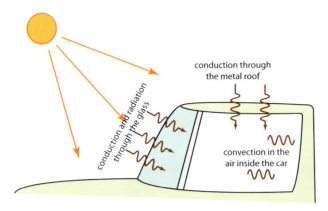

conduction through
the metal roof

conduction and radiation
through the glass

convection in the
air inside the car

Figure 7.5. Heat transfer processes at work heating the interior of a car.

feeling. Your eyes can't see it the way they can visible light, but your skin can feel it. When you feel warm in the sun even on a cool day, it is the IR radiation from the sun that is warming you. When the inside of your car gets hot, it is because of IR radiation from the sun striking the car and transferring into the car by the radiation passing through the glass and by conduction through the metal. Figure 7.5 depicts all three heat transfer processes as each of them contribute to the warming of the interior of a car in the sun.

7.4   The Kinetic Theory of Gases

As we have seen, the internal energy in a gas relates to the temperature of the gas, and this energy is stored in the gas in the form of kinetic energy in each of the molecules of gas. We now extend these concepts to develop a physical explanation for gas pressure. This explanation is called the *kinetic theory of gases.*

According to the kinetic theory of gases, the atoms or molecules of a gas in a container are zooming around inside the container all the time at high speed, as illustrated in Figure 7.6. The hotter the gas is, the more kinetic energy the particles have and the faster they are zooming around. There are an incredible number of particles in even a small container and as they zoom around, the individual particles of this huge population of particles are constantly striking the walls of the container. These collisions between the particles and the container walls cause the effect we call *pressure.* If the temperature goes up, the particles have higher internal energy, meaning higher kinetic energy, so they are moving faster and striking the container walls more frequently and with more momentum—meaning harder hits—creating higher pressure. We can summarize this theory of gas behavior this way:

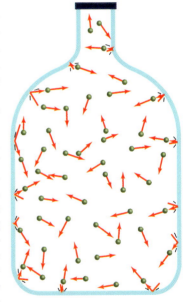

Figure 7.6. Pressure in a container is caused by unimaginable numbers of gas particles continuously striking the walls of the container.

| hotter gas | = | faster moving particles | = | harder hits | = | higher pressure |

A nice example that many of you have seen occurs when you buy a Mylar birthday balloon at the grocery store on a hot day. The balloon vendor fills the balloon with an appropriate amount of helium inside the cool grocery store. When you carry the balloon outside into the blazing heat, the balloon swells up so tightly it appears ready to burst from the high pressure. Then you put it inside a car with the air conditioner blasting away. The air from the car A/C is so cold that the pressure in the balloon decreases significantly, even way below what it is inside the store. This causes the balloon to shrink, collapse, and fall down on the floor of the car. When you arrive home and take the balloon out of the car, the hot outside air warms up the helium. Those atoms pick up speed and begin colliding once again at high speed on the inside of the balloon, causing the pressure to spike back up. All this is caused by the atoms in the helium gas gaining kinetic energy when they warm up and losing kinetic energy when they cool.

7.5   Thermal Properties of Substances

7.5.1 Specific Heat Capacity

Specific heat capacity is a physical property of substances that relates to how a material behaves when it is absorbing or releasing heat energy. The specific heat capacity (or simply, heat capacity) of a substance is defined as the amount of heat energy that must transfer into or out of 1 gram of the substance to change its temperature by 1 degree Celsius.

A substance with a low specific heat capacity changes temperature very easily and quickly because very little energy must be added or removed to change its temperature. Metals such as those shown in Figure 7.7 are like this. A very small amount of heat raises the temperature of a metal a lot, and removing only a small amount of heat causes the temperature of a metal to drop a lot.

A substance with a high specific heat capacity changes temperature only very slowly because it must absorb or release a lot of energy to change its temperature. Water is like this (Figure 7.8). (In fact, water has a great number of very unique properties and this is one of them.) It takes a lot of heat energy to warm up a pan of water even a little bit. And to cool water down requires removing a lot of heat from it just to cool it down a little bit. For comparison, the specific heat capacity of water is about ten times greater than that of copper, and about five times greater than that of aluminum.

Figure 7.7. For low specific heat capacity, think metals. Shown are (L to R) brass, copper, nickel, and aluminum.

Figure 7.8. For high specific heat capacity, think water.

7.5.2 Thermal Conductivity

Like specific heat capacity, *thermal conductivity* is a physical property of materials that relates to how materials behave when they are absorbing or releasing heat energy. The thermal conductivity of a substance is a measure of how well a material transfers

167

heat within its own atomic structure by conduction. Metals have high thermal conductivities, so heat conducts readily through metals. On the other hand, building insulation used inside the walls of buildings and houses has low thermal conductivity, so heat transfers

through it very, very slowly. These materials are called *thermal insulators*. Other materials with a low thermal conductivity are glass and plastic. In fact, high thermal conductivity in materials seems typically to accompany high electrical conductivity. We all know that metals conduct electricity the best. Well, they conduct heat the best, too.

The effect of different thermal conductivities in materials is demonstrated dramatically by watching what happens when ice cubes are placed on blocks of plastic and aluminum. As shown in Figure 7.9, the block on the left is made of some kind of composite wood and plastic material. The block on the right is made of aluminum. The ice cubes were placed on the

Figure 7.9. High thermal conductivity, illustrated by the block on the right, which is made of aluminum.

blocks at the same time and the photo was taken after about 30 seconds. Which material has a higher thermal conductivity? What do you think the two blocks feel like if you touch them? Which direction is the heat flowing?

7.5.3 Heat Capacity vs. Thermal Conductivity

It is important for you to spend enough time on this topic so you can confidently distinguish between the two properties we have been discussing, specific heat capacity and thermal conductivity. Students often get these properties confused. It will help for us to consider the actual values of these quantities for some common materials.

Specific heat capacity has to do with the quantity of heat required to warm a substance or the quantity of heat that must be removed from a substance to cool it. As you can see in Figure 7.10, the heat capacity of water is huge compared to everything else. (In fact, ammonia is the only compound with a higher specific heat capacity than water.) Also notice that the specific heat capacity of metals is quite low compared to water.

The huge heat capacity of water is a major factor in regulating the temperatures on earth. (The atmosphere is another major factor.) About 70% of the earth is covered by water,

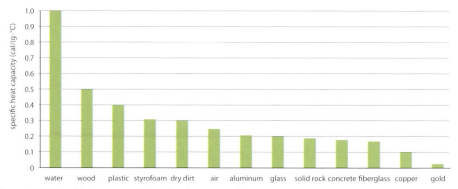

Figure 7.10. Specific heat capacities of various materials.

primarily in the oceans. During the day, the radiation from the sun warms the oceans, but a lot of energy must be added to all this water to warm it. At night, the thermal energy in the water escapes into the atmosphere, cooling the water. But again, a lot of heat energy has to be removed from water for it to cool down. The result is that the ocean water covering the earth does not change temperature very much. Since there is so much water on the earth, the steady temperature of the water helps regulate the temperature of the atmosphere, and thus of the entire planet. Without the oceans, the temperature swings from day to night on earth would be huge, and complex life on earth could not survive.

By contrast, thermal conductivity is about how well heat flows through a substance by conduction. As shown in Figure 7.11a, the thermal conductivity of metals completely dominates everything else. Since the thermal conductivities of all other materials are so small compared to metals, I made a separate chart in Figure 7.11b for the nonmetal substances so that you can compare them. Notice that the thermal conductivities of water and air are minuscule, because heat transfers through fluids mainly by convection, not by conduction, as we discussed previously.

The data in these charts suggest some things you will want to think about when addressing questions about specific heat capacity and thermal conductivity. Since the specific heat capacity of water is so much greater than the heat capacity of anything else, the presence of water is the major factor in questions about how fast things heat up or cool down, or about how much energy must be added to a substance to warm it up or removed from it to cool it down. And since the thermal conductivity of metals is so much greater than the thermal conductivity of anything else, the presence of metal is the major factor in how rapidly heat flows through a substance by conduction.

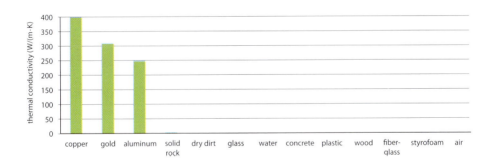

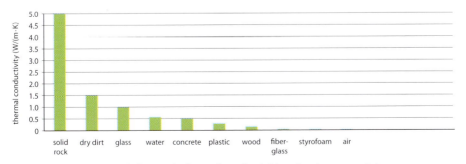

Figures 7.11a (top) and 7.11b (bottom). Thermal conductivities of various materials.

Chapter 7 Exercises

Temperature Unit Conversions

Fill in the missing values in the table. Warning: It is very common for students to think these conversions are easy (they are), complete this exercise, and proceed to get the conversions wrong on quizzes. The errors are typically due to incorrect use of the calculator, so make sure you know how to use your calculator correctly.

	°F	°C	K
1		32.0	
2	56.5		
3			455.0
4	−17.9		
5		−41.6	
6			79.0

Answers

1. 89.6°F, 305.2 K
2. 13.6°C, 286.8 K
3. 359.2°F, 181.8°C
4. −27.7°C, 245.5 K
5. −42.9°F, 231.6 K
6. −317.6°F, −194.2°C

Heat Transfer and Kinetic Theory Study Questions

1. Imagine a closed glass jar (like a Mason jar) full of ice water sitting on a table in the middle of a metal building (say, a tool shed). The metal building is in a clearing in the sun. Starting with the sun, identify every heat transfer process that happens for energy to transfer from the sun where it starts to the ice where it ends up.

2. People who live in places where summer temperatures are much higher than winter temperatures have to adjust the amount of air in their car tires a few times per year. Explain why this is and identify when air must be added (raising the pressure) and when air must be released (lowering the pressure).

3. When driving on a long trip, I like to take along a vacuum bottle (also called a Thermos, which is actually a brand name) full of hot coffee to drink on the road. A vacuum bottle consists of two metal bottles, one inside the other. They are connected together only at the top, and the air is removed from between the bottles, creating a vacuum between them. A thick plastic lid opens to the inside of the inner bottle where the hot coffee is kept.
   a. Explain why the vacuum bottle keeps the coffee hot for a lot longer than just about any other type of container.
   b. Even with a vacuum bottle, the hot coffee inside eventually reaches thermal equilibrium with the outside. Explain the heat transfer mechanisms involved in the heat getting out.

    c.   Explain why the bottle would not work nearly as well if the vacuum seal were to fail, allowing air to fill the space between the inner and outer bottles.

    d.   Explain why the vacuum bottle works just as well if the contents are iced tea instead of hot coffee.

4.   Cans of spray paint include this warning on the label: "CONTENTS UNDER PRESSURE. Avoid prolonged exposure to sunlight or heat sources that may cause bursting. Do not store above 120°F." Explain why heat sources or sunlight could cause the can to burst.

5.   Consider a Styrofoam ice chest full of ice and canned drinks sitting in the sun at the beach on a hot summer day. Explain the heat transfer processes involved in bringing the contents of the ice chest into thermal equilibrium with the environment. In your answer, be sure to consider the fact that initially there is no liquid in the ice chest, but as the ice begins to melt there will be lots of water in there.

6.   How does the internal energy of a gas relate to the kinetic theory of gases?

7.   Explain the cause of the pressure of a gas in a container.

Specific Heat Capacity and Thermal Conductivity Study Questions

In responding to the following questions, you need to consider whether the situation described has to do primarily with specific heat capacity or with thermal conductivity. You may also need to refer to the graphs in Figures 7.10 and 7.11.

1.   Consider a large glass of water and an iron horseshoe that have the same mass, both at room temperature. If these are placed in the sun for five minutes, what happens to the temperatures of the water and the horseshoe? Which one warms up faster and why? After sunset, which one cools off faster and why?

2.   If you were trying to stay warm in an unheated deer blind on a frosty morning, which would be best to have on the floor in front of you—a big bowl of steaming hot water or a chunk of iron with the same mass and temperature as the water? Why?

3.   When a person bakes a casserole in a glass baking dish to take to a party, she often covers the dish with aluminum foil while transporting it in the car to the party. If her purpose is to keep the casserole hot, which is better—to cover the dish with foil or to cover it with a tight fitting plastic lid? Explain your reasoning.

4.   Think about the previous question some more. Despite your answer, people *do* use aluminum foil to cover hot food. Also, thermal blankets for keeping people warm in cold weather emergencies are lined with metal foil. Given the high thermal conductivity of metals, why is this? (The answer must have to do with something other than conduction! See page 212, question 10.)

5.   In northern states, houses are commonly covered in wood siding, but in southern states, brick homes are more common. Assuming that the thermal properties of brick are similar to those of solid rock, why is this? (See Figure 7.11b.)

6.   Suppose you were shopping and saw a bag of aluminum "ice" cubes for sale. The bag says they save water, are reusable, and great for ice chests because they are so light weight. Would such a product be a good idea or a lousy one? Explain your

reasoning.

7. A computer chip generates a fair bit of heat when it is busy working. Try to explain why electronics manufacturers often mount components like computer chips on aluminum blocks called "heat sinks."

8. If you place a brass belt buckle and a cup of water in a freezer, which one reaches 0°C the quickest and why? (Assume they weigh about the same.)

9. Inside a hot car on a summer day, the temperature sometimes get so high that some objects inside the car can burn your skin. Let's say you slide into the hot car wearing shorts. The surface of a cloth seat does not burn your skin, but the buckle of the seat belt does. This seems odd, since they are at thermal equilibrium inside the car, and thus at the same temperature. Try to explain why the metal buckle burns your skin while other materials at the same temperature do not. (Think about the two blocks in Figure 7.9.)

10. Explain why a koozie helps to keep a cold beverage in a can to stay cold longer. (The term "koozie" may not be in your dictionary, but your internet search engine will show you what koozies are if you don't know.)

11. Explain why the same koozie could be used to keep a hot beverage in a can to stay hot longer (as if anyone ever wanted to drink a hot beverage from a can).

12. Explain why a cold beverage stays cold longer in a glass bottle than in an aluminum can.

13. Aluminum cans have a thin plastic coating on the inside. Speculate on how this coating affects a cold beverage in a can.

14. The Gulf Stream is a massive ocean current of warm water that moves from the Gulf of Mexico up the Atlantic seaboard, northeast across the Atlantic Ocean, and right up and past the United Kingdom. When water reaches the UK, its temperature is still about 8°C warmer than surrounding waters. Use this information to explain the fact that even though London is 8 degrees of latitude farther north than Toronto, the average temperature in London in January is 7°C and the average temperature in Toronto in January is −2°C.

15. Referring back to the questions about the vacuum bottle (p. 170, #3), years ago the inner container in vacuum bottles was made of glass. (These fell out of favor for practical reasons. Whenever a kid dropped his lunch box, the glass vacuum bottle inside got smashed.) Explain why the glass vacuum bottle keeps beverages hot even longer than a metal vacuum bottle does. *glass has lower thermal conductivity*

16. If you are going to build a house in a location with an extremely cold climate, does it make more sense (from an energy point of view) to build the walls of concrete or of wood (assuming an equal thickness of either one)? In your response, explain what the building materials are supposed to accomplish as far as heat energy is concerned. *wood has lower conductivity*

17. If you are going to build a house in a location with an extremely hot climate, does it make more sense (from an energy point of view) to build the walls of stones or of mud bricks, which when dried would have similar thermal properties to those of dirt (assuming an equal thickness of either one)? In your response, explain

what the building materials are supposed to accomplish as far as heat energy is concerned.

18. Consider the previous question again, and imagine that you live in the 19th century and have no air conditioning in your house. Explain why the house is more comfortable with a dirt floor than with a concrete floor.

19. Refer again to Figure 7.9. Explain why the ice cube on the right is melting so fast.

20. Referring yet again to Figure 7.9, the water forming from the cube on the right is coming from the bottom of the ice cube, where it is in contact with the aluminum block. For the ice to melt it must first be warmed up to 0°C. What do you think is happening to the temperature in the center or at the top of the ice cube?

21. Figure 7.11b indicates that the thermal conductivity of fiberglass insulation is actually slightly *higher* than the thermal conductivity of air. So why is this material used in walls for insulation? Why not just leave the walls empty except for the air?

22. Explain why the aluminum case of an Apple MacBook or iPad always feels cool when you touch it.

Do You Know ...   *What is the temperature in outer space?*

The temperature in outer space is about 3 K, that is, about 3°C above absolute zero! How can there be a temperature in space where there are no atoms there to have internal energy? The answer is electromagnetic *radiation*. No matter where scientists point their detection equipment into space, the same microwave radiation is found. This microwave radiation is what produces temperature in outer space. In fact, the presence of the *cosmic microwave background* is now understood to be strong evidence for the Big Bang, where the radiation originated. Of course, the universe was a lot hotter and smaller at the Big Bang. But as the universe expands, it cools. The cosmic microwave background appears to be the left-overs of radiation from the original creation event.

A bolometer is a device designed to detect the cosmic microwave background. It uses the principles of both specific heat capacity and thermal conductivity to create a temperature difference that is detected by an electric circuit. The spiderweb bolometer shown was developed by NASA. As radiation strikes the thin web, its temperature becomes higher than that of the surrounding material.

CHAPTER 8
Pressure and Buoyancy

A Floating City

An SH-60F Sea Hawk helicopter flies the national ensign alongside the USS Dwight D. Eisenhower under way in the Mediterranean Sea, July 10, 2012. This Nimitz-class aircraft carrier is enormous. It carries 3,200 personnel and 90 aircraft. The ship is 1,115 feet long (nearly four football fields) and weighs 113,800 tons—over 225 million pounds.

How in the world can a steel vessel this big and this heavy float on water? The answer takes us back to a brilliant engineer in ancient Syracuse.

OBJECTIVES

Memorize and learn how to use these equations:

$$P = \rho g h \qquad P = \frac{F}{A} \qquad P_{abs} = P_{gauge} + P_{atm}$$

After studying this chapter and completing the exercises, students will be able to do each of the following tasks, using supporting terms and principles as necessary:

1. Explain the cause of pressure on objects submerged in a fluid.
2. Explain the cause of air pressure.
3. Distinguish between gauge pressure and absolute pressure.
4. Convert absolute pressure to gauge pressure and vice versa.
5. Calculate the gauge pressure and the absolute pressure at a given depth below the surface of a liquid.
6. Relate Blaise Pascal's explanation of how Torricelli's barometer works, including Pascal's bold new theory about the vacuum.
7. Determine the pressure created by a given force applied to a particular surface area.
8. Explain Archimedes' principle of buoyancy and use it to explain why things float in liquids.
9. Calculate the weight that can be supported by a regular, buoyant solid when the solid is fully or partially submerged in water.

8.1   Pressure Under Liquids and Solids

In the previous chapter, we saw that at the microscopic level pressure is caused by atoms or molecules hitting against the walls of a container of gas. Now we will look at pressure from a different perspective, at the "macro" level, where we consider fluids in bulk rather than as individual particles. Examining fluids in bulk gives us an easy way to calculate the pressure under the surface of a liquid.

Consider the ordinary glass of water illustrated in Figure 8.1. The water at the top of the glass is being attracted to the earth by the earth's gravity, but it is being supported from underneath by the layer of water just beneath it. The third layer of water down has to support both the first two layers. Each layer down is supporting all the layers above it. At a given depth, it is the weight of the liquid above that causes the pressure at that depth. From this simple analysis we see that the pressure at a given depth depends on the earth's gravity and on the depth.

Clearly, the denser the liquid is, the heavier a given volume of the liquid is, and the harder lower layers have to push to support the upper layers. So the pressure at a certain depth also depends on a third factor—the density of the liquid.

In fact, the pressure at a given depth, h, in a liquid is the product of these three factors, or

$$P = \rho g h$$

Figure 8.1. Each layer of water in a glass supports the weight of the water above it.

Figure 8.2. French scientist and mathematician Blaise Pascal (1623–1662).

where P is the pressure in pascals (Pa), ρ is the density of the liquid (kg/m³), g is the acceleration due to gravity (9.80 m/s²), and h is the depth in the liquid (m). Notice this interesting fact about this equation: the pressure beneath a liquid does not depend on the shape or volume of the container; it depends only on the depth. Imagine a vertical soda straw 30 feet tall and full of water. The pressure at the bottom of the straw is the same as the pressure 30 feet under the surface of a lake.

The pascal is another derived unit, the MKS unit for pressure. The pascal is named for Blaise Pascal, an amazing 17th-century French scientist, mathematician, and philosopher (Figure 8.2). (Some day you should read Pascal's famous book *Pensées*. The title means "thoughts." The book consists of Pascal's notes for a book he was writing to defend the Christian faith. Pascal was unable to finish the book, but it is very much worth reading anyway.)

The pascal (Pa) is equivalent to units of newtons per square meter (N/m²), which represents an amount of force per unit area. In the U.S., the more familiar USCS units pounds per square inch (psi) are usually used for pressure. In casual speech, people often simply say so many "pounds" of pressure when they mean pounds per square inch.

▼ Example 8.1

Determine the pressure at the bottom of a typical swimming pool, 9.00 ft in depth. State your result in pounds per square inch (psi).

We do this calculation in MKS units like we always do, converting the result to psi at the end. Writing down the givens and converting to MKS units we have

$$h = 9.00 \text{ ft} \cdot \frac{0.3048 \text{ m}}{\text{ft}} = 2.743 \text{ m}$$

$$\rho = 998 \ \frac{\text{kg}}{\text{m}^3}$$

$$P = ?$$

Recalling that the density of water is approximately 998 kg/m³, and the acceleration due to earth's gravity is 9.80 m/s², we insert these values, along with the given depth, into the equation and compute the pressure:

$$P = \rho g h = 998 \ \frac{\text{kg}}{\text{m}^3} \cdot 9.80 \ \frac{\text{m}}{\text{s}^2} \cdot 2.743 \text{ m} = 26,800 \text{ Pa}$$

Now, using the pressure conversion factor from Appendix A,

$$26,800 \text{ Pa} \cdot \frac{14.7 \text{ psi}}{101,325 \text{ Pa}} = 3.89 \text{ psi}$$

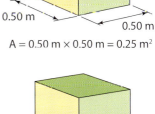

A = 0.50 m × 0.50 m = 0.25 m²

Let's think some more about the MKS unit for pressure, the pascal. As I mentioned just before Example 8.1, one pascal is equivalent to one newton of force per square meter of surface area. This means that pressure can be defined as

$$P = \frac{F}{A}$$

where P is the pressure in pascals (Pa), F is a force in newtons (N), and A is a surface area in square meters (m²). As an illustration of how this formula is applied, consider the two cartons shown in Figure 8.3. Each carton weighs 125 N. The upper carton rests directly on the floor, so the surface area supporting the weight of the carton is the area of its base, which is 0.25 m². The lower carton rests on four feet, each of which has an area of 6.25 cm² at the base, for a total area of 25 cm², or 0.0025 m². The pressure on the floor under the upper carton is

2.5 cm × 2.5 cm

A = 2.5 cm × 2.5 cm × 4 = 25 cm² = 0.0025 m²

Figure 8.3. The pressure under the feet of the lower box is 100 times higher than the pressure under the upper box because the weight is supported by 100 times less area.

$$P = \frac{F}{A} = \frac{125 \text{ N}}{0.25 \text{ m}^2} = 500 \; \frac{\text{N}}{\text{m}^2} = 500 \text{ Pa}$$

But the pressure under each of the feet of the lower carton is

$$P = \frac{F}{A} = \frac{125 \text{ N}}{0.0025 \text{ m}^2} = 50,000 \; \frac{\text{N}}{\text{m}^2} = 50,000 \text{ Pa}$$

▼ Example 8.2

When a certain rifle fires, the gases from the exploding gunpowder apply a force of 18,424 N to the bullet, which is 0.308 inches in diameter. Determine the pressure present inside the rifle barrel when the gun is fired.

Writing the givens and converting to MKS units, we have

$$F = 18,424 \text{ N}$$

$$D = 0.308 \text{ in} \cdot \frac{2.54 \text{ cm}}{\text{in}} \cdot \frac{1 \text{ m}}{100 \text{ cm}} = 0.007823 \text{ m}$$

$$P = ?$$

To compute pressure, we need the force and the area. The force is given, so we just compute the area of the end of the bullet using the formula for the area of a circle.

$D = 0.007823 \text{ m}$

$r = \dfrac{D}{2} = 0.003912 \text{ m}$

$A = \pi r^2 = 3.1416 \cdot (0.003912 \text{ m})^2 = 0.00004808 \text{ m}^2$

Now we compute the pressure.

$P = \dfrac{F}{A} = \dfrac{18,424 \text{ N}}{0.00004808 \text{ m}^2} = 383,200,000 \text{ Pa}$

This value has been rounded to four significant digits, since we are going to need to end up with three significant digits in our result. Finally, we convert this value into psi.

$P = 383,2000,000 \text{ Pa} \cdot \dfrac{14.7 \text{ psi}}{101,325 \text{ Pa}} = 55,600 \text{ psi}$

This result has been rounded to three significant digits, as required.

8.2   Atmospheric Pressure

8.2.1 Air Pressure

We began this chapter by noting that pressure under a liquid is due to the weight of the liquid. This principle applies to any fluid, gases as well as liquids. This means that we can think of the pressure in the atmosphere around us as caused by the weight of the air molecules above us. Of course, we call this pressure *atmospheric pressure*. When weather forecasters refer to the local value of this pressure, they call it *barometric pressure*. A *barometer* is an instrument that measures barometric pressure.

However, there is one big difference between the pressure under the surface of a liquid and the atmospheric pressure we experience. This difference is due to the fact that gases are compressible and liquids basically aren't. So in a liquid like water, the density of the liquid is the same at any depth. But in a gas such as the atmosphere, the weight of the air above a given altitude compresses the air molecules below, causing the air below to be denser than the air above. As a result, the density of air is greatest at the ground and decreases with increasing altitude.

The fact that air density varies with altitude explains why there is less oxygen to breath at high elevations, such as in the mountains. If we lived under water like fishes, the pressure would change a lot as we swam up and down because water is about 830 times denser than air, but the density of the medium we lived in would remain constant. But in air, the density drops as the altitude increases, so there is less air to breathe.

At sea level, atmospheric pressure is about 101,325 Pa, or in USCS units, 14.7 psi. These values are listed in Appendix A for your convenience. You can see that these atmospheric pressure values appear in the pressure conversion factor we used in the last two example problems. There is also a pressure unit called the *atmosphere* (atm), defined in terms of atmospheric pressure. One atmosphere is defined as the atmospheric pressure at sea level, so 1 atm = 101,325 Pa. We are so accustomed to atmospheric pressure that we typically don't notice it, unless we feel it changing during a storm front or while flying in an airplane. We even deliberately ignore it in some applications involving pressure, as explained in the next section.

8.2.2 Barometers

The mercury barometer for measuring air pressure was invented in 1643 by Italian physicist and mathematician Evangelista Torricelli (Figure 8.4). To understand how the barometer works, refer to Figure 8.5. The diagram is a conceptual illustration of how the barometer is constructed, using a vacuum pump to remove the air from a glass tube set into a bowl of mercury. (Torricelli didn't have a vacuum pump, so he accomplished this a different way, but the result is the same.) As the air is removed from the glass tube, the mercury rises into the tube to a height of approximately 760 mm. With the air removed, the tube is capped. The device now registers changes in atmospheric pressure caused by the weather or by transporting the barometer up to the top of a tall building where the air pressure is a wee bit lower.

Figure 8.4. Italian physicist and mathematician Evangelista Torricelli (1608–1647).

As atmospheric pressure changes, the height of the mercury column in the glass tube rises and falls. At normal atmospheric pressure at sea level, the height of the mercury is 760 mm, which is about 29.9 inches. (This is why weather reporters refer to atmospheric pressure as "thirty inches of mercury.") Normal barometric pressure variation in the U. S. is from about 725 mm to 775 mm of mercury (28.5 to 30.5 inches). The higher the atmospheric pressure is, the farther it pushes the mercury up into the barometer tube. So if the local barometric pressure is 30.4 inches of mercury, that is a high-pressure day and the weather is typically clear, sunny, and calm. When the barometric pressure falls, it means weather is headed your way as high-pressure regions push wind and moisture toward your now low-pressure region. You might be interested to know that the world record low baro-

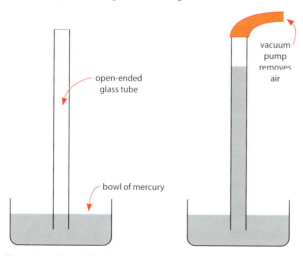
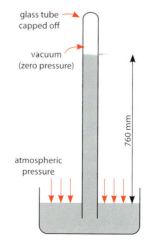

Figure 8.5. Assembly of a mercury barometer.

179

metric pressure (other than the pressure inside a tornado, which is extremely low) is 25.7 inches, measured in 1979 during a typhoon in the Pacific Ocean. The pressure unit *torr* is named after Torricelli, and normal atmospheric pressure is equal to 760 torr.

A vacuum is a complete void in space in which there is no matter of any kind. In 1643, scientists did not believe that a vacuum could exist (an idea dating back to Aristotle), so at the time the barometer was invented no one could formulate a correct explanation for how it worked. The mystery concerned the unidentified force that holds the mercury up in the tube. Blaise Pascal, for whom the SI unit of pressure is named, explained how the barometer worked in 1647, theorizing that a vacuum did exist in the glass tube above the mercury. On the basis of his own experiments, Pascal explained that there was no air pressure above the mercury column in the tube and that the pressure of the atmosphere acting on the surface of the mercury in the bowl *pushes* the mercury up into the tube into the vacuum region. This explanation is correct, but since at the time no one accepted the possibility of a vacuum, scientists did not want to accept Pascal's bold theory. But as we have seen with other theories, a theory gains strength when it repeatedly leads to successful hypotheses, and this one did.

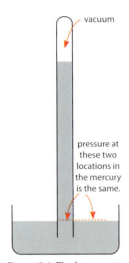

Pascal's explanation is easy to understand if we calculate the pressure in the tube 760 mm below the surface of the mercury. The density of mercury is 13,600 kg/m³. Using our equation $P = \rho g h$, the pressure at a depth of 0.76 m comes out to 101,300 Pa, which agrees with our value for atmospheric pressure to four significant digits. This calculation demonstrates that the pressure at the same height in the mercury inside and outside the tube is the same, as shown in Figure 8.6. I leave it to you to apply this same principle to the operation of a diving bell in the exercises.

Figure 8.6. The barometer illustrates the principle that the pressure at two places at the same depth in any liquid is the same.

8.2.3 Absolute Pressure and Gauge Pressure

Consider the city water supply, typically distributed to homes at a pressure of around is 80 psi (equal to about 550,000 Pa). This value is the net pressure inside the piping, not including atmospheric pressure.

To explain, consider a plumber with a pressure gauge to measure the water pressure in the 80 psi water line. When the plumber's gauge is not connected to anything, it reads 0 psi, even though it is in the air where the atmospheric pressure is 14.7 psi. This pressure reading is illustrated by the gauge on the left side of Figure 8.7, where atmospheric pressure is labeled 0 psi. When the plumber attaches the gauge to the water line, it reads 80 psi. A pressure scale like this that reads 0 at atmospheric pressure is called a *gauge pressure* scale. A gauge pressure reading reports the pressure above atmospheric pressure. When placed in a vacuum, this pressure gauge reads −14.7 psi.

A pressure gauge that reads 0 in a complete vacuum where there is no pressure (such as in a chamber from which all the air molecules have been removed) is reporting the *absolute pressure*. Such a gauge reads 14.7 psi in normal air, and when attached to a water line with 80 psi pressure, the absolute pressure gauge reads 94.7 psi, the sum of the 80 psi water pressure and the 14.7 psi atmospheric pressure.

Although we will not get into any of the more advanced calculations here, I note in passing that most calculations performed in physics and chemistry involving gases require the pressure to be in absolute units. Obtaining the absolute pressure from a gauge pressure

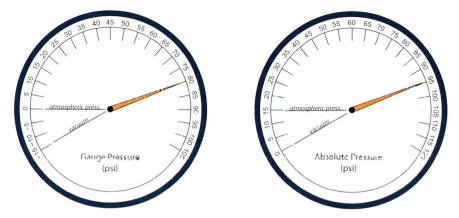

Figure 8.7. Pressure gauges showing a pressure reading of 80 psi (gauge) on a gauge pressure scale (left) and an absolute pressure scale (right). Reference points for atmospheric pressure and vacuum are also shown.

reading is simple; one simply adds atmospheric pressure to any gauge pressure measurement to obtain the absolute pressure. In other words,

$$P_{abs} = P_{gauge} + P_{atm}$$

where P_{abs} is the absolute pressure, P_{gauge} is the gauge pressure, and P_{atm} is the local value of atmospheric pressure. When pressures are measured with ordinary pressure gauges, the pressure measurement is a gauge pressure. To indicate this, the units for gauge pressure measurements are sometimes written as "psig," meaning, "psi gauge."

▼ Example 8.3

To help ensure that his car is as safe as possible, Mr. Washington checks the air pressure in his car tires regularly. He maintains the pressure at 35.0 psig. Determine the absolute pressure in Mr. Washington's tires and state the result in both psi and Pa.

The absolute pressure in psi is found by adding atmospheric pressure to the gauge pressure:

$$P_{abs} = P_{gauge} + P_{atm}$$

Since atmospheric pressure is 14.7 psi, the absolute pressure in the tires is

$$P_{abs} = 14.7 \text{ psi} + 35.0 \text{ psi} = 49.7 \text{ psi}$$

We use the conversion factor from Appendix A on this value to obtain the absolute pressure in pascals.

$$P_{abs} = 49.7 \text{ psi} \cdot \frac{101,325 \text{ Pa}}{14.7 \text{ psi}} = 342,575 \text{ Pa}$$

Rounding this value to three significant digits as the problem requires, we have

$P_{abs} = 343{,}000$ Pa

The pressure formula given in the first section of this chapter determines the gauge pressure at a given depth under a liquid. The absolute pressure at any depth is the pressure from the formula due to the liquid plus atmospheric pressure. A simple calculation shows that in water at a depth of 2.5 meters (8 feet), the absolute pressure is about 125,000 Pa, which is about 125% of atmospheric pressure. This suggests that going four times this deep puts the pressure at about 200,000 Pa, which is about twice atmospheric pressure. In fact, we have here the SCUBA divers' rule of thumb: pressure increases by one atmosphere for every 10 meters (33 feet) of depth.

8.3   Archimedes' Principle of Buoyancy

Archimedes of Syracuse (Figure 8.8), a third-century BC mathematician, inventor, and engineer, is regarded as the greatest scientist of antiquity. His principle of *buoyancy* is the reason huge ships carry their cargo without sinking. *Archimedes' principle* is as follows:

> An object submerged in any fluid experiences a buoyant force equal to the weight of the fluid it displaces.

A *buoyant force* is a force pushing up on an object when the object is partially or fully submerged. Anyone who has ever tried to pull a flotation device under water knows that a strong buoyant force is pushing the object toward the surface. You also know that when you are in a swimming pool, you do not have to exert much force with your legs to hold yourself up. This is because of the buoyant force that is helping to hold you up. Archimedes' principle allows us to calculate the value of this buoyant force.

Applying Archimedes' principle generally involves both the weight equation from Chapter 3, $F_w = mg$, and the density equation from Chapter 6, $\rho = \dfrac{m}{V}$. Applying Archimedes' principle centers on the task of determining the *weight* of the liquid displaced by the submerged object. *This weight is equal to the buoyant force.* For objects in water, we calculate this buoyant force by using the density equation and the density of water first to find the mass of the displaced water. Then we use the weight equation to find the weight of the water. This weight is the buoyant force.

As illustrated in Figure 8.9, in the simple case of a completely submerged flotation device (such as a block of Styrofoam), the volume of the float and the volume of water displaced by the float are the same. We calculate the buoyant force by using this volume and the density of water (998 kg/m³) to get the weight of the displaced water. On the left side of the figure, an additional force is being applied to the float to hold

Figure 8.8. Archimedes of Syracuse (c. 287–212 BC).

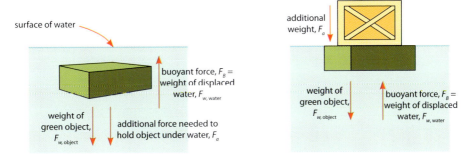

Figure 8.9. Holding a flotation device under water (left) and using the buoyant force to support extra weight (right).

it under water. On the right, additional weight is added on top of the float—just the right amount of weight to submerge the float completely without causing it to sink.

Now, we know from Newton's laws of motion that in both cases if the float remains stationary, the upward forces balance with the downward forces so that no net force is present. This gives us an equation we can use to solve buoyancy problems:

$$F_B = F_{w,\,object} + F_a$$

where F_B is the buoyant force (N), $F_{w,\,object}$ is the weight of the float (N), and F_a is any additional downward force applied to the float or pushing down on top of the float.

Note that there are two weights involved in this calculation. The weight of the displaced water, $F_{w,\,water}$, is equal to the buoyant force, F_B. The other is the weight of the submerged or floating object, $F_{w,\,object}$. As you see in the examples to follow, you should use subscripts to keep these two weights clearly identified so they don't get confused.

▼ Example 8.4

A kid is playing with a small flotation device in a swimming pool. The float weighs 0.0562 N and has a volume of 0.0100 m³. The kid has a cord tied to the float and is pulling the float downward with the cord, attempting to pull the float under water. How much force does the kid have to apply to the cord to pull the float completely under water? State your answer in pounds.

This situation is illustrated by the diagram on the left in Figure 8.9, shown again below. The buoyant force in this problem is equal to the weight of 0.0100 m³ of water (the volume of the float). We begin by using the density and weight equations to determine how much this water weighs.

$$V = 0.0100 \text{ m}^3$$

$$\rho = 998 \ \frac{\text{kg}}{\text{m}^3}$$

$$F_{w,\,water} = ?$$

$$\rho = \frac{m}{V}$$

$$m = \rho V = 998 \ \frac{kg}{m^3} \cdot 0.0100 \ m^3 = 9.98 \ kg$$

$$F_B = F_{w,\,water} = mg = 9.98 \ kg \cdot 9.80 \ \frac{m}{s^2} = 97.80 \ N$$

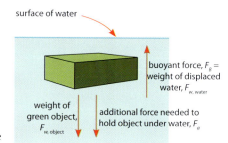

surface of water

buoyant force, $F_B =$ weight of displaced water, $F_{w,\,water}$

weight of green object, $F_{w,\,object}$

additional force needed to hold object under water, F_a

When the float is under water, 97.80 N of water have been displaced by the float, so this is the buoyant force on the float. Now we just use the general force equation to solve for the applied force.

$$F_B = F_{w,\,object} + F_a$$

$$F_a = F_B - F_{w,\,object} = 97.80 \ N - 0.0562 \ N = 97.74 \ N$$

Finally, we convert this value to pounds and round to three significant digits as the problem requires.

$$F_a = 97.74 \ N \cdot \frac{1 \ lb}{4.45 \ N} = 22.0 \ lb$$

▼ Example 8.5

A small boat dock floating on Styrofoam blocks has dimensions 3.0 m wide × 4.0 m long × 0.50 m high. If the density of Styrofoam is 55 kg/m³, how much weight can be piled on the dock before it submerges and sinks?

For simplicity, in this example we will ignore the weight of the boards of the dock, and just consider the Styrofoam blocks. This problem is similar to the previous one, except that we are not given the weight and volume of the floats. We must calculate these from the given dimensions and density of the floats. The situation is just like the right side of Figure 8.9, shown again on the next page.

The volume of the Styrofoam blocks is

$$V = 3.0 \ m \cdot 4.0 \ m \cdot 0.50 \ m = 6.0 \ m^3$$

As before, this volume is also the volume of the displaced water. We now calculate the weight of this water because that weight is equal to the buoyant force. As before, we solve the density equation for mass, and insert the volume and density values to get the mass of the water displaced when the Styrofoam is completely submerged.

$$\rho = \frac{m}{V}$$

$$m = \rho V = 998 \ \frac{kg}{m^3} \cdot 6.0 \ m^3 = 5988 \ kg$$

The weight of this mass of water is

$$F_{w,\,\text{water}} = mg = 5988 \text{ kg} \cdot 9.80\,\frac{\text{m}}{\text{s}^2} = 58{,}682 \text{ N}$$

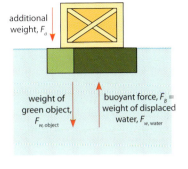

This weight is the buoyant force, F_B, when the Styrofoam is fully submerged.

The weight of the Styrofoam itself is found the same way, using the density of Styrofoam.

$$m = \rho V = 55\,\frac{\text{kg}}{\text{m}^3} \cdot 6.0 \text{ m}^3 = 330 \text{ kg}$$

$$F_{w,\,\text{Styrofoam}} = mg = 330 \text{ kg} \cdot 9.80\,\frac{\text{m}}{\text{s}^2} = 3234 \text{ N}$$

Now we use the same generic force equation to calculate the additional weight these Styrofoam blocks can support, F_a.

$$F_B = F_{w,\,\text{object}} + F_a$$
$$F_a = F_B - F_{w,\,\text{object}} = 58{,}682 \text{ N} - 3234 \text{ N} = 55{,}448 \text{ N}$$

Rounding this value to two significant digits, we have

$$F_a = 55{,}000 \text{ N}$$

The buoyant force must support the Styrofoam itself, which weighs about 3,200 N or 720 lb. But the Styrofoam blocks can support approximately 55,000 N or 12,300 lb of additional stuff, which is over 6 tons!

It is easy to see from this example that a large volume of air space in the hold of a ship gives rise to a huge buoyant force, enough to hold up the heavy, steel-hulled ship and all its cargo. Hopefully, it is also clear that if the ship fills up with water, it sinks.

8.4   Flotation

So long as the weight of an object is less than the buoyant force that would be present if the object were completely submerged, the object floats. If an object is floating, it is only partially submerged, so the object floats deep enough in the water so that the submerged part provides a buoyant force equal to the weight of the entire object. With a buoyant force equal to the weight, the object stays at the surface of the water.

The human body is about 3/4 water. Fat tissue is less dense than water, while muscle and bone tend to be denser than water. Overall, the density of the human body is very close to the density of water, so the buoyant force available from submersion is close to a person's weight. Whether a person floats depends on the person's body fat content. Most people float, even if just barely. Inhaling helps flotation because the overall density of the body decreases when the lungs are expanded and full of air. Floating is easier in sea water than fresh water because the dissolved salts in sea water make it denser than fresh water. And we are told that everyone floats easily in the Dead Sea. Submerging oneself in that water (the saltiest on earth) is next to impossible because the water is so dense.

Archimedes' principle applies to objects in the atmosphere as well as objects in water. However, since a volume of air weighs only about 1/830th as much as the same volume of water, the buoyant force due to air is only 1/830th the buoyant force provided by water. Thus, air can provide the buoyant force needed to hold up a helium balloon, but it won't hold up anything much heavier, such as Wile E. Coyote when he runs off a cliff. (Usually, my students get that reference. But if you haven't seen those cartoons, it just means that I am getting old!)

Hot air balloons work by filling the huge balloon with hot air, which is less dense than air at normal temperature. The density change caused by the heating of the air is not great, so hot air balloons must be quite large to provide enough buoyant force to lift the balloon and the gondola with its passengers. But the balloon works because of Archimedes' principle. The balloon displaces a certain volume of air and the weight of this air is the buoyant

Do You Know ...   What was Archimedes' "eureka" discovery?

Archimedes is famous for his discovery of how to measure the density of an irregularly shaped object. The medieval drawing below shows Archimedes in his bath. The story goes that the king of Syracuse had given a quantity of gold to an artisan with which to make a crown. The crown the artisan delivered weighed the same as the gold the king had supplied, but the king suspected that the artisan had substituted lead for some of the gold, measured out enough lead to make the total weight match the original weight of the gold, and kept the remaining gold for himself. He posed the problem to Archimedes to figure out how to determine if the crown was made of pure gold, without changing it.

Archimedes knew that the key was to measure the density, because the density of gold and lead are different. So if the crown had lead in it, its density would be different from the density of pure gold. Archimedes was pondering this problem as he lowered himself into his bath. He watched the water in the tub rise as he lowered himself in. Suddenly he realized that he knew how to measure the volume of the crown—simply measure out some water, submerge the crown in the water, and measure the new volume. The difference is the volume of the crown. Knowing the volume, he could calculate the density! Archimedes was so excited that he had solved the king's problem

that he leapt from his bath and ran naked down the streets of Syracuse shouting, "Eureka!," which means, "I found it!" Sadly, when Archimedes' discovery was applied to the artisan's crown, it was found that the artisan had indeed kept out some of the gold. Not good.

The method of measuring volume Archimedes discovered is called the *displacement method*. You used this same method of measuring volume in the Density experiment.

force. Inside the balloon is hot air, which is less dense than the air outside the balloon and so weighs less.

Chapter 8 Exercises

Pressure Problems

1. Calculate the pressure 33.5 ft under the surface of a lake. State your result in both kilopascals (kPa) and pounds per square inch (psi).

2. Calculate the pressure 125 ft under the surface of the ocean. Assume that the density of seawater is 1.025 g/cm³. State your result in both kPa and psi.

3. If a barometer is made with water instead of mercury, how tall is the liquid column at normal atmospheric pressure? Perform this calculation in MKS units but state your result in feet.

4. A certain water tank is built in the form of a vertical cylinder 65 ft high and 10.0 ft in diameter. If the water in the tank is 60.6 ft deep, calculate the pressure on the floor of the tank. Do this calculation using the standard equation $P = \rho g h$.

5. Repeat the previous problem, this time solving the problem by calculating the volume and weight of the water and the area of the bottom of the tank. Calculate the pressure as $P = F/A$. See how your two results compare.

6. For your answer to problem 4, compute the absolute pressure at the bottom of the tank. State your result in both kPa and psi.

7. In a real water tank, the air space at the top of the tank is under pressure. This air pressure helps to pressure up the water supply. Determine the absolute pressure at the bottom of the tank in problem 4 if the air pressure in the top of the tank is 55.0 psig. (Hint: One of the pressure laws discovered by Pascal is that if you increase the pressure somewhere in a liquid, the pressure increases everywhere in the liquid by the same amount.)

8. Calculate the absolute pressure in both kPa and psi at the depths specified in problems 1 and 2.

9. The average barometric pressure in Denver, Colorado is 83.4 kPa. How many millimeters does a mercury barometer read in Denver? The density of mercury is 13,600 kg/m³.

10. The concrete floor in a college physics building has a pressure rating of 2,800 psi. Scientists are planning to set up a large, heavy experimental apparatus in the room. The equipment weighs 2,200 pounds and rests on four circular feet, each 0.75 inches in diameter. Can the floor safely support the equipment?

11. The owners of a luxury hotel are planning to put a large aquarium on a concrete floor rated at 3,200 psi. The hotel has an open, high lobby with balconies looking into the lobby from each floor, and the owners want the aquarium to be quite deep so that people can look at it from several different floors in the hotel. If they do not wish for the load on the floor to exceed 75% of the floor rating, how deep can the aquarium be?

12. Assume that on the bottom of a pair of adult men's shoes the contact area with

the floor is 120 cm² per shoe. Calculate the pressure on the floor under the shoes of a 195-pound man. State your result in both Pa and psi.

13. A woman wearing high heels leans back on her heels. The bottoms of the heels of her shoes are square, 0.750 cm on each side. Determine the pressure on the floor under her heels if the woman weighs 130 lb. State your result in both Pa and psi.

14. History tells us that Alexander the Great used a glass diving bell to explore the Mediterranean Sea, as shown in the 16th-century painting to the right. A diving bell is a large vessel, closed on top and open on the bottom. What is the absolute air pressure inside a diving bell lowered to a depth of 20.0 feet? (The increased pressure causes water to rise into the diving bell from the bottom, but ignore this for this problem.) State your result in psi. Use the density of saltwater: 1.025 g/cm³.

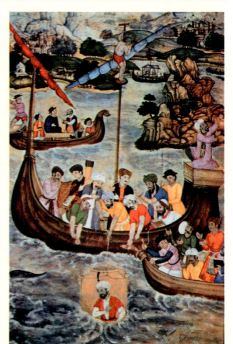

Answers

1. 99.9 kPa, 14.5 psi
2. 383 kPa, 55.5 psi
3. 34.0 ft
4. 181 kPa
5. 181 kPa
6. 282 kPa, 40.9 psi
7. 661 kPa
8. Prob 1: 201.2 kPa, 29.2 psi; Prob 2: 484 kPa, 70.2 psi (The addition rule for significant digits, which we have not covered, allows us to have the fourth digit in the first answer.)
9. 626 mm
10. Yes. Pressure under each foot is 1,200 psi.
11. 1,700 m (This is much taller than the tallest building in the world, so the floor is no concern. However, the glass and joints in the aquarium might be.)
12. 36,000 Pa, 5.2 psi
13. 5,200,000 Pa, 750 psi (You don't want to get stepped on by one of those heels.)
14. 23.57 psi (Four digits; see answer to #8)

Buoyancy Problems

1. Calculate the buoyant force acting on a submerged 2 × 4 if the board's dimensions are 1.5 in × 3.5 in × 12 in. State your result in both N and lb.

2. Calculate the buoyant force on the same piece of wood from the previous problem (in N and lb) if the board is submerged in (a) saltwater and (b) mercury. The densities of saltwater and mercury are 1.025 g/cm^3 and 13.6 g/cm^3, respectively.

3. A standard 55-gallon drum, shown to the right, is 22.5 inches in diameter and 33.5 inches tall. An empty drum is to be submerged under the water of a lake and held there by a cable attached to the bottom. The empty drum weighs 54.0 pounds. Determine the tension in the cable holding it down. (The tension in a cable is the force the cable is applying to the drum.)

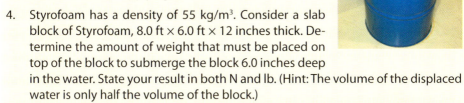

4. Styrofoam has a density of 55 kg/m^3. Consider a slab block of Styrofoam, 8.0 ft × 6.0 ft × 12 inches thick. Determine the amount of weight that must be placed on top of the block to submerge the block 6.0 inches deep in the water. State your result in both N and lb. (Hint: The volume of the displaced water is only half the volume of the block.)

5. In the old days before the invention of Styrofoam, boat docks often floated on horizontal 55-gallon drums, like the one in problem 3 above. If a boat dock built from 450 pounds of lumber rests on eight horizontal 55-gallon drums, how much additional weight can the dock support if the drums are not to be more than 50% submerged? State your result in N and lb. Don't forget about the weight of the drums.

Answers

1. 1.0×10^1 N, 2.3 lb
2. saltwater: 1.0×10^1 N, 2.3 lb, mercury: 140 N, 31 lb
3. 1.90×10^3 N
4. 5,900 N, 1300 lb
5. 4,600 N, 1.0×10^3 lb

Pressure Study Questions

1. Consider a hand-operated well pump for drawing water, such as the one shown to the right. These pumps work by drawing a vacuum on the top of a pipe that extends down to the water table under ground. Explain why a pump like this cannot pump water from a well more than 34 feet deep.

2. City water pressure is typically around 80 psig. A commercial-style tankless toilet requires 25 psi to operate. If the floors in a multi-story office building are 12 feet apart, explain how an architect determines the highest floor where the toilets operate properly without an additional pump to pressure up the water supply.

CHAPTER 9
Waves, Sound and Light

The Antenna that Discovered the Cosmic Microwave Background

From NASA's website:

The horn reflector antenna at Bell Telephone Laboratories in Holmdel, New Jersey was built in 1959 for pioneering work in communication satellites for the NASA ECHO I. The antenna was 50 feet in length and the entire structure weighed about 18 tons. It was comprised of aluminum with a steel base. It was used to detect radio waves that bounced off Project ECHO balloon satellites. The horn was later modified to work with the Telstar Communication Satellite frequencies as a receiver for broadcast signals from the satellite. In 1990 the horn was dedicated to the National Park Service as a National Historic Landmark.

For more on the cosmic microwave background, see page 173.

OBJECTIVES

Learn how to use these equations:

$$v = \lambda f \qquad \tau = \frac{1}{f}$$

After studying this chapter and completing the exercises, students will be able to do each of the following tasks, using supporting terms and principles as necessary:

1. Explain what a wave is.
2. On a graphical representation of a wave, identify the wave parameters and parts: crest, trough, amplitude, wavelength, and period.
3. Define the *frequency* and *period* of a wave.
4. Describe the following five wave interactions, giving examples of each:
   a. Reflection
   b. Refraction
   c. Diffraction
   d. Resonance
   e. Interference
      i. Constructive interference
      ii. Destructive interference
5. Give examples of longitudinal, transverse, and circular waves and the media in which they propagate.
6. Define *infrasonic* and *ultrasonic* and give examples of these types of sounds.
7. Define *infrared* and *ultraviolet* and give examples of these types of radiation.
8. Calculate the velocity, frequency, period, and wavelength of waves from given information.
9. Given the frequency of a wave, determine the period, and vice versa.
10. List at least five separate regions in the electromagnetic spectrum, in order from low frequency to high frequency.
11. State the frequency range of human hearing in hertz (Hz) and the wavelength range of visible light in nanometers (nm).
12. State the six main colors in the visible light spectrum in order from lowest frequency to highest.
13. Identify the relations: frequency and pitch; amplitude and volume.
14. Explain how waves of different frequencies (harmonics) contribute to the timbre of musical instruments.

9.1   Modeling Waves

There are many phenomena in the physical world that can be modeled as waves. Examples are sound, light, earthquakes, electrical signals, radio waves, and even the internal patterns of electrons inside atoms. The great variety of wave phenomena makes the study of waves a topic of high importance in physics.

9.1.1 Describing Waves

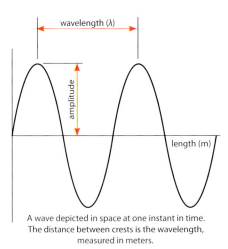

A wave depicted in space at one instant in time. The distance between crests is the wavelength, measured in meters.

Figure 9.1. A graphical representation of a wave in space.

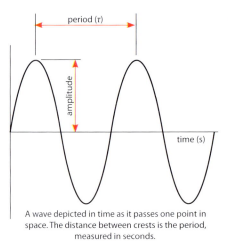

A wave depicted in time as it passes one point in space. The distance between crests is the period, measured in seconds.

Figure 9.2. A graphical representation of a wave in time.

A *wave* is a disturbance in space and time that carries energy from one place to another. A wave can be a single pulse, such as a tsunami caused by an earthquake, or a continuous *wave train*, such as a beam of light or sound waves from a blaring horn. We represent waves graphically as shown in Figures 9.1 and 9.2. The tops of a wave are called *crests* (or *peaks*); the bottoms are called *troughs* (or *valleys*). The height of the wave from the center line to a crest is called the *amplitude*. The amplitude relates to how much energy the wave is carrying and the units depend on what kind of wave it is. (We will not deal with calculations involving the amplitude in this text.) Louder sounds, brighter lights, and more destructive earthquakes are all examples of waves with higher amplitudes.

The waves shown in these two figures are often called *sine waves*, and waves possessing this very typical shape are said to be *sinusoidal*. We will encounter this term again in a later chapter, and you will learn the mathematical details when you study trigonometry.

Since waves are disturbances in both space and time, we can think of the length of a wave in distance (spatial) units or time units. If we are thinking about distance in space, the length of one cycle of a wave is called the *wavelength*, measured in meters. When we consider a wave varying in time, the duration of one cycle of the wave is called the *period*, measured in seconds. The period is the amount of time required for the wave to complete one cycle. You came across this same definition before when you performed the Pendulum Experiment. When you are looking at a graph of a wave, be sure to notice whether the horizontal axis is labeled distance or time because the units on the horizontal axis will either be meters or seconds, respectively. Whether we are speaking of the period or the wavelength, these variables represent the span from one crest of the wave to the next crest.

9.1.2 Categorizing Waves

We could divide all people into two categories, male or female. A different way to categorize all people would be in the three categories of child, adolescent, or adult. Similarly, there are different ways to categorize waves.

One way to categorize waves is by whether the wave needs a *medium* (matter of some kind) in which to propagate. (We use the word *propagate* instead of travel or move when we are talking about waves because waves don't simply travel from place to place. They also spread out as they go and the term propagate connotes this better.) *Mechanical waves* need a medium in which to propagate. Most waves are like this. Sound does not propagate in a vacuum but it does propagate in media such as air and water. By contrast, *electromagnetic waves* propagate in the vacuum of empty space without a medium. This is how energy gets

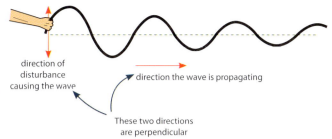

direction of disturbance causing the wave

direction the wave is propagating

These two directions are perpendicular

Figure 9.3. Whipping the end of a cord up and down forms a transverse wave—the direction the wave source is oscillating is perpendicular to the direction of wave propagation.

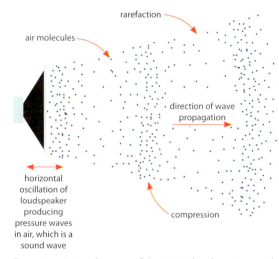

rarefaction

air molecules

direction of wave propagation

horizontal oscillation of loudspeaker producing pressure waves in air, which is a sound wave

compression

Figure 9.4. A visualization of the air molecules in a sound wave emerging from a loudspeaker.

here from the sun, as you know from our study of energy in Chapter 4. Electromagnetic waves also propagate in some media—visible light propagates in air and through glass, and radio waves propagate in air and through wooden walls.

A second way to categorize all waves is according to the relationship between the direction of oscillation that is causing the wave and the direction the wave is propagating. There are three basic possibilities. In *transverse waves*, the oscillation is *perpendicular* to the direction of propagation. Light and waves on strings are examples of transverse waves.

A transverse wave on a string is depicted in Figure 9.3; transverse electromagnetic waves are depicted later in the chapter in Figure 9.19.

With *longitudinal waves*, the oscillating motion causing the wave is *parallel* to the direction of propagation. A sound wave in air is an example, and considering how sound is produced by a loudspeaker pushing back and forth is a helpful way to think about this and remember it. Figure 9.4 depicts a loudspeaker producing longitudinal sound waves. The wave train is actually a succession of pressure fluctuations in the air above and below atmospheric pressure.

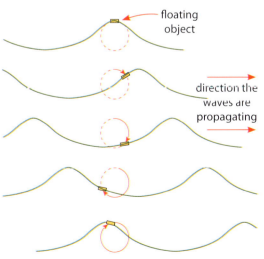

floating object

direction the waves are propagating

Figure 9.5. A water wave causing a floating object to loop around in a circular fashion as the wave goes by.

We address sound waves in more detail, along with the other terms I placed in this diagram, later in this chapter.

The third type of wave is *circular waves*, mainly represented by waves on water. In circular waves, the oscillating action is moving in a circular fashion. Because of this, a floating object on water moves in a circular pattern as the waves pass by underneath it, as illustrated in Figure 9.5.

9.1.3 Modeling Waves Mathematically

There are two important equations for you to learn for our work with waves. Both these equations involve letters from the Greek alphabet, which seems to make these equations challenging for students to remember correctly. Another factor that confuses students is the unfamiliar terms associated with the variables in these equations. So to give students a break, I usually do not require students in this course to memorize these two equations. Instead, you should concentrate on remembering the quantities and units of measure the variables represent.

The first equation relates the velocity at which the wave propagates to the frequency and wavelength of the wave:

$$v = \lambda f$$

where v is the velocity of the wave propagation (m/s), λ ("lambda," the Greek letter *l*) is the wavelength of the wave (m), and f is the frequency of the wave (Hz).

The *frequency* of a wave is the number of cycles the wave completes in one second. The MKS units for frequency are cycles per second. We have a name for this unit, which is *hertz* (Hz). So if a wave completes 5,000 cycles per second, its frequency is 5,000 Hz or 5 kHz.

As you see from the equation, the velocity of a wave varies directly with its wavelength and directly with its frequency. But physically, it is better to think of the wavelength as depending on the other two variables in this equation. This is because the wave velocity actually depends on the medium in which the wave is propagating. Also, the frequency depends on the source of the wave; whatever is causing the wave is oscillating at a certain rate, and this rate determines the frequency of the wave being produced. So since the wave velocity is determined by the medium, and the frequency is determined by the source of the wave, the wavelength is the variable that really depends on the other two. So solving the wave equation for the wavelength, we have

$$\lambda = \frac{v}{f}$$

This equation shows us that wavelength (the dependent variable) varies inversely with frequency (the independent variable). In other words, when one goes up, the other goes down. Knowing this is particularly useful in the study light later in this chapter.

▼ Example 9.1

Determine the wavelength of a 1.500 kHz sound wave. (This frequency is right around the pitch people make when they whistle.) Assume that sound propagates at 342 m/s in air.

Write the givens and do the unit conversions.

$$f = 1.500 \text{ kHz} \cdot \frac{1000 \text{ Hz}}{\text{kHz}} = 1{,}500 \text{ Hz}$$

$$v = 342 \ \frac{\text{m}}{\text{s}}$$

$$\lambda = ?$$

Now complete the problem. Write the wave equation in its standard form first, then do the algebra to solve for the wavelength.

$$v = \lambda f$$

$$\lambda = \frac{v}{f} = \frac{342 \ \frac{\text{m}}{\text{s}}}{1500 \text{ Hz}} = 0.228 \text{ m}$$

Although the frequency is given with four significant digits, the wave velocity is given with only three significant digits, so the result is stated with three significant digits.

 Example 9.2

Determine the frequency of the light produced by a laser operating at a wavelength of 488 nm in air.

Write the givens and do the unit conversions.

$$\lambda = 488 \text{ nm} \cdot \frac{1 \text{ m}}{10^9 \text{ nm}} = 4.88 \times 10^{-7} \text{ m}$$

$$v = 3.00 \times 10^8 \ \frac{\text{m}}{\text{s}}$$

$$f = ?$$

Now complete the problem. Again, write the wave equation in its standard form first, then do the algebra to solve for the frequency.

$$v = \lambda f$$

$$f = \frac{v}{\lambda} = \frac{3.00 \times 10^8 \ \frac{\text{m}}{\text{s}}}{4.88 \times 10^{-7} \text{ m}} = 6.148 \times 10^{14} \text{ Hz}$$

Both the wavelength and the speed of light values have three significant digits, so we round this result to three significant digits.

$$f = 6.15 \times 10^{14} \text{ Hz}$$

195

As I mentioned near the beginning of this chapter, the length of time it takes for any oscillating system, including a wave of some kind, to complete one full cycle is called the period, measured in seconds. The period of a wave relates directly to the frequency. Since the units for frequency are cycles per second, the reciprocal of the frequency has units of seconds per cycle. This time value, the number of seconds in one cycle, is the period of the wave, although when we speak of the period or use it in equations we just call the units "seconds." We use the Greek letter τ ("tau," the t in the Greek alphabet) to represent the period of a wave or of any other oscillating system. Since period and frequency are reciprocals, this gives us our second equation,

$$\tau = \frac{1}{f}$$

You have already been using the speed of light in air or a vacuum, $c = 3.00 \times 10^8$ m/s, in calculations from previous chapters. Since all light is electromagnetic waves, and since our main wave equation has the velocity in it, you will now be using the speed of light even more frequently in calculations. Take special note at this point that radio waves propagate at the speed of light, not at the speed of sound. They are not sound waves (you can't hear them); they are electromagnetic waves. Microwaves and X-rays are also electromagnetic waves, so any problem involving any of these requires you to use the speed of light for the velocity.

▼ Example 9.3

Radio station KUT FM in Austin broadcasts a carrier signal at 90.5 MHz. Determine the wavelength and period of this wave. State the period in nanoseconds.

As I mentioned above and discuss more later, all radio waves (FM, AM, short wave, etc.) are part of the electromagnetic spectrum and propagate at the speed of light. Knowing this, write the givens and do the unit conversions.

$$f = 90.5 \text{ MHz} \cdot \frac{10^6 \text{ Hz}}{\text{MHz}} = 9.05 \times 10^7 \text{ Hz}$$

$$v = 3.00 \times 10^8 \ \frac{\text{m}}{\text{s}}$$

$$\lambda = ?$$

$$\tau = ?$$

The problem requires us to perform two separate calculations. Calculate the wavelength first using the given information.

$$v = \lambda f$$

$$\lambda = \frac{v}{f} = \frac{3.00 \times 18^8 \ \frac{\text{m}}{\text{s}}}{9.05 \times 10^7 \text{ Hz}} = 3.31 \text{ m}$$

All the values in the problem have three significant digits, as does this result. Now compute the period.

$$\tau = \frac{1}{f} = \frac{1}{9.05 \times 10^7 \text{ Hz}} = 1.105 \times 10^{-8} \text{ s} \cdot \frac{10^9 \text{ ns}}{\text{s}} = 11.05 \text{ ns}$$

This value is in nanoseconds, as required, but it has one extra significant digit. Rounding to the required three significant digits gives the final result.

$\tau = 11.1$ ns

9.2   Wave Interactions

There are several phenomena common to all types of waves. These phenomena are also called *interactions*. We are going to look at five of them. For some of these descriptions, it is helpful to think of the wave as a ray, like the thin line of light we are familiar with in a laser beam. In some of the illustrations below, I have shown rays as arrows pointing in the direction of propagation.

9.2.1 Reflection

Waves *reflect* off surfaces or objects. As the wave or ray approaches the surface, it is called the *incident ray*. The ray that reflects off the surface is called the *reflected ray*. The *law of reflection* states:

The angle of incidence equals the angle of reflection.

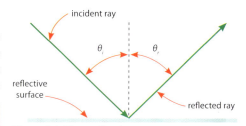

Figure 9.6. Wave reflection.

These two angles are defined with respect to a line perpendicular to the reflecting surface. This perpendicular line is called the *normal line*. Figure 9.6 depicts a reflective surface with the normal line shown (the dashed line), and a ray of light striking the surface at the angle of incidence. Using standard notation, we designate the angle of incidence as θ_i and the angle of reflection as θ_r. (The Greek letter θ, "theta," which is the "th" sound, is used throughout mathematics and physics to represent measures of angles.) So we can write the law of reflection as $\theta_i = \theta_r$.

9.2.2 Refraction

All waves refract but we typically only notice it with light waves. The speed of light in a plastic prism, glass, or water is slower than the speed of light in air. When the light enters the new medium, it changes speed instantly. This change of speed causes the light to change direction, or *refract*. Figure 9.7 shows a ray of red laser light coming almost straight down in the air and refracting as it crosses

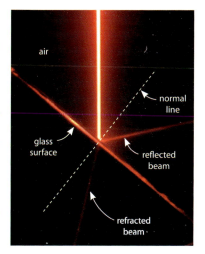

Figure 9.7. Refraction, illustrated by a 633-nm laser passing from air into a glass prism. Some light can be seen reflecting off the upper side of the prism.

197

the boundary from air into a glass prism. Some of the light also reflects off the prism, obeying the law of reflection, and heads off to the right in the photograph. The amount the light bends depends on both the media in which the wave is propagating. In the case of the photo in Figure 9.7, these media are air and glass. The amount of refraction also depends slightly on the color of the light, which depends on the wavelength. Sunlight is a combination of all different colors, and when it passes through a prism the different colors refract different amounts. This causes the sunlight to split into different colors as in a rainbow, an effect known as *dispersion*.

The familiar image of a straw in a glass of water, and the way it appears to be broken, is because of the refraction of light as it passes from air to water, reflects off the straw under the water, and refracts again as it passes from water to air. The light we see reflected from the underwater part of the straw has refracted twice and is propagating in a direction different from the light reflecting from the upper part of the straw. The light we see reflecting from the upper part never went under water and never refracted.

9.2.3 Diffraction

Diffraction occurs when a wave bends around the corner of some obstruction. A familiar example of diffraction is that a person's voice can be heard around the edge of a barrier like the corner of a building because of the sound waves bending around the corner. Without diffraction, it would be very difficult to hear what people say because sound waves would propagate straight out of our mouths without spreading out, like the beam of a flashlight. And just as the flashlight creates a spot of light, the sound wave would create only a

Figure 9.8. Combined red and green laser beams are split apart and bent by a diffraction grating.

spot of sound. (The flashlight beam diffracts too, but the wavelengths of light are so short that diffraction of light is more difficult to perceive.) As a result of diffraction, a laser beam can be arranged to create a so-called *interference pattern*. This effect is described in the section on wave interference below.

A really cool example of the bending of waves from diffraction occurs when laser light is beamed through a *diffraction grating*, as shown in Figure 9.8. In this figure, a yellow laser beam composed of the light from 633-nm red and 543-nm green lasers hits a diffraction grating. A diffraction grating is a glass plate with very fine grooves etched into its surface (70 grooves per mm for the grating used to make the image shown, which is actually not very many for a diffraction grating). The red and green laser beams, visible at the bottom of the image, are combined in the plastic cube. The yellow beam that emerges strikes the grating in the center of the image and the red and green beams fan out separately beyond the grating. The grating splits the light into many beams, each of them bending more and more away from the center. You also see that the two colors are bent different amounts, just as with refraction. Many of the separated beams are hitting the white screen I placed in the beam path, but you also see distant red and green dots where other beams are hitting the wall in the back of the room.

9.2.4 Resonance

When an original wave interacts with its reflection in a medium, a *standing wave* is formed in the medium. When a standing wave is present in a medium, the nodes, peaks, and troughs of the wave remain in the same position—unlike a propagating wave in which the nodes and peaks propagate at the wave velocity. Standing waves also form when a wave propagates in a moving medium, if the wave and medium are moving in opposite directions.

Under certain conditions, standing waves give rise to the phenomenon of *resonance*. To illustrate, consider a wave train continuously propagating in a medium such as sound waves propagating in a room or waves propagating on a string. As the waves reach a boundary such as a wall or the end of the string, they reflect back into the region where the original waves are, so the original waves and the reflected waves are mingling together in the same space, creating standing waves in the medium. If the dimensions of the medium (e.g., room or string) correspond to an integral multiple of half the wavelength, the crests and troughs of the original and reflected waves line up with each other and remain that way, producing a standing wave with a much greater amplitude. This is the condition of resonance. Figure 9.9 shows such a standing wave giving rise to resonance on a string stretched between two posts and vibrated at 60 Hz on one end.

Figure 9.9. A standing wave on a vibrating string strung across a classroom.

In most media most of the time, the energy in waves is absorbed. Sound dies out in a room because of absorption by the surfaces in the room and the air in the room. Light does the same thing. But when resonance occurs, the energy in the original and reflected waves is not dampened much by the medium because the medium itself naturally tends to oscillate at that frequency. Thus the energy of the original and reflected waves in the medium adds together without significant dampening, creating the resonance. The amplitude of the standing wave shown in Figure 9.9 is about 6 cm, but the device used to vibrate the string only moves up and down with an amplitude of about 1 cm. The resonance occurs when the tension in the string is set such that the string naturally tends to vibrate at 60 Hz. At other tensions, the amplitude of the standing waves on the string is approximately equal to the amplitude of the device causing the string to vibrate.

Most of the time, such as with sound or light waves, we cannot see standing waves, but on a vibrating string we can. The wave on the string is not really standing still, it just appears to be because the nodes and peaks of the wave and its reflection produce the standing wave that is standing still. That the wave is not actually stationary is demonstrated by the high-speed photograph shown in Figure 9.10. The device vibrating the string is in the foreground and the waves are propagating away from the wave source, just as they are in Figure 9.3.

As mentioned above, for resonance to occur, the dimension of the medium in which the wave is propagating (the length of the string in this case) must be equal to an integral

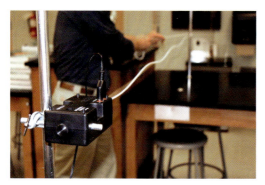

Figure 9.10. Transverse waves being formed on a string by the up and down motion of a string vibration device, and propagating down the string.

multiple of half the wavelength of the wave. For the wave on the string, if the wave is 1 m long, resonance occurs if the length of the string is any multiple of 0.5 m. For a resonance to occur with a sound wave in a room, at least one of the dimensions of the room (height, width, or depth) must equal an integral multiple of half the wavelength of the sound wave. Resonance is what is happening when someone plays a wind instrument, when you blow across the top of a bottle, or when you find the sweet notes that resound more loudly than others while you are humming in a tile and glass shower stall. (You should try it.)

Another example of resonance arising from standing waves is the pattern formed on the surface of a liquid when the container is vibrating, as in the vibrating coffee in a coffee cup shown in Figure 9.11. To make this pattern for the photograph, I had to make the table vibrate at just the right frequency so that the coffee cup sitting on the table vibrated the coffee, producing resonance in the waves on the surface of the coffee.

All musical instruments depend on resonance to produce their sounds, and patterns of resonances enable us to distinguish one instrument from another. To illustrate, let's say you want to play the note known as "middle C" on a flute. To do this, the air in the flute resonates at the frequency of middle C, also called C4, which is 278.4375 Hz. Now let's have

Figure 9.11. Standing waves on the surface of coffee in a vibrating coffee cup.

a clarinet play this same note. To do this, the clarinet also resonates at 278.4375 Hz. The two instruments are producing sound from resonances at the same frequency, but they sound quite different.

One of the reasons they sound different is because while both instruments are resonating at the frequency of middle C, they are also both simultaneously resonating at many other frequencies that are multiples of middle C. The main resonant frequency is called the *fundamental* (middle C in this case) and the other frequencies resonating in the instrument are called *harmonics*. Sometimes the fundamental is called the first harmonic. The second harmonic is usually a frequency that is twice the frequency of the fundamental. The third harmonic is three times the fundamental, and so on. In some instruments, the harmonics increase by odd or fractional multiples but in general, harmonics are multiples of the fundamental frequency. When writing about the different harmonic frequencies present in a sound wave, the fundamental is denoted as f_1, and the second, third, and higher harmonics are denoted f_2, f_3, and so on.

Figure 9.12 depicts the first three harmonics that resonate on a string. I drew them on a string because they are easy to visualize (exactly like the standing wave shown in the photo of Figure 9.9), but the same basic idea applies to waves resonating in any medium, such as sound waves in a room or in a wind instrument. The horizontal line in the middle of the figure represents the string when it is not vibrating. As you see, the string length is equal to one half-wavelength of the fundamental, f_1, shown in blue. Two half-wavelengths, or one full wave, of the second harmonic (in red) fit on the same string. Three half wavelengths of the third harmonic fit on the string. This pattern continues for the higher harmonics, f_4, f_5, and higher.

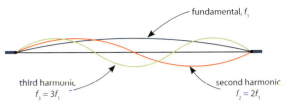

Figure 9.12. The first three harmonics that resonate on a string.

The various harmonic frequencies resonate with different intensities (that is, different amounts of energy) in different instruments. Generally, the fundamental has the greatest amplitude (i.e., it is the loudest) and the amplitudes diminish for higher and higher harmonics. The amplitudes of the various harmonics of every violin are similar, which is an important reason why every violin sounds like a violin. The harmonic content of every flute is similar, so they all sound like flutes. But even among flutes, there are subtle differences in harmonic content, depending on the materials from which they are made and the details of their construction. So a particular instrument has a little more energy at this harmonic, and a little less energy at that harmonic. These differences allow us to distinguish one flute from another.

A high-quality instrument is one that has a very desirable blend of harmonics, making the instrument sound smooth or mellow. Lower-quality instruments have a less desirable blend of harmonics, making the instrument sound harsh, muddy, or wooden. (Lower-quality instruments are also typically more difficult to play, not as loud, and so on, but these are separate issues.) The particular combination of harmonics (and some other factors we aren't going into) give each instrument a unique sound quality that makes the instrument identifiable as itself. This unique sound quality is called the instrument's *timbre* (pronounced "tamber").

A graphical representation of the relative loudness (or intensity) of the

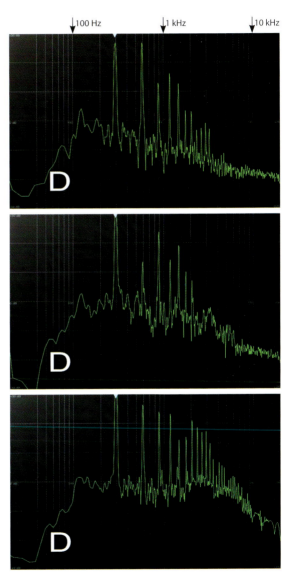

Figure 9.13. Harmonic spectrum of a flute (top), clarinet (middle), and saxophone (bottom), each playing 293.665 Hz (D4).

fundamental and other harmonics in a particular instrument is called the *harmonic spectrum* of the instrument. Figure 9.13 shows actual harmonic spectra for three different wind instruments playing the same note, D4, the D just above middle C. The fundamental frequency of this pitch is 293.665 Hz. On the graphs, the vertical scale is the loudness in a unit called *decibels*. The horizontal scale is the frequency on a *logarithmic scale*—a scale where equal increments represent increases of an order of magnitude. The vertical line just to the right of the capital D is at 100 Hz, and the lines to the right go up by 100 Hz each until 1,000 Hz. They go up by 1,000 Hz each to the right of that until 10 kHz, and the right edge of the graph is at 20 kHz. The lines at 100 Hz, 1,000 Hz, and 10,000 Hz are spaced at equal increments.

The peaks in the graph show the energy intensity at each resonant frequency (each harmonic). You can easily see the fundamental in each graph, represented by the left-most large peak, right around 300 Hz. The second harmonic at about 600 Hz and third harmonic at around 900 Hz are clearly visible, as are other higher harmonics at multiples of 300 Hz. The jagged curve across the bottom of all the spectra is simply the low-level background noise in the room where these images were captured.

Examining these spectra in light of what we know about the way these instruments sound is instructive. In the flute spectrum, you can see that the intensity of the first and second harmonics (which are the fundamental D pitch and the D one octave up, D5) is the same. This is unique to the flute spectrum and is the chief reason for the velvety sound quality flutes have. There are 11 significant harmonics in the harmonic spectrum of this flute.

By contrast, the clarinet spectrum has only seven significant harmonics in it. The harmonic spectrum is much simpler than the other two instruments, giving the clarinet a pure, mellow sound.

Finally, notice the very rich harmonic spectrum of the saxophone. The higher-order harmonics in the sax are significantly more intense than the corresponding harmonics in the other two spectra, including pronounced harmonics at f_8 and higher. There are so many strong, high-frequency harmonics that the harmonic spectrum looks like a comb and it is difficult to tell where they end. These strong higher harmonics help to explain the reedy sound quality the sax is known for.

9.2.5 Interference

One of the most historically significant wave phenomena is *interference*. It was interference patterns produced by light in 1801 that first gave scientists strong evidence that light was a wave phenomenon. (Before that everyone accepted Isaac Newton's hypothesis that light was made of particles.) Interference occurs when two different waves arrive at the same place at the same time. Two such waves always add together, but the resulting sum depends on how the two waves are aligned with each other. As illustrated in Figure 9.14, if two waves arrive at the same place *in phase*, with their crests and troughs lined up, the energies of the waves add together, producing an effect stronger than either of the individual waves. This effect is called *constructive interference*. Constructive interference with sound waves produces a loud spot. With light waves, constructive interference causes a bright spot of light. By contrast, if one wave has to propagate far-

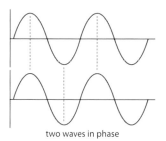
two waves in phase

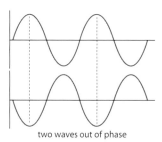
two waves out of phase

Figure 9.14. Pairs of waves in phase and out of phase.

Do You Know ...   *Do skyscrapers have resonant frequencies?*

Sometimes resonances can be harmful or dangerous, and technologies are developed to prevent resonances from occurring. An amazing example of this is seen in the Taipei 101 tower in Taiwan, completed in 2004. The tower was ranked as the world's tallest building for six years.

Every building has a resonant frequency, a natural frequency at which it sways back and forth. Engineers design skyscrapers to be flexible in order to bend with the wind, but if a flexible building begins swaying at its resonant frequency due to high winds, oscillation can get started that leads to disaster if the building sways too far.

The engineers of the Taipei 101 designed a stupendous system to counteract the possibility of high winds causing the building to oscillate due to resonance. They designed a 728-ton pendulum to hang in the building, suspended from the 92nd floor down to the 87th floor. As the building sways, the pendulum's inertia pulls it back the other way. *Dampening* is the term used to described the effect of mechanical or electrical systems installed to restrain a system and prevent oscillation. The mass on the huge damper pendulum in the Taipei 101 is a sphere 18 feet in diameter. It is made of steel plates five inches thick and is the largest damper sphere in the world.

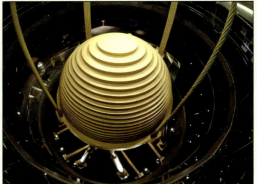

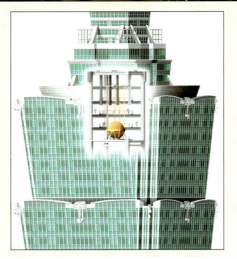

Figure 9.15. A green light spot on a dark screen with a shadow in the center.

ther than the other one it arrives later and can be *out of phase* with the first one so that the crests of one wave line up with the troughs of the other. This is called *destructive interference* because when the waves add together they partially or completely cancel each other out. With sound waves, destructive interference creates a dead spot where the sound cannot be heard at all. With light waves, a dark spot is created where there is no light.

Before we look at an example of interference, consider this thought experiment. Imagine we have a flashlight with a green light bulb. Let's point the flashlight beam at a screen in a dark room and hold up a thin broom handle in the beam so it makes a shadow on the wall. We expect to see an image on the wall that looks like Figure 9.15.

Now let's do the same thing with a green laser beam. Since the beam is only about 1 mm in diameter, we place a very thin obstruction in this beam, about the thickness of the lead in a mechanical pencil. This time we do not see a spot of light with a shadow in it. Instead, we see an interesting pattern of alternating light and dark spots called an *interference pattern*, shown in the actual photograph I took in Figure 9.16. Not only do we have a bright spot in the center instead of a shadow, but the light from the diffracted beam is now spread out into a wide strip of alternating light and dark spots! How does this happen?

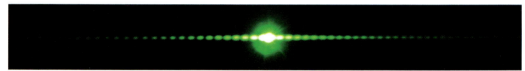

Figure 9.16. The interference pattern resulting when a thin obstruction is placed in a laser beam.

The key is that unlike ordinary sources of light, in laser light all the light waves in the laser beam are perfectly in phase with each other. Light like this, which is characteristic of lasers, is said to be *coherent*. When this coherent light hits a thin obstruction (it has to be thin because of the very small wavelength of visible light), the light *diffracts*, just like sound waves diffracting around the corner of a building. In other words, the two light beams passing on either side of the obstruction bend and spread out.

Look what happens as this diffracting light continues past the obstruction and heads for the screen. As illustrated in Figure 9.17, the two beams formed by the original beam splitting and going around each side of the obstruction each spread out and strike the screen. When the two beams hit the screen, there are places where the two light beams arrive in phase, producing constructive interference (the bright spots), and other places where they arrive out of phase, producing destructive interference (the dark spots). In the center, the two beams propagate the exact same

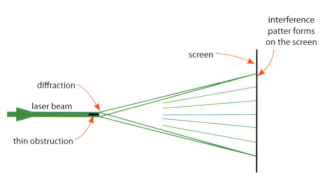

Figure 9.17. Placing a thin obstruction in the path of a laser beam causes diffraction and interference.

distance. They arrive at the screen in phase and constructive interference occurs, creating the center bright spot. To the left or right of the center, the two beams arrive out of phase when one beam is exactly one half wavelength behind the other, and the result is a dark spot from destructive interference. Moving further from the center, we reach the point where one beam is exactly one full wavelength behind the other and the waves are in phase again. This in-phase/out-of-phase pattern continues further to the left or right of the center.

9.3   Sound Waves

9.3.1 Pressure Variations in Air

As I mentioned previously, sound waves are mechanical waves. That is, they propagate in a medium (matter) of some kind. We most often think of sound waves as propagating in air, but we have all heard sound waves under water, too. Sound waves also propagate in other media, which is why you can hear sounds through closed doors or walls. Sound waves cannot propagate in a vacuum, which means all the explosions and screeching space fighter planes we see in science fiction movies are absurd because there is no sound in space. (Sorry.)

In air, sound waves are created by causing the pressure in the air to fluctuate above and below atmospheric pressure, as illustrated back in Figure 9.4. The regions of higher pressure are called *compressions* because in these regions the air molecules are compressed more closely together than they are at atmospheric pressure. Regions where the pressure is a bit lower than atmospheric pressure are called *rarefactions* (pronounced "RARE-uh-faction"), and in these regions the molecules are more spread out than normal. A sound wave in air is a continuous train of successive compressions and rarefactions. In a 1,000 Hz sound wave (like a low-pitched whistle), the pressure goes up and down at a rate of 1,000 cycles per second.

In air, sound waves propagate at a velocity of around 342 m/s, depending on temperature, humidity, and pressure.

9.3.2 Frequencies of Sound Waves

Humans can hear sound waves in the frequency range of about 20 Hz to 20 kHz. Frequencies lower that this are said to be *infrasonic*. (The Latin prefix *infra* means below.) Frequencies higher than this range are said to be *ultrasonic*. (The Latin prefix *ultra* means above or beyond.) You should be careful not to confuse these terms with the terms subsonic and supersonic, which refer to the velocity of a moving object like a jet airplane, and not to the frequencies of sound waves. (Subsonic refers to a velocity slower than the speed of sound in air; supersonic refers to a velocity greater than the speed of sound.) Table 9.1 lists a few reference points for frequencies of common musical notes and sounds.

Perfect human hearing extends up to around 20 kHz, but by middle age few people can hear sounds above 15 kHz. This is natural as our ear drums stiffen over time and do not respond as well to high frequencies. If you listen to a lot of loud music, especially with ear buds or headphones, your hearing degrades more rapidly. Although human hearing does not extend above 20 kHz, dogs up to nearly 30 kHz. For this reason, dog whistles emit a frequency of around 27 kHz that dogs hear but humans don't.

A *frequency response curve* is a graph of loudness vs. frequency across the spectrum of human hearing. Such curves are used to assess how evenly a sound system produces different frequencies. To check the frequency response of a sound system, the sound engineer

27.5 Hz	lowest note (A) on a piano keyboard (i.e., its fundamental)
41.2 Hz	lowest note (E) on a bass guitar (fundamental)
82.4 Hz	lowest note (E) on a guitar (fundamental)
440 Hz	A above middle C on a keyboard (fundamental), called "A 440" and used as a reference for tuning pianos and other concert instruments
100 Hz–3000 Hz	primary range of frequencies in the human voice
1000 Hz	frequency of a low whistle
4186 Hz	highest note (C) on a piano keyboard (fundamental)
6 kHz–8 kHz	the range for the "s" sounds we make between our teeth when we say a word like "Susan" (this sound is called *sibilance*); also the range of the sound from cymbals

Table 9.1. Reference frequencies of some well-known sounds.

plays *pink noise* through the system. Pink noise has equal sound energy in every octave of the audible spectrum, 20 Hz to 20 kHz, and sounds like a roaring waterfall. If the sound system reproduces all the different frequencies equally, the frequency response curve is flat on average across the audible spectrum.

Figure 9.18 shows the frequency response of the small studio monitor speakers on my desk while playing pink noise. Relative to the loudness at 1 kHz, the speakers' output is a bit down at 1.3 kHz, and quite a bit up in the mid-bass region of 250–400 Hz. There is essentially no bass response below 100 Hz, which cuts out the bottom two octaves of sound that might be present in the music program. This is always the case with small speakers and is the reason sound systems these days are often equipped with a subwoofer.

For a pair of small speakers, the performance represented by the curve in Figure 9.18 is actually pretty good. However, note that the frequency response shown in the graph was not measured in a proper sound chamber that eliminates sound reflections off the walls (an *anechoic* chamber), and I did not use a microphone whose own frequency response had been taken into account. Thus, the curve shown actually represents the combined frequency response of the speakers, room, and microphone. Resonances in the room are probably the cause of the emphasis in the 250–400 Hz range. Also included in the mix is the frequency response of the elec-

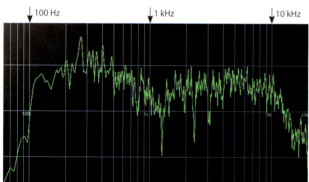

Figure 9.18. Frequency response curve of small, high-quality speakers.

tronic system used to play the pink noise, although these days the frequency response of electronic systems is very flat. The big contributors to the frequency response of an audio system are the speakers and the room.

0 dB	This level is called the *threshold of human hearing* because it is the quietest sound human ears are capable of hearing. You will probably never be in a place this quiet in your entire life. (If you were, you would have the undesirable sensation of being completely deaf.)
15 dB	This is about the loudness of faint rustling leaves on a quiet day.
65 dB	Quiet conversation in a quiet room.
85 dB	A moderately noisy environment. Legally, OSHA requires that if the noise level in a workplace (like a factory) is consistently above 85 dB, the workers must be supplied with hearing protection. This is because studies have shown that being exposed to noise levels of 85 dB or higher for 8 hours a day over many years results in permanent hearing loss. Headphone music listeners beware!
95 dB	Loud music in a movie theater.
106 dB	The upper limit of a medium-quality stereo system.
110 dB	An excruciatingly loud rock concert.
130 dB	This is usually called the *threshold of pain*, for obvious reasons.
150 dB	A jet engine heard at a distance of 1 m. This is why the ground crew at airports wear hearing protection.
170 dB	A .30–06 rifle heard at a distance of 1 m. This is why sensible people also use hearing protection when at the rifle range. It is also why movies showing people whispering to one another right after firing a bunch of guns at bad guys are not realistic. Temporary hearing loss after a lot of gun fire makes it hard to hear whispers.

Table 9.2. Reference loudness of well-known sounds.

9.3.3 Loudness of Sound

We measure the loudness of sound waves with a variable called *Sound Pressure Level* (SPL), measured in decibels (dB). In this text, we aren't doing any calculations involving decibels, but you might be interested in some common reference points for the dB scale listed in Table 9.2. The decibel values shown are approximate. I guess those of you who are into heavy metal (I am not) will enjoy the term for a loudness of 130 dB.

9.3.4 Connections Between Scientific and Musical Terms

It is really handy to make the connections between what we are studying in this chapter and other terms you probably often hear. If you play an instrument, you should know that what musicians refer to as the *pitch* of a musical note is what scientists, and this chapter, call the frequency of its fundamental. If you listen to music, like everyone else on the planet, you will appreciate knowing that what music lovers refer to as *volume* or *loudness* corresponds to the amplitude of the sound wave.

9.4   The Electromagnetic Spectrum and Light

It is difficult to understand exactly what electromagnetic waves are, but don't let this bother you. Scientists have been puzzling over electromagnetic radiation in one way or another for many centuries. We now understand that light, and all other electromagnetic radiation, is composed of transverse waves of oscillating electric and magnetic fields in space. (This description may not help much, but we will address fields again briefly later on when we get to magnetism, so hang in there!) When we speak of wavelengths and frequencies of electromagnetic radiation, what we are talking about are the wavelengths and frequencies of the electric and magnetic field waves. These oscillating fields are pretty hard to visualize, but Figure 9.19 is an attempt at a 3-D representation of a propagating electromagnetic wave. The heights of the waves represent the strengths of the electric and magnetic fields at any point in space as the wave train passes by.

To get another view of propagating electromagnetic waves, consider this. Back in Chapter 6 we learned that energy is quantized. Electromagnetic waves carry energy, and thus *light is quantized, too*. After physicists had understood light to be a wave phenomenon for nearly 200 years, in 1905, Albert Einstein proposed that light was quantized, and in 1926 chemist Gilbert Lewis coined the term *photon* that we now use to describe a single quantum of electromagnetic energy. Photons behave like particles, but they have phase like waves, and they exhibit all the other wave interactions we have studied in this chapter.

Figure 9.20 depicts a group of propagating photons in a beam of light. Since light has wave properties, but is also quantized into lumps of energy, physicists sometimes call pho-

Do You Know ...   What causes sonic booms?

If you are ever around when a jet flying faster than the speed of sound flies over, it is likely that you will hear a *sonic boom*. Sonic booms were heard often in the northeast U.S. from 1976 to 2003 when the supersonic *Concorde* jet was providing high speed passenger service from New York and Washington DC to Europe.

A sonic boom sounds like its name—a big, loud *boom!* For an explanation of what causes the boom, take a look at the photograph below. The dark silhouette is a T-38 Talon training jet traveling faster than the speed of sound. The lines spreading out from the jet are *shock waves* in the air. These shock waves are massive density changes in the air, spreading out from the jet like the wake of a boat. Changes in density cause changes in pressure. Pressure waves in air are sound waves and travel at the speed of sound.

So as the jet flies over, it creates a massive pressure fluctuation—or shock wave—

in the air as a wake which travels out from the jet at the speed of sound. When the shock wave arrives at the listener's location, the listener hears a *boom* as the shock wave passes by.

The photograph was taken in 1993 by Leonard Weinstein, using a technique called Schlieren photography. Weinstein invented the first Schlieren camera that can photograph shock waves from aircraft in flight.

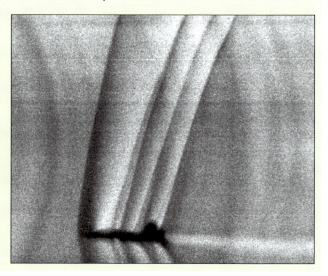

tons *wave packets*, so in the diagram I have shown the photons as little, individual wavy things. The way I have drawn this diagram, there are three particular things to notice about this beam of light. First, every photon is penetrating the blue reference plane at the same place in the wave—at the crest of a wave. This means all these photons are in phase with each other. When all the photons in a beam of light are in phase, the light is said to be *coherent*. The special thing about laser light is that it is like this—laser light is

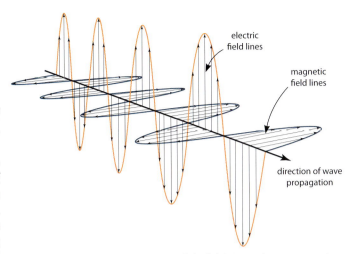

Figure 9.19. A visual representation of the fields in an electromagnetic wave.

coherent. Light from light bulbs or the sun is not like this. Instead, there is no relationship between the phases of the photons; they are random.

Second, I drew the electric fields of all the photons vertically—the waves all oscillate up and down. Normal light is not like this, either. The orientation of any of the photons could be vertical, horizontal, or at any angle. When the electric fields are all aligned in the same direction, the light is said to be *polarized*. Polarizing sunglasses only let through the photons whose electric fields are vertical (like those in the diagram), so photons oriented differently do not pass through the glasses. (This description is a wee bit oversimplified, but it will do.)

Finally, when the phase of all the photons is the same in a given reference plane, as they are in the figure, the photons together form what we call a *plane wave*. Plane waves are very useful in the study and application of optics.

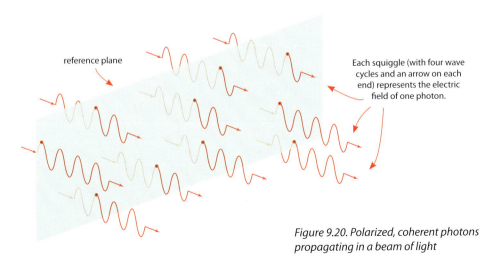

Figure 9.20. Polarized, coherent photons propagating in a beam of light

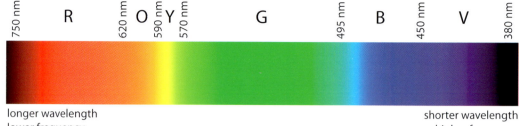

Figure 9.21. The visible light portion of the electromagnetic spectrum showing approximate reference wavelengths.

Returning to our wave discussion, visible light is only one small piece of the *electromagnetic spectrum*, the entire frequency range of electromagnetic radiation. Different frequencies of visible light are perceived as different colors by our eyes. Red light has the lowest frequency in the *visible spectrum*. Violet light has the highest frequency in the visible spectrum. As every school child knows, the sequence of colors from low frequency to high frequency is red—orange—yellow—green—blue—violet, as shown in Figure 9.21.

When we speak of electromagnetic radiation, sometimes we speak in terms of frequencies, as in the previous paragraph. In other cases, we speak in terms of wavelengths. As we saw earlier in this chapter, frequency and wavelength are inversely proportional, so we could just as easily state the difference between red light and violet light in terms of wavelengths this way: red light has the longest wavelength in the visible spectrum; violet light has the shortest wavelength in the visible spectrum. In fact, it is pretty easy to remember the wavelength range of the visible spectrum, which is about 700 nm to 400 nm (red to violet). (Some reference sources have the visible spectrum starting at 750 or 800 nm.) And as you see from the approximate reference wavelengths in Figure 9.21, orange light consists of wavelengths from around 590 to 620 nm, yellow is in the 570 to 590 nm range, green is in the 495 to 570 nm range, and blue light is in the 450 to 495 nm range.

Electromagnetic radiation with frequencies too low for humans to see is called *infrared radiation (IR)* since red light has the lowest frequency of the visible light spectrum. (Remember, *infra* means below.) Electromagnetic radiation with frequencies too high for us to see is called *ultraviolet radiation (UV)* since violet light is the highest frequency in the spectrum of visible light.

The major bands of the electromagnetic spectrum and their approximate wavelengths are shown in Figure 9.22. It is not necessary to memorize all the values, but you do need to know the names for these different regions in the electromagnetic spectrum, and you need to be able to write them in order from lowest frequency to highest frequency.

As additional examples, the electric and magnetic waves of green light with a wavelength of 532 nm oscillate 566,000,000,000,000 times each second! The microwaves your microwave oven uses complete 2,450,000,000 cycles each second.

It is interesting to consider how the different wavelengths of electromagnetic radiation relate to phenomena we are familiar with. You probably know that the reception of an AM radio in a car blanks out when the car drives under a bridge. If you calculate the wavelength of the carrier frequency of an AM radio station, you will see why. Radio station KLBJ here in Austin, Texas, broadcasts at 590 kHz. Using the wave equation, we calculate the wavelength of the radio waves propagating out from the KLBJ tower, and it turns out to be 508 m—nearly a third of a mile. Imagine a bunch of waves this long reflecting around

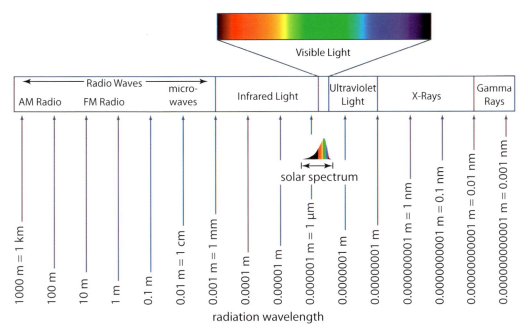

Figure 9.22. The electromagnetic spectrum.

every which way in the air. It is likely that none of the waves is aimed directly at the bridge opening by the broadcast tower, so they all arrive at the bridge opening at various angles, being reflected there by buildings and hills. Thus, very few of the waves have just the right orientation to fit under the bridge and pass through. Most of them reflect off the structure around the bridge, so the signal under the bridge is so weak that the car radio loses it. The wavelengths in the FM range are much shorter, only about 3 m (10 ft). Waves this short fit easily into the large space under the bridge and fly right under it, so FM radios don't blank out as we drive through. The digital radio channels launched about 15 years ago operate in the range of 2.3 GHz! These wavelengths are small enough that the signal is usable just about everywhere within the broadcast region of the satellites.

How small are they? Well, as another example, consider the metal screen with holes in it on the front door of a microwave oven. With a frequency of 2.45 GHz (not far from where the digital radio signals are), the wavelength of this radiation is 12.2 cm. The holes in the door screen are about forty times smaller than this, at 3 mm or so, so the waves reflect around inside the oven where they can do their job heating up food, instead of escaping out into the room and harming us.

Chapter 9 Exercises

Wave Computations

For problems involving sound waves in air, use $v = 342$ m/s.

1. What is the period of a 60.0 Hz tone?
2. What is the frequency of a sine wave with a period of 2.155×10^{-5} s?

3. If a dog whistle emits an ultrasonic tone at 26.0 kHz, what is the period of the wave?

4. For problem 3, what is the wavelength of the wave? State your answer in cm.

5. The Austin FM radio station KMFA broadcasts an electromagnetic radio wave at 89.5 MHz. Determine the wavelength and period in μs of this wave.

6. The San Antonio AM radio station KAHL broadcasts an electromagnetic wave at 1310 kHz. Determine the wavelength and period in μs of this wave.

7. Determine the period and frequency of the light emitted by a 542-nm green laser. State the period in ns and the frequency in GHz.

8. A carbon dioxide laser used for cutting steel has a wavelength of 10.6 μm. Determine the period and frequency of the wave. State the period in μs and the frequency in MHz.

9. Ultrasonic devices are available that drive pests out of homes and buildings by emitting sound waves at 33 kHz. Calculate the period and wavelength of the sound wave these devices emit. State the period in ms and the wavelength in mm.

10. Certain rooms where electronic equipment is tested are completely enclosed by copper screening to keep out electromagnetic waves. This works because electromagnetic waves do not travel through electrically conductive materials such as copper. If a room is surrounded with screening that has holes 2.00 mm across, what is the approximate frequency range of waves that are kept out by the screen or that pass through the screen?

11. In the spectrum of visible light, blue light has a wavelength of 470 nm, green light has a wavelength of 550 nm, and red light has a wavelength of 680 nm. Calculate the frequencies for each of these colors of light.

12. The lowest frequency humans can hear is about 20 Hz. Let's call it 20.00 Hz. What is the wavelength of this wave in air? Assume sound propagates at 342.0 m/s in air.

13. Calculate the wavelengths in nm for the following types of electromagnetic radiation.
    a. gamma rays, $f = 4.67 \times 10^{20}$ Hz
    b. X-rays, $f = 9.9876 \times 10^{18}$ Hz
    c. microwaves, $f = 2.555 \times 10^{10}$ Hz
    d. ultraviolet light, $f = 1.172 \times 10^{15}$ Hz
    e. infrared light, $f = 2.83 \times 10^{13}$ Hz

14. A microwave oven radiates electromagnetic energy into the cooking compartment at a frequency of 2.45 GHz. How does this compare to the 3-mm diameter holes in the screen of the door of a microwave oven? Why are the holes the size they are? (When you are asked to compare two values in science and math it means you should determine their ratio. That is, determine how many times bigger one is than the other, or what percentage the smaller one is of the larger.)

15. Calculate the wavelength of a 1.00 kHz tone propagating as a sound wave in each of these different media.

medium	speed of sound		medium	speed of sound
air	342 m/s		water	1,402 m/s
steel	5,130 m/s		helium	965 m/s

Answers

1. 0.0167 s
2. 46.40 kHz
3. 3.85×10^{-5} s
4. 1.32 cm
5. $\lambda = 3.35$ m, $\tau = 0.0112$ μs
6. $\lambda = 229$ m, $\tau = 0.763$ μs
7. $f = 5.54 \times 10^5$ GHz, $\tau = 1.81 \times 10^{-6}$ ns
8. $f = 2.83 \times 10^7$ MHz, $\tau = 3.53 \times 10^{-8}$ μs
9. 0.030 ms, 1.0×10^1 mm
10. Waves with $f > 1.50 \times 10^{11}$ Hz pass through the screen.
11. blue 6.4×10^{14} Hz, green 5.5×10^{14} Hz, red 4.4×10^{14} Hz
12. 17.10 m
13. gamma 0.000642 nm, X-ray 0.0300 nm, microwave 11,700,000 nm, UV 256 nm, IR 10,600 nm
14. $\lambda/D \approx 40$
15. air 0.342 m, water 1.40 m, steel 5.13 m, helium 0.965 m

Study Questions

1. What is a wave?

2. Write paragraphs explaining reflection, refraction, diffraction, resonance, and interference.

3. Explain why one guitar sounds different from another, according to the discussion in Chapter 9.

4. Distinguish between mechanical and electromagnetic waves.

5. Look at your answers for problems 5 and 6 above and try to explain the fact that when a car drives under a bridge, FM radio signals come through fine while AM radio signals do not.

6. What does it mean to say that two waves are in phase?

7. Distinguish between transverse, longitudinal, and circular waves.

8. Explain why you cannot hear radio stations in the air right now (or any other time).

9. Think of as many ways to distinguish sound waves from light waves as you can.

10. Using words and not equations, distinguish between the period and frequency of a wave.

11. Explain why laser light is coherent.

CHAPTER 10
Introduction to Electricity

Static Man

This kid picked up a handful of static electricity on his trip down the plastic slide. Some kids pick up so much charge on a plastic slide that they shock one another. Friction is the most familiar of three ways static accumulations of electrical charge form.

OBJECTIVES

After studying this chapter and completing the exercises, students will be able to do each of the following tasks, using supporting terms and principles as necessary:

1. Describe the roles of Alessandro Volta and James Clerk Maxwell in the development of our knowledge of electricity.
2. Define and distinguish between static electricity and electric current.
3. Explain what static electricity and static discharges are.
4. Describe three ways static electricity forms and apply them to explain the operation of the Van de Graaff generator and the electroscope.
5. Describe the general way the atoms are arranged in metals.
6. Use the idea of the "electron sea" to explain why electric current flows so easily in metals.

Electricity and electric circuits are all around us and our lives are massively affected by electrical appliances and gadgets. For this reason, it is important and valuable for you to learn the basics of electrical theory and circuit calculations. We begin in this chapter with a historical survey of Western discoveries in the area of electricity, moving on to the details of static electricity and electric current. In the next chapter, we address the specifics of how electric circuits work and the computations associated with them.

10.1   The Amazing History of Electricity

Being an electrical engineer, I like to tell the story of electricity and magnetism. Included here is a short account of this interesting saga. As always, the things you need to know and remember are listed in the Objectives List and the Scientists List in Appendix D.

10.1.1 Greeks to Gilbert

The earliest recorded observations of electrical phenomena were by the ancient Greeks around 600 BC. They observed that amber rubbed with fur attracts certain materials such as feathers. This is an effect of static electricity, which we address in some detail later in this chapter.

The modern story of our scientific knowledge about electricity and magnetism begins with William Gilbert (Figure 10.1), Queen Elizabeth's personal physician, who spent a lot of time investigating static electricity and magnetism. In 1600 (the same year Kepler moved to Prague to work with Tycho Brahe), Gilbert proposed that the earth acts like a large magnet. Gilbert's work was the first major scientific work published in England. His new attitude toward science emphasized experiment and scientific observations. (Remember Kepler's keen interest in comparing theory with *data*.) He was one of the first scientists to embrace this new attitude, which, of course, was to become the

Figure 10.1. English scientist William Gilbert (1544–1603).

very bedrock principle of science. Gilbert discovered many new materials with electrical properties and coined the word *electric* from the Greek and Latin words for amber. He also invented an early form of a device called an *electroscope*, described later in the chapter.

10.1.2 18th-Century Discoveries

Figure 10.2. Leyden jars.

About 140 years later, the *Leyden jar* was invented, named after the city in the Netherlands where one of the inventors lived. A Leyden jar, pictured in Figure 10.2, has a metal lining in it that allows it to hold electric charge. About 10 years after that, a new, more sensitive version of the electroscope was invented in England. Both these gadgets store electrical charge, meaning an excess of negatively charged electrons. Of course, everyone has heard the story from back in 1752 of how Benjamin Franklin proved that lightning was electrical by charging up a Leyden jar connected to his kite flying in an electrical storm. (You know better than to try this yourself, right? If not, have you considered the possibilities of what happens if the kite string you are holding in your hand gets struck by lightning? Mr. Franklin was not aware that his life was at risk.)

Franklin also started the convention of denoting the two types of electric charge as positive and negative, and he established the *law of conservation of charge* (yet another of the important conservation laws in physics). Before Franklin, "positive" and "negative" were called "vitreous" and "resinous" because electricity was believed to consist of two kinds of "fluid." (Back then lots of things were considered to be fluids—fire, electricity, heat, even an "ether"

Figure 10.3. English scientist Joseph Priestly (1733–1804).

in space that people thought was necessary for light to travel in. All of these notions were found to be incorrect, causing the theories to change yet again!)

In 1767, Joseph Priestly (Figure 10.3) found that the force of attraction between two electric charges varies inversely as the square of the distance between them, just like the gravitational force between masses in Newton's law of universal gravitation. An accurate mathematical model for the electrical attraction between two charges was formulated in 1785 by French physicist Charles-Augustin de Coulomb, and this formula is now known as *Coulomb's law*. Have a look at the box on the next page to read about some fascinating parallels between electric forces and gravity.

In 1791, Luigi Galvani (Figure 10.4), professor of anatomy in Bologna (Italy), was studying muscle contractions in frog legs. He noticed that frog legs hanging on copper hooks from an iron balustrade (or railing) twitched as if alive when they touched the iron.

Intriguing Similarities between Gravity and Electricity

Even though we won't be doing any calculations with Coulomb's law, it is still fascinating to consider it, especially because of its similarity to the law of universal gravitation. Let's compare these laws side by side, Newton's on the left, Coulomb's on the right:

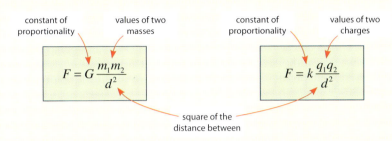

These two equations are both about forces. Newton's equation gives the force of attraction between two masses. Coulomb's equation gives the force between two charges, which could be attractive or repulsive, depending on whether the charges are both positive, both negative, or one of each. The thing I want you to notice here is that these two equations look just alike. They both have the square of the distance between the objects in the denominator. Newton's law of force between two masses has the two masses multiplied in the numerator and Coulomb's law of force between charges has the two charges multiplied together in the numerator. Why is the equation that describes gravitational force just like the equation that describes electrical force? How is it that these forces can be modeled with such simple equations in the first place? Remember what I wrote when we were looking at Kepler's third law of planetary motion? The wise hand of a purposeful creator is definitely in evidence here!

He made a fork with iron and copper prongs and reproduced this result in his lab, discovering that electric charge causes the muscles in a frog's leg to contract. When Galvani's friend, *Count Alessandro Volta* (Figure 10.5) heard about this, he was fascinated. After nine years of thought and experimentation he invented the *voltaic pile* in 1800, the predecessor to the modern battery.

An example of a voltaic pile from the early 19th century is shown in Figure 10.6 and a diagram of how the pile is assembled is shown in Figure 10.7. In Volta's original design, the cells are made of metal disks about four inches in diameter and half an inch thick. The stack of cells is about one and a half feet tall and encased in a tall glass jar or wooden frame. The pile consists of a stack of many individual *cells*. Each cell consists of a layer of copper, an *electrolyte* layer, and a layer of zinc. An electrolyte is a salt solution that conducts electricity. In Volta's pile, the electrolyte layer consisted of cloth or cardboard soaked in brine (salt-

Figure 10.4. Italian anatomy professor Luigi Galvani (1737–1798).

Figure 10.5. Italian scientist Alessandro Volta (1745–1827).

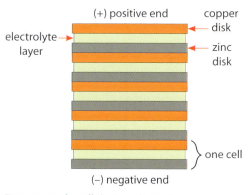

(+) positive end — copper disk

electrolyte layer

zinc disk

one cell

(–) negative end

Figure 10.7. The cells in the voltaic pile.

Figure 10.6. The voltaic pile.

water). The invention of the voltaic pile was huge because for the first time scientists had a reliable source of electricity they could use in the lab so that electricity could be studied further.

10.1.3 19th-Century Breakthroughs

By 1820, scientists knew that electric current and static electricity were different manifestations of the same physical phenomenon, but no one knew how these connected to magnetism. Then a funny thing happened one day in 1820 to Hans Ørsted (Figure 10.8), professor at the University of Copenhagen. While walking to the college one morning, he suddenly realized the connection between electricity and magnetism. He went into his lecture hall and right in front of his students, he connected a circuit of wire to a voltaic pile and placed a magnetic compass near the wire. When he switched on the current, the compass needle deflected. This was the first demonstration of the connection between electricity and magnetism. His students did not realize how significant this was and were unimpressed. (I love the irony of this.) Ørsted discovered that a compass near a current-carrying conductor deflects and deduced that electric current produces its own magnetic field. He called this *electromagnetism* and immediately published his results. The physics world went crazy!

When André-Marie Ampère (Figure 10.9) heard that news, he began experimenting and within a few weeks showed that two parallel current-carrying wires attract or repel each other just like magnets. He then figured out the equation for the strength of the magnetic field as a function of the current, a relationship now known as *Ampère's law*. Ampère's experiments with magnetic fields in coils of current-carrying wire led him to suggest that natural magnetism was the result of tiny circulating currents inside the atoms in magnetic materials. This stunning theoretical leap is now considered to be correct!

Now consider Michael Faraday (Figure 10.10) in England. He had been apprenticed for seven years as a bookbinder and got into physics and chemistry by reading the books he was binding. When he completed his apprenticeship at age 20, he began attending lectures by Sir Humphry Davy, one of the most well-known scientists in London (see page 19). Faraday sent Sir Davy a 300-page book of the notes he had taken from Davy's lectures and asked for a position working in Davy's lab. Faraday was given his chance, and my oh my did he make good on it! He started out working on the chemistry of gases, but his interests moved toward electricity and magnetism.

In 1831, Faraday demonstrated the opposite effect from Ampère's Law: that a magnetic field can induce an electrical current in a nearby conductor. This is now known as *Faraday's law of magnetic induction*. This discovery led to the development of generators, motors, and transformers. Your life would be utterly different, and much more difficult, if we did not have these devices. Faraday had almost no mathematical training, yet he brilliantly theorized the existence of "fields" to explain magnetic phenomena. We address some of the details of Ampère's and Faraday's laws in Chapter 12.

Michael Faraday was a devout Christian. His biographers have written that "a strong sense of the unity of God and nature pervaded Faraday's life and work." He is also considered by many to be one of the greatest experimentalists of all time.

The story of electricity reaches its conclusion (at least, the pre-quantum era conclusion) with the amazing Scotsman James Clerk Maxwell (Figure 10.11). Like Faraday, Maxwell was a devoted Christian. Building on Faraday's experimental results, in 1864 Maxwell published what is perhaps the greatest achievement in theoretical physics of all time, the four equations now known as *Maxwell's equa-*

Figure 10.8. Danish scientist and professor Hans Ørsted (1777–1851).

Figure 10.9. French scientist André Ampère (1775–1836).

219

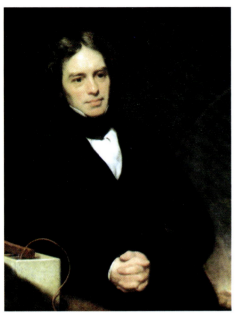

Figure 10.10. English scientist Michael Faraday (1791–1867).

Figure 10.11. Scottish theoretical physicist James Clerk Maxwell (1831–1879).

tions. These four equations constitute a complete description of all known electric and magnetic phenomena in the universe! In addition to his supreme achievement with electromagnetic theory, Maxwell also contributed to the pioneering work on the kinetic theory of gases, which we addressed back in Chapter 7. Maxwell's theoretical work is ranked at the same level as that of Newton and Einstein.

Maxwell's equations are so elegant and so simple (at least to those who know vector calculus) that I must show them to you. Only those who study physics, mathematics, or engineering in college learn what these equations and their symbols mean, but just have a *look* at them and consider how so much can be said with so little. The four equations can be written several different ways, but Figure 10.12 shows them in their so-called *differential form*.

$$\nabla \cdot D = \rho_f$$

$$\nabla \cdot B = 0$$

$$\nabla \times E = -\frac{\partial B}{\partial t}$$

$$\nabla \times H = J_f + \frac{\partial D}{\partial t}$$

Figure 10.12. Maxwell's four beautiful equations.

Back in Chapter 2, we looked with amazement at how a simple equation like Kepler's third law of planetary motion models the real physical world so elegantly. Maxwell's equations, universally regarded as one of the greatest and most beautiful discoveries in the history of physics, are even more astonishing. These equations first proved that light was an electromagnetic wave. From the study of these equations came the idea that we could produce radio waves and transmit information with them, which is without doubt one of the most influential ideas in the history of human technology. And while you are pausing here to consider Maxwell's discovery, consider the profound truth that the universe God made, a universe of apparently infinite complexity, can be modeled with four short equations like these. And while you are considering it, give thanks for it. Creation is a wonderful gift.

Do You Know ...   *Who made the first color photograph?*

James Clerk Maxwell made the first color photograph in 1861. The photograph is an image of a knot tied in a piece of tartan ribbon. Maxwell constructed the image from three separate photographs, each taken with a different color filter. When composited, the three images reconstructed the original color article with its full variety of colors.

10.2   Charge and Static Electricity

10.2.1 Electric Charge

In the context of physics and chemistry, when we refer to *charge* we mean charged subatomic particles. As you already know, there are two subatomic particles that have charge, protons with positive charge and electrons with negative charge. All electrical phenomena are due to these charges. All matter is made of atoms and all atoms have protons and electrons in them, so charged particles—charges for short   are everywhere. (Atoms also contain neutrons, which have no charge.)

Most of the time, the colossal numbers of positive and negative charges that matter is made of are so evenly distributed in a substance or object they essentially cancel each other out, so there is no noticeable net electrical charge and we aren't even aware of the charges all around us (and in us). But there are two special circumstances when charges do things that make us notice them—when charges are in motion and when charges accumulate somewhere.

As I discuss later in the chapter, electrons in metals are free to move around, which is why metals conduct electricity so well. The protons in metals are locked in the *nuclei* (plural of nucleus) of the atoms and are not free to move around. For this reason, when we talk about charges in motion or charges accumulating in or on metals, we almost always mean *electrons*, not protons.

When electric charge is moving or flowing it is called an *electric current*. We'll get to that in the next chapter. Here we are going to examine charges that accumulate together and are at rest. The word physicists use to describe something that is not moving is "static," which comes from the Greek word *stasis*, meaning "standing still." *Static electricity* is an accumulation of charge that is stationary. An easy way to accumulate static electricity is by combing your hair with a rubber comb. As we will see in the next section, the friction between the comb and the hair causes static electricity to build up. Then when the comb is held near some small bits of paper, the paper jumps onto the comb from the electrical attraction, as shown in Figure 10.13.

Isn't it funny how so many of the things we have encountered come in threes? Well, you will not be surprised to learn that there are three ways static electricity forms. In the description that follows, I refer to a device called an *electroscope* to illustrate these three ways. Figure 10.14 shows an electroscope when it has no static electricity on it (left) and when it is charged with static electricity (right). The device consists of a metal rod with a metal sphere on top and a pair of metal hooks on the bottom. From the hooks are suspended two strips of aluminum foil called *leaves*. The glass flask and red rubber stopper aren't part of the action; they are only there to support the metal parts.

Figure 10.13. Static electricity causes small bits of paper to be attracted to a rubber comb.

Now consider what happens if an accumulation of electrons—static electricity—is placed on the metal sphere. Since the electrons all have the same negative charge, they repel each other, so they spread themselves out on the metal parts, pushing as far from each other as possible. Electrons move easily in metals because metals are good conductors of electricity and electricity *is* moving electrons. So the electrons spread out (*conduct*) up and down the metal parts, including down into the leaves. Since the leaves are free to move, they swing out due to the electrons on each leaf repelling the electrons on the other leaf. The photograph on the right in Figure 10.14 shows the electroscope with its leaves forced outward due to the static electricity on the metal parts of the electroscope. Just to drive the point home with more familiar materials, Figure 10.15 shows a simple home-made electroscope made of a clamp, a length of copper wire, and two strips of aluminum foil. Since the leaves are standing out from each other, we know that this electroscope has a charge of static electricity on it.

10.2.2 How Static Electricity Forms

The first way static electricity forms is by *friction*. When you walk across a nylon carpet in the winter, your feet are actually scraping loose some of the electrons from the atoms in the carpet. These electrons accumulate on you and spread out all over your body! Then when you reach out to touch a door knob (or your sister), the excess charge accumulated on you jumps to the door knob (or your sister), causing the exciting *arc* and snapping sound you are no doubt familiar with. The same thing happens when you slide your backside into a car with nylon upholstery. The friction causes electrons to get rubbed onto your body and stay there. Then when you are ready to open the door, the static electricity on you *discharges* when you touch the metal door handle to get out of the car.

foil
leaves

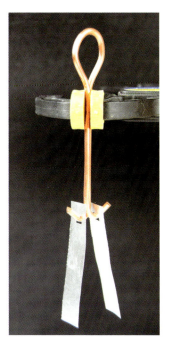

Figure 10.14. The electroscope uncharged (left) and charged with static electricity (right).

The term discharge is used to denote the arc or spark that occurs when a large accumulation of static electricity is suddenly released. Electrons are constantly trying to get away from each other because they all have the same negative charge. This is especially true when many of them have been forced to accumulate. So when an object comes close enough for them to jump to, they do! They literally jump through the air and their violent path through the air creates a *plasma* that releases energy in the form of light and sound, as shown in Figure 10.16.

Figure 10.15. A home-made electroscope. Electrons conducted onto the two leaves push the leaves apart.

The biggest arc any of us ever see is a lightning bolt, which is a whopping big discharge of the static electricity that builds up in clouds and discharges to the ground.

Interestingly, the friction involved when gasoline is flowing in a rubber hose causes quite a bit of static electricity to build up on a car during fueling. This is obviously quite dangerous, so gasoline hoses have special conductors inside the rubber to provide a path for the accumulating charge to "drain off," thus preventing a sparking hazard with gasoline vapors around.

Let's go back to the electroscopes in Figures 10.14 and 10.15. Friction comes into the electroscope demonstration in a couple of ways. First, we typically start this demonstration by rubbing, using friction to create an accumulation of charge which we can then use for the rest of the experiment. The way I usually do it is to rub a Styrofoam cup on my hair because this friction causes electrons from my hair to accumulate on the cup. (You have probably done this yourself at some point with a balloon, which you can then stick to the wall.) Another way to do it is

Figure 10.16. The violet streak of light is an arc from a static discharge as electrons on the large metal sphere jump through the air to the small sphere.

Do You Know ...   *Why are plasmas conductive?*

In Chapter 6, we saw that one of the phases of matter is the plasma. A plasma is an *ionized* gas. An *ion* is an atom that has lost or gained one or more electrons so that it has a net electrical charge. (Michael Faraday coined the term ion, from the Greek word for *wanderer*.) An ionized gas, or plasma, is formed when the atoms in the gas are ionized. Being full of ions, plasmas are electrically conductive. As a result, the properties of plasmas are so different from the other phases of matter that the plasma is considered to be a separate phase of matter.

Plasmas form when the presence of a very strong electric field strips away the electrons from the atoms in a gas. As it happens, such conditions exist throughout the universe and plasma is actually the most prevalent phase of matter in the universe!

The electric fields that cause plasma to form can also impart energy to the other electrons in the atoms of the gas, causing the atoms to emit photons of visible light. The familiar neon signs work this way, so we could accurately refer to these lights as "plasma signs." The electric field that produces the neon plasma is provided by a transformer with electric leads inserted into the sealed tube containing the gas.

Essentially the same process occurs during a lightning strike. A high electric field between clouds and the ground causes molecules in the air to ionize, creating an electrically conductive channel between the cloud and the ground. The ionized air molecules emit light the same way the neon lights do. As soon as the electrically conducting channel opens up between the cloud and the ground, the electrical charge in the cloud (which caused the electric field in the first place) instantly flows to ground— a bolt of lightning.

to rub a glass rod with some silk, which causes electrons to build up on the glass rod. Or you can do it the way the Greeks did by rubbing a piece of amber with some fur. However you do it, the friction from rubbing is what causes the electrical charge to accumulate.

In the electroscope demonstration, friction is used again to get the electric charges off the Styrofoam cup and onto the metal sphere on the top of the electroscope. Just touching the cup to the metal sphere won't do it because Styrofoam doesn't conduct electricity. Styrofoam is an *electrical insulator*. The electrons on the cup cannot make their way to the metal sphere; they are stranded on the cup. But if I rub the metal sphere with the cup, the electrons on the cup are literally scraped off the cup and onto the metal. You cannot see any of this happening, of course. All you see is that the leaves of the electroscope swing out because the charge placed on the electroscope spreads out as much as it can, since like charges repel.

The second way static electricity forms is by *conduction*. Now, before we go any further I would like for you to pause and notice that you saw this exact term before when we were talking about heat transfer. The word means something completely different here, so please take note so you don't get these two different uses of the word confused.

Everyone knows that certain substances, like metals, conduct electricity. That is, electrons flow easily in these substances. If you create a static build-up by friction and then touch the object that has the static electricity on it with a metal rod, the electric charge flows, that is *conducts*, in the metal rod and flows onto whatever the rod is touching. This effect is demonstrated in the electroscope when the charged object (Styrofoam cup, glass rod, or whatever) touches the metal sphere. When this happens, the electric charge conducts down the metal rod in the electroscope and into the metal foil leaves. And now that there is excess electrical charge in the leaves, they push away from each other. Conduction occurs again when I touch the metal sphere atop the electroscope with my hand, allowing the extra charge to drain off the electroscope

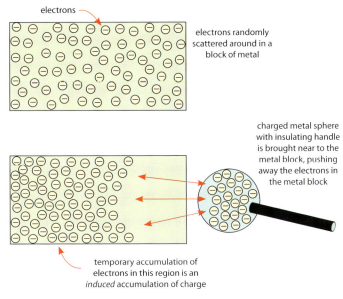

electrons

electrons randomly scattered around in a block of metal

charged metal sphere with insulating handle is brought near to the metal block, pushing away the electrons in the metal block

temporary accumulation of electrons in this region is an *induced* accumulation of charge

Figure 10.17. Charge accumulation by induction.

by flowing onto me. (Touching the electroscope like this does not produce a shock because the amount of charge built up on the electroscope is quite small.)

The third way static electricity forms is by a process called *induction*. (Warning—This term also appears in Chapter 12 in our discussion of Faraday's law of induction. You must keep that concept from getting confused with this one. It's just the way it is.) Let's pause for a brief aside on the word induce. This word basically means "force to happen." If you induce someone to confess a crime, you put them in a position where they figure that is their only option. When labor is induced in an expectant mother, it is brought about by drugs that force it to happen. So, back to electricity. When static electricity—an accumulation of electric charges—is formed by induction, it is somehow forced to happen.

Now, consider a block of metal, as illustrated in Figure 10.17. It is chock full of electrons that are completely free to move around. What do all those electrons do if an object, such as a metal sphere, is charged up with excess electrons and brought near to the metal block? As shown in the figure, they move away from the charged object because negative charges always move away from each other if they can. Even if the charged sphere never touches the metal block, the electrons in the metal block crowd up together on the opposite side of the block from where the charged object is, producing an accumulation of charge (static electricity). Of course, the charges in the static electricity on the sphere bunch up more closely together as well.

When a static accumulation is created this way (by induction), the static charge accumulation is temporary. It only stays there as long as the charged object that is causing (inducing) it is present. Pull the charged object away and all those electrons in the metal block relax and spread back out in the block.

This effect is demonstrated in the electroscope when the charged Styrofoam cup is brought near the metal sphere but not allowed to touch it. The leaves move apart because of the charge induced in them, or forced down into them, by the nearby presence of the charged Styrofoam cup. But the induced accumulation is temporary and as soon as the

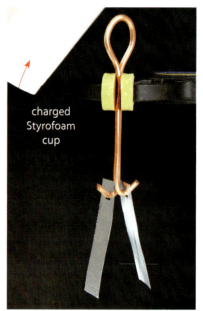

charged
Styrofoam
cup

Figure 10.18. The charge on the cup induces a static accumulation of charge in the leaves of the electroscope.

Figure 10.19. A physics teacher having fun with the Van de Graaff generator.

cup is pulled away the charges in the metal parts of the electroscope relax and spread back out again. Figure 10.18 shows the home-made electroscope again. In this case, no excess charge is on the electroscope, but there is excess charge on the Styrofoam cup. When the cup is brought near the top of the electroscope, electrons are forced down into the leaves and they repel.

When I demonstrate induction with the electroscope, students sometimes get the mistaken notion that electric charges jump from the cup to the electroscope. Well, they don't. If the cup does not touch the metal sphere, no charges actually transfer from the cup to the metal. Induction occurs without any charges transferring from the cup to the metal sphere and it is a temporary effect that goes away as soon as the cup is withdrawn.

A well-known device that produces static electricity is the Van de Graaff generator. These machines are loads of fun, as you can see in Figure 10.19. Figure 10.20 shows photos of the Van de Graaff generator outside and inside. The generator uses an electric motor to run a rubber belt in a loop from the base of the machine up through the plastic tube to the dome at the top. At the bottom, there is a wire screen, called a comb, that deposits electrons onto the belt as it runs. In the dome, there is another comb that collects the electrons off the belt and allows them to conduct out onto the metal dome, where they cause all kinds of mischief for students who like getting shocked.

The lower photo in Figure 10.20 shows a close-up of the upper comb and how it is positioned. In the photo, you see the amber-colored roller on top of the vertical plastic tube. The rubber belt, which is missing in this photo, wraps over this roller. The metal strands on the

Figure 10.20. The Van de Graaff generator, and a close up of the pulley and comb on the inside of the dome at the top.

lower edge of the comb are very close to the belt without touching it. As the belt turns, the electrons hop off the belt and onto the comb, which is attached with metal legs to the dome. Thus the electrons accumulate on the dome in large quantities. This machine produces quite a large accumulation of static charge on the dome, so much in fact that the dome can produce an impressive arc in the air to a nearby metal object (Figure 10.16) or to the body of a nearby student!

10.3   Electric Current

10.3.1 Flowing Charge

Electric current is flowing or moving charge. In principle, this flowing charge can be either positive or negative: flowing positive charges (protons) and flowing negative charges (electrons) both qualify as electrical currents. But, as stated earlier, in ordinary circumstances the protons in a solid substance are locked in the atomic nuclei of atoms that are held in place. So although a current of protons can be created in a laboratory under special circumstances (such as in the LHC, see page 84), that is not what normally happens.

On the other hand, electrons in metals are free to move around. We have already seen how electrons accumulate to form static electricity. In metals, the electrons are made to flow in a wire like water in a pipe. This is what ordinary electric current is—flowing electrons.

10.3.2 Why Electricity Flows So Easily in Metals

In our study of static electricity, I noted that one way for static electricity to accumulate is by conduction, when electrons flow in a metal. An interesting thing to think about is *why* electrons flow so easily in metals. In other words, why is it that metals are such good electrical conductors?

Recall from back in Chapter 6 that there are two basic ways the atoms in solid substances can be arranged. In many solids, the atoms are combined in molecules—tiny clusters of atoms. The other basic form is a crystal structure, in which the atoms in the solid are arranged in an orderly, geometric crystal lattice. As it turns out, metals are like this—*metals are crystals.*

In addition to their crystal structure, there is another important thing about the way atoms in metals are arranged. Every atom has a certain number of electrons. The electrons reside in regions around the nucleus called *orbitals*. In non-metallic substances, the electrons of an atom are held in their orbitals around the atomic nucleus by the force of electrical attraction between the positive protons in the nucleus and the negative electrons. However, in metal atoms, things are a bit different. In metal atoms, there are so many orbitals available for electrons to be in that the outer orbitals of the atoms in the crystal lattice overlap. Electrons in orbitals near the nucleus remain there. But electrons in the outer orbitals are able to move freely through the overlapping orbitals and are not bound to any particular atomic nucleus. These electrons that are free to move around are called *conduction electrons*. When an electrical force called a "voltage" is present, the conduction electrons take off and start flowing through the crystal lattice. We will address the topic of voltage in the next chapter as part of our study of DC circuits. For now, we will just focus on what is going on with these conduction electrons in metals.

In metals, the conduction electrons have no idea what atom they came from or belong to and it doesn't matter. What matters is that the conduction electrons move easily through the crystal lattice when a voltage is applied to the metal. It also matters that there is an in-

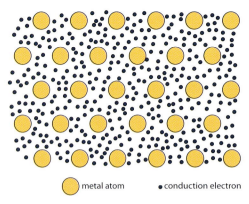

metal atom   • conduction electron

Figure 10.21. Metallic crystal with the "electron sea."

credibly large number of these free electrons. In fact, there are so many free conduction electrons in a metal that scientists actually refer to this ocean of electrons as the *electron sea*.

Figure 10.21 depicts the crystal structure in the atoms of a metal, with the conduction electrons in the electron sea scattered throughout the crystal lattice.

Chapter 10 Exercises

Historical Questions

1. Describe the origin of the voltaic pile.
2. Describe the design and operation of the voltaic pile.
3. Why was the invention of the voltaic pile so important?
4. What are Maxwell's equations?
5. Why are Maxwell's equations recognized as such a great achievement in theoretical physics?

Static Electricity Study Questions

1. Explain what static electricity is and how it forms.
2. Describe two examples of friction causing a build-up of static electricity.
3. Explain what electrical conduction is and describe how it is involved in the operation of an electroscope.
4. Explain what induction is and describe how it is involved in the operation of an electroscope.
5. What is a static discharge?
6. Why do we say that an accumulation of static electricity by induction is a temporary effect?

Electric Current Study Questions

1. Explain what electric current is.
2. Explain what the "electron sea" in a metal is.
3. Why is it that the electric current in everyday electrical circuits consists of moving electrons and not moving protons?
4. Why do metals conduct electricity so well?

Do You Know ...   *Whose pictures did Einstein have on his walls?*

Like you, Albert Einstein liked to have pictures of people who inspired him. Einstein kept pictures of Isaac Newton, Michael Faraday, and James Clerk Maxwell on his study wall.

Isaac Newton

Albert Einstein

Michael Faraday, holding a glass bar from an experiment with magnetism

James Maxwell

CHAPTER 11
DC Circuits

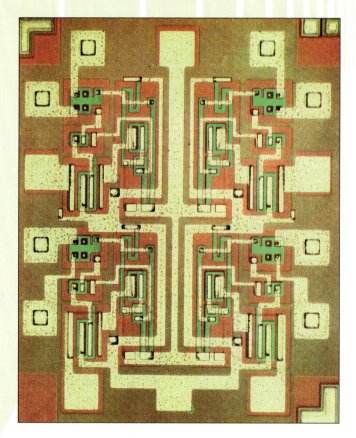

DC Circuits Go Future

In the early days of electronics, circuits were made from individual parts connected together with wires and solder. Then in the 1960s, researchers began studying how to make electronic circuits by etching the circuit into thin layers of conducting and insulating material deposited on top of one another. The image above is from a small circuit made this way back in 1968. Nowadays, technologies allow millions of circuits like the one above to be placed on a single chip smaller than a fingernail.

OBJECTIVES

Memorize and learn how to use these equations:

$$V = IR \qquad P = VI$$

After studying this chapter and completing the exercises, students will be able to do each of the following tasks, using supporting terms and principles as necessary:

1. Using the analogy of water being pumped through a filter, give definitions by analogy for *voltage*, *current*, *resistance*, and *potential difference*.
2. Explain what electric current is and what causes it.
3. State Kirchhoff's two circuit laws.
4. Calculate the equivalent resistance of resistors connected in series, in parallel, or in combination.
5. Use Ohm's law and Kirchhoff's laws to calculate voltages, currents, and powers in DC circuits with up to four resistors.

11.1 Understanding Currents

11.1.1 Electric Current

In the previous chapter, we discussed static electricity, which is a stationary accumulation of charged particles. Now we are going to discuss charges in motion. This is what electric *current* is, moving electric charge. Since there are two types of charged particles, positive protons and negative electrons, we have an electric current if a group of either of them is in motion. However, as before with static electricity, in all practical electric circuits made with wires or other metal conductors, the protons are fixed in place in the metal and the electrons are the charges that flow in the wires. As we saw at the end of the previous chapter, in a metal the conduction electrons are completely free to move around. An electric circuit is just an arrangement that forces them to flow in a certain direction so they can do some valuable work for us.

11.1.2 The Water Analogy

The best way to understand how electricity works is to consider it as analogous to water flowing in a pipe. We cannot see electricity flowing, and for most people electricity is some mysterious force that makes modern life convenient but causes serious fires or injuries if it gets out of hand. But electricity is just electrons flowing in wires and the way they do it is a lot like water molecules flowing in a pipe. So, we will do well to begin this study by understanding the physics of flowing water. After that, you will have a much easier time understanding the physics of electricity.

If we pump water in a closed circuit of piping through, say, a water filter, we have a system that is analogous to an electric circuit in nearly every respect. The sketch in Figure 11.1 shows the water circuit.

Imagine that we have a little water pump, like what we would see in an aquarium, and we are pumping the water through the filter to keep it clean. Let's think for a minute about what causes water to flow in a pipe. Imagine you and a friend are going to play a water game. You are each holding one end of a garden hose and the hose is filled to the brim with water.

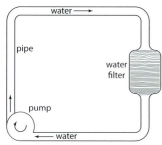

Figure 11.1. The water circuit as an analogy for the electric circuit.

Now you both put the end of the hose in your mouth and blow hard. Who is going to lose? That is, whose mouth gets full of water? As we can all easily guess, the person who blows with the lowest pressure will get the mouth full of water. The reason is that water always flows from high pressure to low pressure. In fact, it is the *difference* in pressure that makes it flow at all. If you both put the hoses in your mouths but do not blow, or blow with the same pressure, the water does not go anywhere. But when a difference in pressure is created, the water always flows toward the lower pressure. So in summary, no pressure difference, no flow; and when there is a pressure difference, the water flows toward the low pressure.

With this in mind, let's think more about the water circuit and what the pressures must be like at different points around the circuit in order for it to work. Figure 11.2 is another sketch of the water circuit in which I have added pressure gauges to the pipes at seven places around the loop. Let us imagine that we are reading the gauges to determine what the pressure is at any point around the circuit. If we make a graph of what the "pressure profile" around the circuit looks like based on the pressures all the gauges are indicating, we find that the pressures around the circuit vary as shown in Figure 11.3.

Notice that the highest pressure in the circuit is at the discharge (outlet) of the pump, and the lowest pressure in the circuit is at the suction (inlet) of the pump. It has to be this way because outside the pump water always flows toward low pressure. (Inside the pump are blades like those on a fan that push the water toward the discharge, forcing the water to go where the higher pressure is. Outside the pump, the water flows by itself, always toward the lower pressure.)

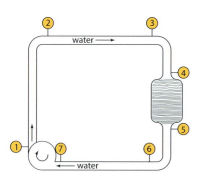

Figure 11.2. Water system with numbered pressure gauges added to the pipes.

Notice also that except for the drop between gauges 4 and 5, the pressure decreases steadily from place to place around the circuit. Again, this has to be the case because water only flows from point A to point B if the pressure at point B is lower than the pressure at point A. Finally, notice that the largest pressure drop in the system is across the water filter, between gauges 4 and 5. This is because the pipes are unobstructed and only a small difference in pressure between two points causes the water to flow easily. But the water filter is packed with sand and charcoal and what not, so to make the water flow through it the pressure drop across it has to be very large, just as a person has to blow hard to push water through a clogged hose. In fact, *almost all* the pressure drop from the pump discharge to

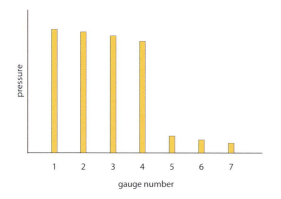

Figure 11.3. The pressure profile around the water circuit.

Component	Role
Pump	Makes the water flow. It does this by creating a difference in pressure. Outside the pump, water always flows toward lower pressure.
Pipe	Provides a contained pathway in which the water flows.
Water	Flows in the pipes. Flowing water actually consists of many individual molecules moving along in the pipe.
Water Filter	Provides resistance to the flow of water. The more material the filter has packed into it, the harder it is for the pump to make the water flow. The filter is doing a practical job (cleaning the water), and it is the reason we have the water circuit in the first place.

Table 11.1. Components in the water circuit.

the pump suction occurs across the water filter. All this behavior is exactly analogous to the way electricity flows in wires.

As a last step of analysis about the water circuit before we make the analogy to electricity, read through Table 11.1, which summarizes the important points I have made (and a few other details) about the different components in the water circuit and the roles they each play.

11.2   DC Circuit Basics

11.2.1 AC and DC Currents

Electricity flowing in a wire works exactly the same way as water flowing in a pipe. We have the same four basic components and the way they work is analogous to the way the components in the water circuit work. Before we proceed, you should know the difference between *DC* and *AC* circuits. You have probably heard of these terms. "DC" means *direct current*. It is a bit of a strange term, but it basically means the currents and voltages in the circuit, which you will understand soon, are constant, steady values. This is the case with any circuit powered by a battery. The internal electronics inside virtually every electronic gadget are DC circuits. (The AC from the power receptacle is converted to DC right inside the device, or by an adapter built into the plug.) The math for dealing with DC circuits is our topic in this chapter. "AC" means *alternating current*. We are not going to discuss AC circuits very much in this book, except for a bit in the next chapter. But in an AC circuit, the current and voltage are constantly changing, increasing and decreasing back and forth, going from positive to negative and back again. This kind of electric system is used for the entire electrical power distribution system that carries electricity to all the homes, offices, and factories. So the electricity that comes from a wall receptacle is AC.

11.2.2 DC Circuits and Schematic Diagrams

Let's now consider a DC electric circuit consisting of a battery, some wire, and a resistor, as shown in Figure 11.4. This type of drawing is called a *schematic diagram*. Schematic diagrams use symbols to show the components in the circuit and how they are connected, but they do not show what the circuit actually looks like physically. The wires may be a mile long, bending along on power poles along a roadside, or they may be tiny copper connections inside a computer. The schematic only shows the components and how they connect to each other.

Before we begin analyzing this circuit, a few words about the symbols are necessary. The flowing current is indicated by the arrows labeled with a capital I. The letter I is used as the symbol for electric current. (I suppose this is because the letter C is used for another electrical variable, capacitance, which we will not get into.) The symbol for the battery in the circuit in Figure 11.4 is actually the symbol for a single-cell battery, like a flashlight battery. The long bar on the end of the battery symbol indicates the positive end of the battery. The current comes out of this end of the battery and flows in the wire toward the resistor. Often the (+) and (−) symbols are not shown. The student is supposed to know that the long bar is the positive end. As you know, all batteries produce a certain voltage, which I will explain further below. But the (+/−) symbols indicate that the voltage at the (+) end of the battery is higher than the voltage at the (−) end, the difference being the voltage of the battery. Note that the (−) symbol does not mean the voltage is actually *negative*, just as the pressure at suction of a pump is not negative. The (+/−) symbols simply tell you which end of the battery has the highest voltage (+). Finally, the universal symbol for a *resistor* is the zigzag line segment.

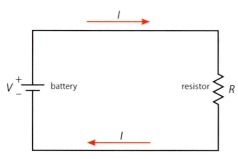

Figure 11.4. The basic DC circuit.

The small batteries used in flashlights, calculators, and other gadgets are single-cell batteries. The voltage a battery produces is fixed by its cell chemistry, so to produce a higher voltage we stack cells on top of one another, just as Volta's pile is a stack of individual cells. For example, car batteries have six cells, an arrangement indicated as shown in Figure 11.5. However, it is common when studying DC circuits to just use the battery symbol for a single-cell battery, regardless of the battery voltage, and this is what I do in my diagrams.

Now we are ready to study the electric circuit the same way we studied the water circuit, referring to Figure 11.4. Just like the water circuit, there are four components in the circuit. The battery is analogous to the water pump. Its job is to make the current flow. Water flows when a difference in pressure is present and the water always flows toward the lower pressure. The electrical analog to pressure is *voltage*. A battery is a chemical device that produces a difference in voltage and this difference in voltage forces an electric current to flow in the wire. Electric current always flows from high voltage to low. And just like the water, if there is no voltage drop no current flows.

Figure 11.5. The electrical symbol for a six-cell battery.

I will have more to say about resistors later, but here we must note that this electrical device has the same effect on an electrical circuit as the water filter has on the water circuit. It resists the flow of current and causes a large voltage drop. The resistor is the reason we have the circuit in the first place because the resistor is able to do some kind of work for us, just as the water filter does.

The roles the circuit components play are summarized in Table 11.2. I have also shown the water circuit components in the table so you can compare them side by side.

Component	Role	Component	Role
Pump	Makes the water flow. It does this by creating a difference in pressure. Outside the pump, water always flows toward lower pressure.	Battery	Makes the current flow. It does this by creating a difference in voltage. Outside the battery, current always flows toward lower voltage.
Pipe	Provides a contained pathway in which the water flows.	Wire	Provides a contained pathway in which the electric current flows.
Water	Flows in the pipes. Flowing water actually consists of many individual molecules moving along in the pipe.	Current	Flows in the wires. Flowing electric current actually consists of many individual electrons moving along in the wire.
Water Filter	Provides resistance to the flow of water. The more material the filter has packed into it, the harder it is for the pump to make the water flow. The filter is doing a practical job (cleaning the water), and it is the reason we have the water circuit in the first place.	Resistor	Provides resistance to the flow of current. The more resistance the resistor has, the harder it is for the battery to make the current flow. The resistor is doing a practical job and it is the reason we have the electric circuit in the first place. The resistor can be an actual resistor or some other device represented by a resistor.

Table 11.2. Comparison between the water circuit and the electrical circuit.

11.2.3 Two Secrets

I don't want to add unnecessary confusion to this discussion, but there are two little important details from all this explanation and analogy that need to be said. You won't really need to think about them at all when doing your circuit calculations, but I need to tell you about them so you won't get a wrong impression.

Everything I said above about current flowing from positive to negative and all that is valid only within the convention adopted a hundred years ago that we do the math with circuits by assuming the flow of *positive* charge. (As you know, protons have positive charge.) This convention has the advantage of eliminating a ton of negative signs from our calculations that don't do anything but get in the way, so this is a good convention. (The arrows labeled I in Figure 11.4 show the direction of flow for these fictitious positive charges.) In reality, as I have explained, it is not positive charge that flows in the wires of the circuit but negatively charged electrons, as illustrated in Figure 11.6. So here is secret number one: do believe everything I told you above about the direction the current is flowing and all that, and do the calculations the way I will explain below. Just remember that we define current as the flow of fictitious positive charges. In reality, the electrons are actually flowing in the wire in the opposite direction!

And now the second item. Remember how we said that for current to flow there must be a voltage difference, just like there must be a pressure difference for water to flow? Well, strictly speaking this is absolutely correct. However, in practical circuit analysis, the voltage

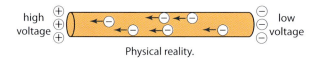

Physical reality.

The mathematical game we play.

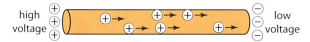

Figure 11.6. In electric circuits, electrons flow in the wire toward the positive terminal of the battery. In our math, we pretend positive charges are flowing the other direction.

difference along a copper wire—between the end of the battery and the near end of the resistor or light bulb or whatever in the circuit—is typically so small that we can completely ignore it in our circuit calculations. So we will! This means we can assume that any two points connected together by solid wire are at the same voltage. So I have drawn the electric circuit again in Figure 11.7 and indicated the voltage at a few places with this simplification in mind. I drew the

circuit with a 9-V battery.

Remember, there actually is an extremely minute voltage drop along the wire from the end of the battery to the resistor, just as there is a small pressure drop from gauge to gauge in Figure 11.3. If there weren't, current would not flow. But this is our second secret: the voltage difference is so small we are going to just forget about it and pretend that along any continuously connected wire the voltage is the same.

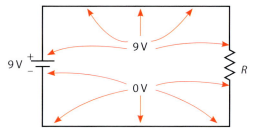

Figure 11.7. Circuit showing voltages (for practical purposes) on the wires.

11.2.4 Electrical Variables and Units

Here we will get our symbols straight for the variables and units of measure used in circuit analysis. The variable we use to describe the force with which the battery is pushing the current is *voltage*, with symbol V, and the unit we use is the *volt*, symbol V. So we measure voltage with volts. These words kind of look alike and their symbols are the same. But the other variables don't look at all like their units of measure and their symbols are not the same. The easiest way to lay it out is in another nice table, as I have done in Table 11.3. The symbols are all self-explanatory except the symbol for the *ohm*, the unit of electrical resistance, Ω. This symbol is the capital Greek letter omega, the last letter in the Greek alphabet. The unit *ohm* is named after the German physicist Georg Ohm (Figure 11.8). Ohm

Variable	Symbol	Units of Measure	Symbol for Units
voltage	V	volt	V
current	I	ampere, or amp	A
resistance	R	ohm	Ω
power	P	watt	W

Table 11.3. Electric variables and units of measure, and their symbols.

discovered the famous equation relating current, voltage, and resistance together, Ohm's law, that we address in the next section.

I will mention a couple of points here about units. First, note that when using the metric prefixes with resistances in ohms, there are two cases in which irregular spellings are commonly used: kilohm and megohm. In both of these cases, the –o or –a on the metric prefix is commonly dropped. Finally, as you know by now, I usually like to work out the MKS units we are using when we encounter new derived units. I assure you that the units of measure in the table above are all MKS units and that they all boil down to combinations of base units. (Current, in amperes, *is* one of the seven base units. Of course, nobody really says amperes anymore; we all just say amps now for short.) But working out the units for volts and ohms is beyond what we need to get into. I will only explain one of them—power.

Figure 11.8. German physicist, mathematician, and teacher Georg Simon Ohm (1789–1854).

Power is the rate at which energy is delivered or used, and is measured in watts (W). We will get to the equation for power in a few pages. A watt of power is an energy rate of one joule per second, or

$$1 \text{ W} = 1 \ \frac{\text{J}}{\text{s}}$$

For example, a traditional 60-watt light bulb, such as we all had in our houses before we replaced the bulbs with compact fluorescents (and then replaced some of them again with LEDs), draws 60 joules of energy every second from the city electrical distribution system. When your parents pay the electric bill, they pay for those joules. (There is a meter on the side of your house that measures how many joules you use, although the units they use aren't the nice joules (J) we use; they are kilowatt-hours (KWH). This weird unit is still just a unit for energy.)

11.2.5 Ohm's Law

The basic equation relating voltage, current, and resistance, called *Ohm's law*, is:

$$V = IR$$

This is one of the most famous equations in physics. One handy way to remember it is that it spells the word *man* in Latin (*vir*). (At least it's handy if you've studied Latin.) As I mentioned above, we are assuming that the voltage on a single piece of wire is the same everywhere on it. If we apply this rule and Ohm's law to a simple circuit with one resistor like the one in Figure 11.4, this means that the voltage of the battery is equal to the voltage drop across the resistor. So if you know the value of the resistance, you can determine how much current is flowing through the resistor. You simply solve Ohm's law for the current *I*, giving

$$I = \frac{V}{R}$$

Then put in the voltage and resistor values and calculate the current that is flowing in that circuit. Let's look at a couple of examples.

 Example 11.1

A battery with a voltage of 2.80 V is connected to a light bulb with a resistance of 9.33 Ω. Determine the amount of current flowing in the wires from the battery to the light bulb. State your answer in milliamps.

As always, begin by writing down the givens.

$V = 2.80$ V

$R = 9.33$ Ω

$I = ?$

In this case, all the givens are in MKS units already. So we proceed to write down Ohm's law, solve the equation for the current, and calculate the result.

$V = IR$

$$I = \frac{V}{R} = \frac{2.80 \text{ V}}{9.33 \text{ Ω}} = 0.300 \text{ A}$$

Finally, we convert this value to milliamps as required.

$$I = 0.300 \text{ A} \cdot \frac{1000 \text{ mA}}{1 \text{ A}} = 300 \text{ mA}$$

We are finished at this point, except that our result does not have the correct number of significant digits. From the given information, you see that we need to have three significant digits in the result. Recall that with a value like 300, the only way to write this so that the zeros are shown to be significant is to express the value in scientific notation. Doing so, we have our final result,

$$I = 3.00 \times 10^2 \text{ mA}$$

▲

▼ Example 11.2

Natural aluminum is found in a mineral ore called bauxite. One process of refining and smelting the ore to get the aluminum out of it uses a DC circuit with a very high current of 235 kA. Assume a voltage of 1,152 V is used to supply the current to the process, which can be modeled as a single resistor. Determine the resistance in the DC circuit.

Begin by writing the givens and performing the necessary unit conversions.

$$I = 235 \text{ kA} \cdot \frac{1000 \text{ A}}{1 \text{ kA}} = 235,000 \text{ A}$$

$$V = 1152 \text{ V}$$

$$R = ?$$

Now write the Ohm's law equation, solve for R, insert the values, and compute.

$$V = IR$$

$$R = \frac{V}{I} = \frac{1152 \text{ V}}{235,000 \text{ A}} = 0.00490 \ \Omega$$

Since the least precise value in the given information has three significant digits, this result has been rounded to three significant digits.

11.2.6 What Exactly Are Resistors and Why Do We Have Them?

There are little devices called resistors that are used in electronic circuits to regulate the voltages and currents throughout the circuit. A complicated electronic gadget like a computer has hundreds of resistors in its circuitry. Figure 11.9 is an image of some older-style, low-power resistors, the way they look just out of the package. The colored bands are coded and represent the resistance, in ohms, of the resistor. You will learn more about the resistor color code when you do the DC circuits experiment. The upper part of Figure 11.10 shows the way these look when installed on a circuit board in an electronic gadget. The lower photo shows the newer-style "surface mount" resistors. Surface-mount resistors are common now in electronics manufacturing, but for experimenting and do-it-yourself electronic projects the old-style resistors are still used.

Now you may or may not end up being an electrical engineer designing circuits with actual resistors. You may relate more immediately to the electronic devices in your house than you do to electronic parts. But at a basic level, most of the electrical devices we use—electric driers, lighting, hedge trimmers, computers, phone chargers, and on and on—act like resistors in the electric circuits that power everything around us. (AC circuits are more complicated than that, but this is an adequate simplification for our purposes here.) Light bulbs, electric heaters, and toasters are actually quite similar to actual resistors, so we use resistors to model those devices in our circuit calculations. In summary, when we do electric circuit calculations, we use the generic symbol for a resistor. But the resistors in our drawing could represent just about any device that needs electric power to operate.

Figure 11.9. Older-style, low-power resistors.

older style electronic resistors

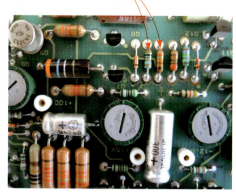

newer style electronic resistors

Figure 11.10. Older and newer styles of electronic resistors such as are found in electronics components.

11.2.7 Through? Across? In?

The prepositions and adverbs we use to describe what is going on in an electric circuit are important, so I want to take a moment here to make these clear. Current flows *in* a wire, just like water flows *in* a pipe. The voltage drops *across* a resistor, just like the pressure drops *across* the water filter. What we mean by this is that the voltage "upstream" of the resistor is higher than the voltage "downstream" of the resistor, so the voltage drops *across* it. We never speak of the voltage *through* a resistor because the voltage is not what is going through it. The current is what is going through it. So currents flow *in* wires and through *resistors*. Voltages drop *across* resistors.

While we are discussing word issues, you should also know that for historical reasons, voltage is also often called *potential*, and a voltage difference is sometimes called a *potential difference*. The phrase "potential difference" does not mean, "a possible difference." It means "difference in voltage" or "voltage drop." Finally, another term for voltage from the old days that you still see from time to time is *electromotive force*, or EMF. I actually like this term, because it reminds me of how the battery is providing the force that moves the current, just like a pump providing the force that moves the water. (I would like it even more if the term were electromotive *pressure* since pressure is a better analogy to voltage, as we saw with the water circuit.)

11.2.8 Voltages Are Relative

Consider for a moment our discussion back in Chapter 4 about energy. The gravitational potential energy an object possesses depends on how high up it is. But an object's height depends on where the zero reference is for height. Where is zero height? The table top? The floor? The ground outside the building? Sea level? The answer is that it doesn't matter. The only reason we ever calculate the E_G in an object is so we can predict how much work it takes to get it up there or how fast it goes if it falls down. These calculations do not depend on the absolute height, which is hard to even define. The calculations really depend only on the *difference* in height from some reference point like the floor or the table top to where the object is. It is only the difference that matters.

The same situation holds for voltages. Defining voltage in an absolute sense is rather abstract and must wait for a future physics course. For now, all we need concern ourselves with are voltage differences, or voltage drops. We simply use the lowest voltage in the circuit as our reference point. Remember where this is? The lowest pressure in the water circuit is at the suction of the pump. Analogously, the lowest voltage in an electric circuit is at the negative end of the battery. Since the only thing that matters about a voltage is how high it is relative to the reference, we just say that our reference voltage is the wire connected to the negative end of the battery. Further, we might as well call this voltage zero volts.

In fact, we might as well point out that in the great electrical systems of our noble nation, the earth itself is used as the zero voltage reference. In our power distribution systems, we realize this by physically bolting part of the circuit to the ground. This is done by driving a long copper *ground rod* into the ground (at your house this rod is about eight feet long) and connecting the power system

Figure 11.11. The symbol for electrical ground.

to it with a hefty copper wire. This is the electrical "ground" that you have probably heard of plenty of times but didn't understand. The electrical people call it ground because the reference point for zero volts in the circuit is the ground, and the circuit is connected to the ground—literally.

In an electric circuit diagram, an electrical ground connection is indicated by the symbol shown in Figure 11.11. Wherever ground is in an electric circuit, that is where the zero volt reference is. Now, I am not going to draw the ground symbol on every circuit example in this book. This is mainly because a lot of common electric circuits, such as flashlights, laptops, cell phones, and so on aren't actually connected to the ground unless they are being charged up. So we usually don't show a ground symbol on DC circuits. But I will show it once just to illustrate.

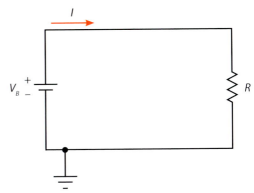

Figure 11.12. An electric circuit with ground connection shown.

Figure 11.12 shows our one-resistor circuit with the ground symbol shown. I have also labeled the voltage, current, and resistance using the standard symbols. V_B stands for the voltage of the battery. From now on, we will simply say that the negative end of the battery is where zero volts is. All the voltages in the circuit go up from there.

11.2.9 Power in Electrical Circuits

We encountered power briefly back in Section 11.2.4 when discussing units. In an electric circuit, the battery (or whatever the power supply actually is, even if it is not a battery) supplies the energy to the devices that need it. The energy is supplied at a certain rate, a certain number of joules per second. This energy rate is called the *power* and we measure it in watts (W). The basic equation for power is

$$P = VI$$

where P is the power (watts, W, that is, J/s). So, if a battery has a voltage V and is supplying a current I, it is supplying power to the circuit in the amount $P = VI$. Or, if a resistor has a voltage drop V across it and a current I flowing through it, then the power consumed by the resistor is $P = VI$.

While we are on the subject of power, let's consider how conservation of energy works in electrical circuits. The battery supplies energy to the circuit and this energy is "consumed" (as we say) by the devices in the circuit. If we are talking about the energy supplied by the battery in one second, then this is the power supplied to the circuit, and this power supplied to the circuit equals the power consumed by all the devices in the circuit. For resistors, the power is *dissipated* (that is, given off) as heat. For other devices, such as motors, the power is used to do mechanical work ($W = Fd$), as with a fan pushing air or a garage door opener lifting the door.

▼ Example 11.3

For the circuit shown in Figure 11.4, assume that the battery voltage is 5.60 V and the resistor value is 7.7 kΩ. Determine the current flowing in the circuit and the power consumed by the resistor. State your results in mA and mW.

Begin by writing the givens and converting to MKS units.

$V = 5.60$ V

$R = 7.7 \text{ k}\Omega \cdot \dfrac{1000\ \Omega}{1\ \text{k}\Omega} = 7700\ \Omega$

$I = ?$

$P = ?$

To solve for the current flowing in the circuit, we use Ohm's law.

$V = IR$

$I = \dfrac{V}{R} = \dfrac{5.60\ \text{V}}{7700\ \Omega} = 0.0007273\ \text{A}$

I will have something to say in the next section about ways to simplify the units of measure we use in circuit calculations so we don't have all these zeros to write down. For now, let's go ahead and convert this value to mA as required by the problem and round off to the required two significant digits.

$I = 0.0007273\ \text{A} \cdot \dfrac{1000\ \text{mA}}{1\ \text{A}} = 0.7273\ \text{mA}$

Rounding to two significant digits we have

$I = 0.73$ mA

To compute the power consumed by the resistor, we use the power equation. The voltage drop across the resistor is equal to the battery voltage and we have just computed the current in the circuit, so we can dive right in.

$P = VI = 5.60\ \text{V} \cdot 0.0007273\ \text{A} = 0.00407\ \text{W}$

Converting this result to milliwatts gives

$$P = 0.00407 \text{ W} \cdot \frac{1000 \text{ mW}}{1 \text{ W}} = 4.07 \text{ mW}$$

And finally, rounding to the required two significant digits gives

$$P = 4.1 \text{ mW}$$

11.2.10 Tips on Using Metric Prefixes in Circuit Problems

Bear with me here for a few paragraphs. My goal in this section is to simplify your life a bit.

All the metric prefixes you have learned and have been using in this course apply to the variables we use in electric circuits, just as they apply to other variables. However, for a couple of reasons, we will not be using all the prefixes for our circuit calculations. For starters, very few people ever bump in to circuits like the example a few pages back that had a current of 235,000 amps. Although that is an actual application in industry, most science students are never going to deal with currents that large. Another factor is the sheer volume of electronic devices in our time. When I was your age (which does not seem that long ago!) a home might have a TV and a radio. Now, as you well know, electronic circuits appear everywhere from your mobile phone and computer all the way down to your car, refrigerator, clothes dryer, and toothbrush.

Because of these two factors I am going to limit the circuit applications that appear henceforth to the voltages and currents we typically see in electronic applications or common household devices. There is actually a pretty good chance that many of you will encounter these kinds of circuits again, either in your future science courses, your hobbies, or your career.

Now, if we stick to low-voltage applications like this, there are some very common ranges of values that come up for the voltages, resistors, currents, and powers involved. The voltages used in electronic devices are typically just a few volts, anywhere from two or three volts up to 15 or 20 volts. Around the house, we have 120-V circuits everywhere, and although these are AC circuits rather than DC, the calculations are the same so we will do some problems with them. None of the voltage values in this range require any metric prefixes, so in our work we will stick to voltages in this range. To be sure, in the electronics world there are specialized circuits that use millivolts or even microvolts, but we will leave them for some other course of study.

When it comes to common resistor values one sees in household circuits or in electronics, resistances in ohms, kilohms and megohms are quite common, so we are going to stick to those. Just as with voltages, there are other prefixes that are used (giving resistances in milliohms or microohms), but we will leave them for another time.

These simplifications on the voltage and resistance values we will use result in some simplifications for the prefixes we will use for currents and powers. My goal here is to save you a bunch of time writing down zeros in your answers to computations, and time performing unit conversions. So let me use some examples to show you what I mean. There are three cases to consider, involving resistance values in ohms, kilohms (kΩ) and megohms (MΩ). For reasons I will explain in a few pages, I am not going to get too hung up on tracking the significant digits in these next few examples. You will see why soon; for now we will keep it simple.

Case 1: Resistances in ohms

With a voltage of a few volts, and a resistance of a few ohms or even a few hundred ohms, it is easiest to use the standard MKS units of amps (A) for the currents and watts (W) for the powers. For example, if we had a voltage drop of 6.55 V across a resistance of 28.3 Ω, we compute the current flowing through the resistor using the standard Ohm's law equation and the standard MKS units, this way:

$V = 6.55$ V

$R = 28.3$ Ω

$I = ?$

$V = IR$

$$I = \frac{V}{R} = \frac{6.55 \text{ V}}{28.3 \text{ } \Omega} = 0.231 \text{ A}$$

And as you have already learned, the power consumed by the resistor is calculated as follows:

$$P = VI = 6.55 \text{ V} \cdot 0.231 \text{ A} = 1.51 \text{ W}$$

Case 2: Resistances in kilohms

What if we have the same voltage as before—a few volts—but with a resistance in kilohms? Let's see how this changes things. Let's use the same voltage as in the previous example, but with a resistance of 28.3 kΩ.

$V = 6.55$ V

$$R = 28.3 \text{ k}\Omega \cdot \frac{1000 \text{ } \Omega}{1 \text{ k}\Omega} = 28,300 \text{ } \Omega$$

$I = ?$

$V = IR$

$$I = \frac{V}{R} = \frac{6.55 \text{ V}}{28,300 \text{ } \Omega} = 0.000231 \text{ A}$$

Now look at these results. Since the resistance was in kilohms instead of ohms, we ended up with three extra zeros in front of the significant digits for the current. In effect, we divided by 1,000 when we converted the resistance to ohms, so our answer is 1,000 times smaller. What if we convert this result to milliamps?

$$I = 0.000231 \text{ A} \cdot \frac{1000 \text{ mA}}{1 \text{ A}} = 0.231 \text{ mA}$$

These are the same digits we had before, except the units are milliamps instead of amps. This happens every time, so we might as well save a bunch of unit conversion work by remembering this mathematical principle: if the resistance is in kΩ, we can leave it that way, put it in the Ohm's law equation as it is, and the current comes out in mA. And what if we go ahead and use this current in milliamps to calculate the power consumed by the resistor?

$$P = VI = 6.55 \text{ V} \cdot 0.231 \text{ mA} = 1.51 \text{ mW}$$

All else remaining the same, if the only change we make is that the current is in milliamps, then the power units are milliwatts. So if the resistance is in thousands of ohms (kilohms) the current comes out in thousandths of amps (milliamps) and the power comes out in thousandths of watts (milliwatts). Remember from the water analogy the effect of the resistance on the circuit: higher resistance reduces the flow of current. In this example we had 1,000 times more resistance because the resistance was in kilohms. This reduces the current and the power by a factor of 1,000 giving us current in milliamps and power in milliwatts.

Case 3: Resistances in megohms

Everything I wrote in the previous case about kilohms giving milliamps and milliwatts applies in a similar fashion when the resistance is in megohms, except the currents and powers have the metric prefix *micro–* (μ). Since "mega" means millions, the resistance in this case is in millions of ohms so the currents and powers come out in millionths of amps and millionths of watts, which are μA and μW. Here are the details of how the math works:

$$V = 6.55 \text{ V}$$

$$R = 28.3 \text{ M}\Omega \cdot \frac{1,000,000 \ \Omega}{1 \text{ M}\Omega} = 28,300,000 \ \Omega$$

$$I = ?$$

$$V = IR$$

$$I = \frac{V}{R} = \frac{6.55 \text{ V}}{28,300,000 \ \Omega} = 0.000000231 \text{ A} \cdot \frac{1,000,000 \ \mu\text{A}}{1 \text{ A}} = 0.231 \ \mu\text{A}$$

And just as before, the metric prefix in the current is there in the power, too:

$$P = VI = 6.55 \text{ V} \cdot 0.231 \ \mu\text{A} = 1.51 \ \mu\text{W}$$

This entire discussion is summarized in Table 11.4. We will always use voltages with units in volts (V). The units for the currents and powers depends on which of the three possible units for resistance is used in the problem.

Voltage Units	Resistance Units	Current Units	Power Units
V (volts)	Ω (ohms)	A (amps)	W (watts)
V (volts)	kΩ (kilohms)	mA (milliamps)	mW (milliwatts)
V (volts)	MΩ (megohms)	μA (microamps)	μW (microwatts)

Table 11.4. The simplified unit system we will use for circuit calculations.

Study these examples and the table until you are comfortable using the units as I have described. I really encourage you to give this some effort. After you work a few problems, you will feel more comfortable with the units. However, when I have taught this to students, I have known some students who never did get comfortable with dealing with the units this way and preferred to handle all the units the standard way, by converting to MKS units. Technically, there is nothing wrong with doing this. It's more cumbersome, but it is not incorrect. So if my discussion of how to simplify the units doesn't seem to simplify things in your mind, you can always fall back to good old MKS units.

11.3   Multi-Resistor DC Circuits

11.3.1 Two-Resistor Networks

So far you have seen only circuits with one battery and one resistor. We have now covered all of the basics for such circuits and are ready to learn how to handle circuits with more than one resistor. In the real electrical world, circuits usually have many resistors, not just one. The bundle of resistors in a circuit, all wired together but not connected to the battery, is called a resistor *network*. Our first task is to learn the different ways resistors can be connected together in a network.

Resistors in circuits can be connected together in two main ways. The first kind of connection is shown in Figure 11.13. The key feature here is that all the current that enters resistor R_1 also enters resistor R_2. There are no other branches or pathways in the circuit, so the current, represented by the red arrows, has no option but to go through both resistors, one after the other. This kind of connection is called a *series* connection and we say

Figure 11.13. Two resistors connected in series.

that the two resistors are connected "in series." We could also wire up three or four resistors in series, or as many as we like. It is important to note that in a series connection, there is nothing between the resistors, no other devices, no connections, nothing except the single wire connecting the series resistors together. If there is anything else there, the resistors are not connected in series. An important principle to note for series connections is this: *when resistors are connected in series, the same current passes through each of them.* (It has no where else to go).

The second way of wiring resistors together is shown in Figure 11.14. Don't freak out at that diagram; the connections shown in the three sketches are the same! I drew it three different ways to show different ways this type of resistor network can be drawn. This kind of connection is called a *parallel* connection and we say that the two resistors are connected "in parallel."

An important warning is critical here. Just because this is called a parallel connection does not mean that when drawn on the paper in a schematic diagram the resistors appear to be parallel to each other on the page. Remember, a schematic diagram tells us nothing about what the circuit looks like, only how the devices are connected. This is why the three networks shown in Figure 11.14 are identical. The way they are drawn on the page is simply for convenience; the important thing is how they are connected.

The key feature to note about a parallel connection like this is that if two resistors are connected together in parallel, then they are electrically connected together at both ends. If

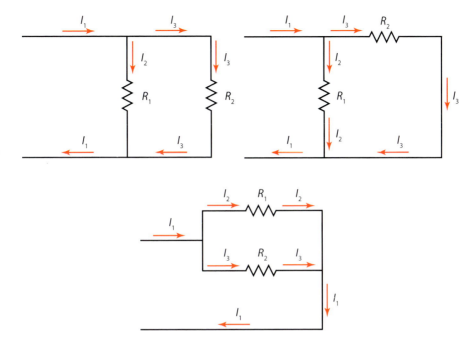

Figure 11.14. Three identical resistor networks showing two resistors connected in parallel.

they are not connected together at both ends, or if there is more than one resistor in one of the branches, they are not connected in parallel.

Notice that in a parallel network the current heading towards the resistors, I_1, divides at the *junction*, or *node*, in the wires. Some of the current, which I show as I_2, goes through resistor R_1, and the rest of the current, which I have labeled I_3, goes through the other resistor, R_2. The amount of current that goes through each branch depends on the resistance in each branch. Keep thinking about the water analogy. If water is flowing in a pipe and comes to a place where the pipe has two branches, the water divides, some going one way and some the other.

Now let's see what these two basic types of networks look like when they are connected to a battery. The sketches in Figure 11.15 each show circuits with two resistors in series. These two circuits are identical. The sketches in Figure 11.16 each show circuits with two resistors in parallel. These two circuits are also identical. They are drawn differently, but they are electrically identical.

As you can see in the series circuit of Figure 11.15, there is only one current and it passes in turn through each of the resistors in the circuit. Referring again to the water circuit, this is analogous to having two water filters, one after the other, in the pipe. The parallel circuit in Figure 11.16 is quite different. In this parallel circuit, at the first junction (or node) the first current I_1 splits into two parts, currents I_2 and I_3. Just like a split or branching in a water pipe, some of the current goes one way and some goes the other way. Downstream of the two resistors, the two currents I_2 and I_3 join up again and combine back to the total current they are before the split, namely, I_1.

As a further illustration of series and parallel circuits, I wired up one of each using a battery and a couple of LEDs. LEDs (light emitting diodes) are small devices that emit light and act like resistors in a DC circuit. In these two circuits, I used one green LED and one red LED. The photos in Figure 11.17 show the two circuits.

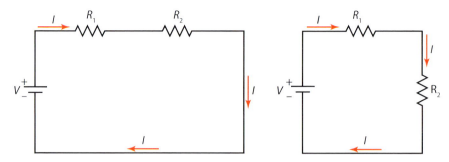

Figure 11.15. Identical DC circuits with two resistors connected in series.

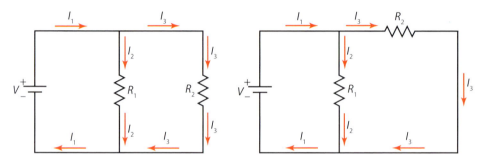

Figure 11.16. Identical DC circuits with two resistors connected in parallel.

Compare the series circuit schematics in Figure 11.15 to the LED series circuit on the left of Figure 11.17. Either of the schematic diagrams can be used to represent the LED circuit when it is wired this way. Notice that the current coming out of the battery (the red wire) must go through the red LED first (R_1), then through the green one (R_2), and finally back to the battery through the black wire to complete the circuit.

Now compare the parallel circuit schematics in Figure 11.16 to the LEDs connected in parallel on the right of Figure 11.17. Again, either schematic can be used to represent the LEDs when they are connected together this way. The current exits the battery through the red wire and splits into two parts at the junction of the red wires. Part of the current passes through the green LED (R_1) and part passes through the red LED (R_2). These two currents

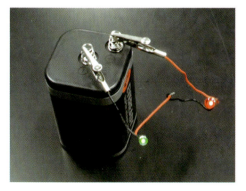
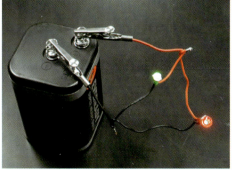

Figure 11.17. Two LEDs connected in a series circuit (left) and a parallel circuit (right).

come together again at the junction of the black wires and then the current returns to the battery to complete the circuit.

11.3.2 Equivalent Resistance

When we perform calculations with circuits, our ultimate goal is to determine the voltages, currents, and powers in every part of the circuit. To calculate the voltages and currents, we must first find out how much current is flowing into the resistor network from the battery. Until we do this, we generally aren't able to figure out much else.

Here is an important first principle to keep in mind: no matter what the resistor network looks like, the battery is connected to the network with just two wires. This means the battery has no way at all of knowing how many resistors there are or how they are wired together. As far as the battery is concerned, there may be one resistor or a hundred. To the battery, regardless of how many resistors there are, the network *seems* like just one resistor. The value of the apparent resistance that the battery perceives is called the *equivalent resistance*, for which we use the symbol R_{EQ}.

Consider this analogy. A horse pulling a cart doesn't know the difference between having 20 bales of hay on the cart each weighing 50 pounds, and 40 bales of hay each weighing 25 pounds. To the horse, it feels like a 1,000 pound load. This is because it *is* a 1,000 pound load. We could even remove all the hay bales and put a single machine weighing 1,000 pounds on the cart and the horse would not know the difference. Similarly, to the battery, the entire resistor network feels like one "load." (The term *load* is even used sometimes when discussing electric circuits to describe the resistor network that the power source is connected to.) Regardless of how the resistors are connected and how many of them there are, the battery responds as if there were only a single resistor connected with a resistance value equal to the R_{EQ} of the resistor network. I will now discuss how we determine the value of R_{EQ} for different types of resistor networks. This presentation is lengthy, not because this topic is particularly difficult but because there are a number of details to review and I want to explain them very carefully. So read on.

We will take this one step at a time. First, the equation for calculating the equivalent resistance for resistors in series is

$$R_{EQ} = R_1 + R_2 + R_3 + ...$$

This equation works for any number of resistors, so long as they are all connected in series. For resistors in parallel, the easiest equation to use for calculating R_{EQ} works only for two resistors. The equation for calculating the equivalent resistance for two resistors in parallel, including the special symbol we use to indicate parallel resistances, is

$$R_{EQ} = R_1 \| R_2 = \frac{R_1 R_2}{R_1 + R_2}$$

There is an equation that can handle more than two resistors in parallel, but we will not use it for our calculations.[1]

---

[1] This equation has an interesting mathematical symmetry with the series equation. Because of this, you might want to at least see the equation, so here it is: $\dfrac{1}{R_{EQ}} = \dfrac{1}{R_1} + \dfrac{1}{R_2} + \dfrac{1}{R_3} + ...$

When using the parallel equation, remember these two tips to help check yourself as you go.

> **Tip 1: The resistance of two resistors in parallel is always less than either of the single values that go into the calculation.**

So your R_{EQ} calculation for two resistors R_1 and R_2 in parallel always gives a result that is less than the value of either of the resistors.

> **Tip 2: When two identical resistors are in parallel, the equivalent resistance is half the value of one of the resistors.**

Thus, two 8 kΩ resistors in parallel has an R_{EQ} of 4 kΩ. This is a time-saving tip.

▼ Example 11.4

The sketches in Figure 11.18 show one series network and one parallel network. I drew each network inside an imaginary box (dashed lines) with two terminals (the little circles) to show how the battery connects (red arrows) to the network. The black arrow is there to show that the R_{EQ} value is the single apparent resistor value that the battery thinks is out there when connected to the network. (Remember, the battery does not know what is in the imaginary box.) Determine R_{EQ} for each network.

For the series network, we just add the resistor values. The R_{EQ} value for the series network is

$$R_{EQ} = 40 \ \Omega + 530 \ \Omega = 570 \ \Omega$$

To the battery, this resistor network seems like a single 570 Ω resistor. In the parallel network, the resistor values are all in kilohms. This is fine as long as all of them are in the same units. Since both resistors are in kilohms, we do the calculation in kilohms. The R_{EQ} value for the parallel network is

$$R_{EQ} = 14 \ \text{k}\Omega \parallel 27 \ \text{k}\Omega = \frac{14 \ \text{k}\Omega \cdot 27 \ \text{k}\Omega}{14 \ \text{k}\Omega + 27 \ \text{k}\Omega} = 9.2195 \ \text{k}\Omega$$

To the battery, this parallel resistor network seems like a single resistor with a resistance of 9.2195 kΩ. Notice here how easy this calculation is when we just keep the resistors in

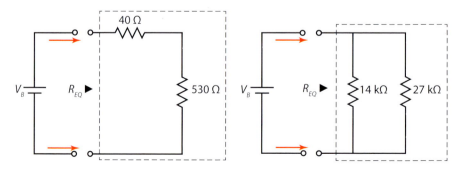

Figure 11.18. Series and parallel resistor networks connecting to a battery.

kilohms. We don't have to write down or key in any of the zeros. We'll discuss significant digits in these calculations next.

11.3.3 Significant Digits in Circuit Calculations

As you probably noticed immediately from the parallel resistor example, my calculation of the equivalent resistance has several more digits in it than we normally permit according to the significant digit rules we have been using since the beginning of this book. You are right! Here is what is going on.

In electric circuit calculations for circuits with more than one resistor, there is often more than one way to do the calculations, and there are usually many steps in the calculations. The ordinary significant digit rules do apply to these calculations, as always. But if we use them, it makes it difficult for you, the student, to make sure you are doing the steps correctly. If you make a small error somewhere, your answer can look correct but be incorrect because the rounding can cover up the error. On the teacher's end, scoring papers can be a nightmare if the teacher cannot tell whether an answer is correct, and it can easily happen that a student's answer is calculated correctly but looks different from the answer key because of rounding repeatedly over several steps of calculations. By the same logic, incorrect answers can look correct.

In this text we are going to handle this problem by using a special digits rule for circuit calculations: *For circuits with more than one resistor, ignore the usual significant digit rules and show four decimal places in every value for every calculation.* This includes equivalent resistances, voltages, currents, and powers. Four decimal places is enough to tell whether your calculation has been done correctly. When comparing your answers to the answer key, the first three decimal places should agree nearly every time. The last place may differ from the key, but if the first three agree you probably solved the problem correctly.

You know that we used a special rule like this once before: with temperature unit conversions our special rule is to show one decimal place. Circuit calculations is the only other kind of computation where we will use a special rule.

11.3.4 Larger Resistor Networks

We have seen how to compute R_{EQ} for series and parallel networks with two resistors. The next step is to apply these techniques to larger networks. Consider the three-resistor network shown in Figure 11.19. For resistor networks with more than two resistors, we always begin the process of calculating R_{EQ} by *starting at the right side of the network*. We work our way from right to left, identifying small groups of resistors that are connected either in series or in parallel. Each time we identify a little group, we calculate a sort of sub-R_{EQ} for that group. Then replace the group with the R_{EQ} for the group. Do this repeatedly, working from right to left, until every resistor in the network has been combined into a single resistor value, the R_{EQ} for the entire network. Hopefully, working the following two examples will make this process clear.

▼ Example 11.5

To begin calculating R_{EQ} for the three-resistor network in Figure 11.19, first note that one of the resistors is in kilohms and the other two are in megohms. This will not do. All resistors need to be in the same units before we begin combining their values together. In these calculations, don't trouble yourself to convert all of the resistor values to MKS units (ohms). You don't want to write down dozens of zeros all the time. Instead, pick the metric prefix that occurs most frequently in the resistor values in the circuit and rewrite all the other resistor values so they have the same prefix. So we convert the units for the 830 kΩ resistor into megohms and do the calculation of R_{EQ} in MΩ.

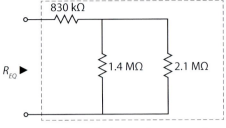

Figure 11.19. Example three-resistor network.

$$830 \text{ k}\Omega \cdot \frac{1,000 \ \Omega}{1 \text{ k}\Omega} \cdot \frac{1 \text{ M}\Omega}{1,000,000 \ \Omega} = 0.83 \text{ M}\Omega$$

Now beginning on the right hand side of the circuit, we see that the 1.4 MΩ and 2.1 MΩ resistors are in parallel. So we calculate an equivalent value of resistance that we can substitute for them.

$$1.4 \text{ M} \| 2.1 \text{ M} = \frac{1.4 \text{ M} \cdot 2.1 \text{ M}}{1.4 \text{ M} + 2.1 \text{ M}} = 0.8400 \text{ M}\Omega$$

Imagine at this point that in the network, the 1.4 MΩ and 2.1 MΩ resistors have been replaced by their 0.8400 MΩ equivalent. The network may now be represented as shown in Figure 11.20.

Now we see that the 0.83 MΩ resistor is in series with the 0.84 MΩ group, so we use the series resistance calculation to compute R_{EQ} for the entire network as

$$R_{EQ} = 0.8300 \text{ M}\Omega + 0.8400 \text{ M}\Omega = 1.6700 \text{ M}\Omega$$

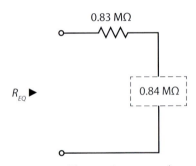

Figure 11.20. Three-resistor network with two of the resistors replaced by their equivalent resistance.

▲

▼ Example 11.6

Consider the four-resistor network shown in Figure 11.21. Notice that I didn't put the Ω symbol by the resistor values. It is common to see circuits that just show the numerical value and the metric prefix, and since all resistor values are in ohms or –ohms with a prefix, the Ω symbol is assumed. Notice also that all the resistors except R_4 are in kilohms. Converting R_4 to kilohms gives 0.87 kΩ, and we use this value in the calculation of R_{EQ}.

To calculate R_{EQ}, begin as before on the right side of the network and work from right to left. Doing so, we see that the first two resistors we encounter are R_3 and R_4, which are in series. Combining these gives

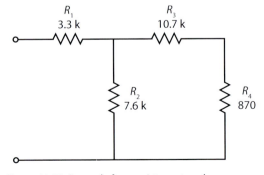

Figure 11.21. Example four-resistor network.

$R_3 + R_4 = 10.7 \text{ k}\Omega + 0.87 \text{ k}\Omega = 11.57 \text{ k}\Omega$

Next, this combined value for R_3 and R_4 is in parallel with R_2, so we execute a parallel calculation between the 11.57 kΩ and R_2.

$$R_2 \parallel 11.57 \text{ k} = \frac{7.6 \text{ k} \cdot 11.57 \text{ k}}{7.6 \text{ k} + 11.57 \text{ k}} = 4.5870 \text{ k}\Omega$$

Finally, the combination of the three resistors we have calculated so far is in series with R_1. Combining them gives

$$R_{EQ} = R_1 + 4.5870 \text{ k} = 3.3 \text{ k} + 4.5870 \text{ k} = 7.8870 \text{ k}\Omega$$

We will work one more example R_{EQ} calculation and then move on to calculating voltages and currents.

▼ Example 11.7

For this example, we use the resistor network shown in Figure 11.22. Since two of the resistors are in ohms and two are in kilohms, it doesn't matter which units we use. I will just convert everything to kilohms and use kilohms for the calculation. This gives us $R_3 = 0.55 \text{ k}\Omega$ and $R_4 = 0.84 \text{ k}\Omega$.

Beginning on the right side of the network, R_2 and R_3 are in parallel, giving

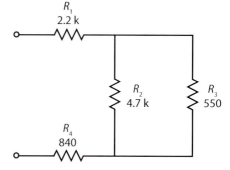

Figure 11.22. Second example four-resistor network.

$$R_2 \parallel R_3 = \frac{4.7 \text{ k} \cdot 0.55 \text{ k}}{4.7 \text{ k} + 0.55 \text{ k}} = 0.4924 \text{ k}$$

This combination is in series with both R_1 and R_4, so we just add these all together to get R_{EQ}.

$$R_{EQ} = R_1 + 0.4924 \text{ k} + R_4 = 2.2 \text{ k} + 0.4924 \text{ k} + 0.84 \text{ k} = 3.5324 \text{ k}\Omega$$

In summary, here are the steps of our procedure for calculating R_{EQ}.

- Make sure all the resistor values in the network use the same units.

- Work from right to left, identifying small groups (usually two) of resistors to combine into a single value.

- When you reach the left side and have included every resistor in the calculation, you have your value for R_{EQ}.

11.3.5 Kirchhoff's Laws

We are almost ready to put a battery on a resistor network and start calculating the voltages and currents. In addition to the equations for combining resistors in series or parallel circuits, there are three equations we use to do the circuit calculations. You already know one of them, Ohm's law, or $V = IR$. The other two are called Kirchhoff's laws. These two laws are named after Gustav Kirchhoff (Figure 11.23), another nineteenth-century German physicist who contributed a lot to our understanding of how electricity works in circuits.

Kirchhoff's junction law says:

> At any junction, or node, in a circuit, the sum of the currents entering the node equals the sum of the currents exiting the node.

This makes perfect sense. Think again about the water analogy. If three pipes are feeding water into some point in a plumbing network and two pipes are where the water exits, then the sum of the three water flows coming into the junction must equal the sum of the two water flows going out. The water has nowhere else to go.

As a quick illustration, assume you have a section of an electrical circuit as shown in Figure 11.24. If you know I_1 and I_2, then I_3 is easy to determine, because $I_1 = I_2 + I_3$, so I_3 is just $I_1 - I_2$. Using the values shown in the sketch, I_3 has to be

Figure 11.23. German physicist Gustav Robert Kirchhoff (1824–1887).

$$I_3 = I_1 - I_2 = 14.5 \text{ A} - 6.9 \text{ A} = 7.6 \text{ A}$$

The other law is *Kirchhoff's voltage drop law*. This law states:

> If you scan around any loop in a circuit, the sum of the voltage rises equals the sum of the voltage drops.

$I_1 = 14.5$ A I_3

$I_2 = 6.9$ A

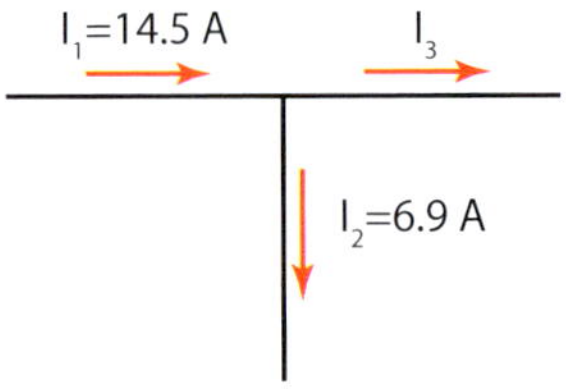

Figure 11.24. A junction, or node, in a circuit.

Sometimes it takes students a while to get their minds around the terminology in this law, so I am going to expand on this with an example and with a new analogy. Hopefully this will enable you to grasp this concept, which, like everything else in the chapter, is relatively simple but tricky to put into words. First, the example.

▼ Example 11.8

Figure 11.25 shows a sketch of a circuit in which I have labeled the voltage drops. I have shown three of the voltage drops as unknowns, V_1, V_2, and V_3. The other values you may take as givens, although soon you will be able to calculate them for yourself.

Pick any loop you like in this circuit. Scan around the loop, marking down the voltage rises and drops. Kirchhoff's voltage drop law says the sum of the rises equals the sum of the drops. Begin in the first loop next to the battery, starting in the bottom left corner of the loop. As you know, this is the zero-volt point in the circuit. As we move around the first loop in a clockwise circle, we first pass a rise of 8 volts as we cross the battery. This voltage rise is followed by a voltage

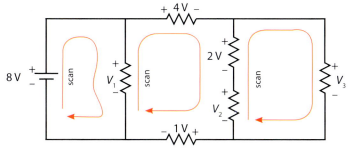

Figure 11.25. Scanning around the loops in a circuit.

drop of V_1 to complete the loop. Applying the voltage drop law, we get

$$8 \text{ V} = V_1$$

$$V_1 = 8 \text{ volts}$$

This first loop illustrates a general principle you should make note of: *devices in parallel always have the same voltage drop across them*. Here, the resistor is in parallel with the battery because the two ends of resistor are connected directly to the two ends of the battery. So the voltage across them must be the same. We use this principle often with resistors, or, as in this case, with a resistor in parallel with the battery.

Now scan the middle loop. Beginning in the lower left corner and scanning around the circuit in a clockwise direction, we first have a rise of 8 volts ($V_1 = 8$ V), then a drop of 4 volts, a drop of 2 volts, a drop of V_2 volts, and a drop of 1 volt. Using Kirchhoff's law, we have

$$8 \text{ V} = 4 \text{ V} + 2 \text{ V} + V_2 + 1 \text{ V}$$

$$V_2 = 8 \text{ V} - 4 \text{ V} - 2 \text{ V} - 1 \text{ V} = 1 \text{ volt}.$$

Finally, scanning the third loop we begin with two rises, 1 volt and 2 volts, followed by the drop V_3. Using Kirchhoff's law, we have

$$1 \text{ V} + 2 \text{ V} = V_3$$

$$V_3 = 3 \text{ volts}$$

Note that we can also solve for V_3 by scanning around the large outer loop of the circuit, where we have a rise of 8 volts followed by drops of 4 volts, V_3, and 1 volt, giving

$$8 \text{ V} = 4 \text{ V} + V_3 + 1 \text{ V}$$

$$V_3 = 8 \text{ V} - 4 \text{ V} - 1 \text{ V} = 3 \text{ volts}$$

I also promised a new analogy to aid in understanding the voltage drop law. Imagine a huge old mansion with many different staircases. It is designed as a sort of crazy fun house so that there is one main staircase for going up to the top floor, like the voltage of a battery,

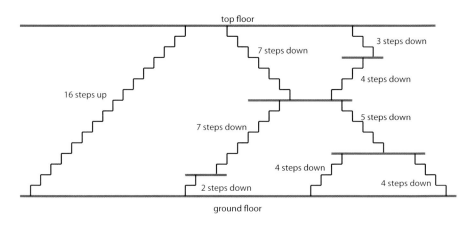

Figure 11.26. Steps in a house as an analogy for voltage rises and drops in a circuit.

and many little ones for going down, like the voltage drops across resistors. These directions correspond to the direction current flows in a circuit. The different routes one can take to get to the top floor and then back down to the ground are like the loops we scanned in the previous example. The number of steps in each staircase corresponds to the value of a voltage rise or voltage drop. The different floors correspond to different voltages above the zero-voltage wire of a circuit.

So consider the sketch in Figure 11.26. The horizontal lines represent the different floors in the house. Notice that no matter what route you take to get down from the top floor, you have to go down 16 steps, which is equal to the number of steps in the up staircase. Also, between any two floors in the house there are the same number of steps, no matter which staircase you take. This means that this house is like the voltage drop law. You can scan around any loop you want involving any staircases you like, but the total steps you go up has to equal the total steps you go down to get back to where you started. And you may also notice that those two staircases at the bottom that have four steps are like resistors in parallel. As I mentioned before, any two resistors in parallel have the same voltage drop, just as these two staircases have to have the same number of steps, since they start and end at the same floors.

11.3.6 Putting it All Together to Solve DC Circuits

The final step in this study is to put everything to use in solving DC circuits. Let's begin with an outline of the general procedure. The steps to solve a circuit are as follows:

1. Number all the resistors (if they are not already numbered) on the circuit diagram. Label the current leaving the positive end of the battery as I_1. Label the voltage drop across each resistor using the same number as the resistor number. Label the current flowing in each of the other branches of the circuit.

2. Calculate the equivalent resistance of the resistor network, R_{EQ}. Do this by working from right to left in the circuit. If R_{EQ} is not a whole number, write down the value of R_{EQ} with four decimal places.

3. Calculate the first current, I_1, using the battery voltage, the equivalent resistance, and Ohm's law, $V = IR$. Record I_1 with four decimal places. Solving Ohm's law for I, we have

$I = V/R$. Applying this to the battery voltage and equivalent resistance of a circuit, we have

$$I_1 = \frac{V_B}{R_{EQ}}$$

4. Working left to right, use Ohm's law and Kirchhoff's two laws to calculate the individual currents and voltage drops for the resistors, writing down each value with four decimal places. You always apply one of these three laws. For any resistor, if you know two out of the three variables in Ohm's law you can calculate the third. In any loop, if you know all the voltage drops but one, you can use Kirchhoff's voltage drop law to determine the unknown voltage drop. If at any junction you know the values of all the currents entering and leaving the junction but one, you can use Kirchhoff's junction law to determine the one unknown current.

5. When required, calculate the powers consumed by the individual resistors using their individual voltage and current values.

6. Assuming that the battery voltage is a few volts (very common), use this rule of thumb to expedite dealing with the metric prefixes on currents and powers: if all the resistors are in Ω, the currents are in A and powers are in W. If all the resistors are in kΩ, the currents are in mA and powers are in mW. If all resistors are in MΩ, the currents are in μA and powers are in μW.

In the examples below, I solve each of the five circuits given as examples above for calculating R_{EQ}. For each of these, the first step of calculating the equivalent resistance is already complete. For each circuit, I add a battery with a certain voltage value and label all the currents and voltage drops. Then I solve for all the voltages and currents. Finally, I compute the power consumed by each of the resistors.

▼ Example 11.9

Our first example is to compute the voltages, currents, and powers for the circuit shown in Figure 11.27.

From the R_{EQ} calculation in Example 11.4 above, we have $R_{EQ} = 570\ \Omega$. Next we calculate the current, I_1, as

$$I_1 = \frac{V_B}{R_{EQ}} = \frac{13.2\ \text{V}}{570\ \Omega} = 0.0232\ \text{A}$$

With this current, we solve for the two voltage drops using Ohm's law:

$$V_1 = I_1 R_1 = 0.0232\ \text{A} \cdot 40\ \Omega - 0.9280\ \text{V}$$

and

$$V_2 = I_1 R_2 = 0.0232\ \text{A} \cdot 530\ \Omega = 12.2960\ \text{V}$$

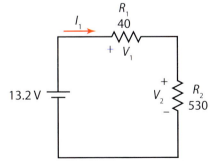

Figure 11.27. Example two-resistor series circuit.

According to Kirchhoff's voltage drop law, voltages V_1 and V_2 add up to the battery voltage, 13.2 V. They actually add up to a teeny bit more because of the rounding in our I_1 calculation.

Now for the power consumed by the resistors. The power consumed by R_1 is

$$P_{R1} = VI = 0.9280 \text{ V} \cdot 0.0232 \text{ A} = 0.0215 \text{ W}$$

and the power consumed by R_2 is

$$P_{R2} = VI = 12.2960 \text{ V} \cdot 0.0232 \text{ A} = 0.2853 \text{ W}$$

▼ Example 11.10

The next example is for the circuit in Figure 11.28.

From the R_{EQ} calculation in Example 11.4 above, we have R_{EQ} = 9.2195 kΩ. However, this circuit is an example of the *one time* that R_{EQ} is not needed for computing the voltages and currents of the resistors! As I have mentioned a couple of times before, devices in parallel always have the same voltage drop across them. This means that in any circuit where every resistor is in parallel with the battery, the voltage drop across every resistor is the battery voltage. In this circuit, both resistors are in parallel with the battery, so

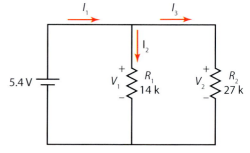

Figure 11.28. Example two-resistor parallel circuit.

$$V_1 = V_2 = 5.4 \text{ V}$$

With these voltages, we can calculate the currents using Ohm's law as

$$I_2 = \frac{V_1}{R_1} = \frac{5.4 \text{ V}}{14 \text{ k}\Omega} = 0.3857 \text{ mA}$$

and

$$I_3 = \frac{V_2}{R_2} = \frac{5.4 \text{ V}}{27 \text{ k}\Omega} = 0.2000 \text{ mA}$$

Notice that I left the resistances in kilohms, which means the current units are in milliamps, as discussed in Section 11.2.10.

Now we calculate the power consumed by each of the resistors.

$$P_{R1} = V_1 I_2 = 5.4 \text{ V} \cdot 0.3857 \text{ mA} = 2.0828 \text{ mW}$$

and

$$P_{R2} = V_1 I_3 = 5.4 \text{ V} \cdot 0.2000 \text{ mA} = 1.0800 \text{ mW}$$

Notice that since the current was in milliamps, the power is in milliwatts.

▼ Example 11.11

The third example uses the three-resistor circuit in Figure 11.29.

From the R_{EQ} calculation in Example 11.5 above, we have $R_{EQ} = 1.6700$ MΩ. We begin by calculating the first current, which is the current coming out of the battery.

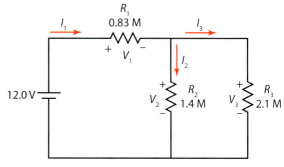

Figure 11.29. Example three-resistor circuit.

$$I_1 = \frac{V_B}{R_{EQ}} = \frac{12.0 \text{ V}}{1.67 \text{ M}\Omega} = 7.1856 \text{ μA}$$

Since the resistor value is in megohms, the current is in microamps. Knowing I_1, we next ask what we can calculate using one of the three laws (remembering that we are working from left to right). The answer is that since we now know the current going through R_1 we can calculate V_1 using Ohm's law. This gives

$$V_1 = I_1 R_1 = 7.1856 \text{ μA} \cdot 0.83 \text{ M}\Omega = 5.9640 \text{ V}$$

Notice what happens to the metric prefixes in this calculation—they cancel out. Since micro– means millionths and mega– means millions, multiplying them eliminates the prefixes entirely, giving a voltage in volts.

Now again we ask what we can calculate next. We are working our way from left to right. As we move to the right of R_1 we have a junction and we also have R_2 nearby. Can we use Kirchhoff's junction law at the node? No, because although we know I_1, we don't know either of the other two currents. Can we make use of Kirchhoff's voltage drop law? Yes. In the first loop we know every voltage except V_2. Writing the voltage drop equation for this loop gives

$$12.0 \text{ V} = 5.9640 \text{ V} + V_2$$
$$V_2 = 12.0 \text{ V} - 5.9640 \text{ V} = 6.0360 \text{ V}$$

Now that we know V_2 we can calculate current I_2 using Ohm's law.

$$I_2 = \frac{V_2}{R_2} = \frac{6.0360 \text{ V}}{1.4 \text{ M}\Omega} = 4.3114 \text{ μA}$$

Still working our way to the right, we can either look at the node to the right of R_1 or we can simply note that R_3 is in parallel with R_2. (Observe that because of the position of R_1, neither R_2 nor R_3 is in parallel with the battery!) Since R_3 is in parallel with R_2, these two resistors have the same voltage drop. The simplest thing to do is the latter approach, but just for its instructional value, let's look at the node. We now know two out of the three currents at the node at the top of the circuit. Applying the junction law here, we have

$$I_1 = I_2 + I_3$$
$$I_3 = I_1 - I_2 = 7.1856 \ \mu A - 4.3114 \ \mu A = 2.8742 \ \mu A$$

Knowing I_3, we can use Ohm's law to determine V_3.

$$V_3 = I_3 R_3 = 2.8742 \ \mu A \cdot 2.1 \ M\Omega = 6.0358 \ V$$

You see that this value differs only a tiny bit from the value we have for V_2. And as you know by now, this is because of rounding and is the reason we are using four decimal places for these calculations.

Finally, we need to calculate the power consumed by each of the resistors.

$$P_{R1} = V_1 I_1 = 5.9640 \ V \cdot 7.1856 \ \mu A = 42.8549 \ \mu W$$
$$P_{R2} = V_2 I_2 = 6.0360 \ V \cdot 4.3114 \ \mu A = 26.0236 \ \mu W$$
$$P_{R3} = V_3 I_3 = 6.0358 \ V \cdot 2.8742 \ \mu A = 17.3481 \ \mu W$$

Again, since each of the currents is expressed in microamps, each of the powers is in microwatts.

Our first three examples involve fairly simple circuits with two or three resistors. In the final two examples we, will look at four-resistor circuits. These will be the most complex circuits we analyze in this course. Solving these circuits involves nothing but the repeated use of the three laws we have to work with, Ohm's law and Kirchhoff's two laws.

▼ Example 11.12

For our fourth example, we use the circuit shown in Figure 11.30. Just as a reminder, our R_{EQ} calculation for this circuit is performed using $k\Omega$ for resistance, so we need to continue to use the individual resistance values in $k\Omega$ in our calculations. From the R_{EQ} calculation in Example 11.6 above, we have $R_{EQ} = 7.8870 \ k\Omega$. Using this value, we calculate the first current, I_1, as usual.

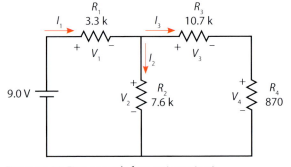

Figure 11.30. First example four-resistor circuit.

$$I_1 = \frac{V_B}{R_{EQ}} = \frac{9.0 \ V}{7.8870 \ k\Omega} = 1.1411 \ mA$$

With this current, we calculate the voltage drop V_1 using Ohm's law.

$$V_1 = I_1 R_1 = 1.1411 \ mA \cdot 3.3 \ k\Omega = 3.7656 \ V$$

Now we move again to the right and consider what we can do next. We cannot apply the junction rule to the node to the right of R_1 because we only know one of the three currents entering and exiting that node. However, we can apply the voltage drop law to the first loop because we know all the voltages in that loop except for V_2. Accordingly,

$$9.0 \text{ V} = V_1 + V_2$$
$$9.0 \text{ V} = 3.7656 \text{ V} + V_2$$
$$V_2 = 9.0 \text{ V} - 3.7656 \text{ V} = 5.2344 \text{ V}$$

Now that we know V_2, we use Ohm's law to calculate I_2.

$$I_2 = \frac{V_2}{R_2} = \frac{5.2344 \text{ V}}{7.6 \text{ k}\Omega} = 0.6887 \text{ mA}$$

Now we are getting there. We know two out of the three currents at the upper node and this allows us to use the junction rule to calculate I_3. Once we know that, we can use Ohm's law to calculate each of the last two voltage drops since I_3 goes through both R_3 and R_4. First, we determine I_3.

$$I_1 = I_2 + I_3$$
$$I_3 = I_1 - I_2 = 1.1411 \text{ mA} - 0.6887 \text{ mA} = 0.4524 \text{ mA}$$

Now we calculate V_3 and V_4.

$$V_3 = I_3 R_3 = 0.4524 \text{ mA} \cdot 10.7 \text{ k}\Omega = 4.8407 \text{ V}$$
$$V_4 = I_3 R_4 = 0.4524 \text{ mA} \cdot 0.870 \text{ k}\Omega = 0.3936 \text{ V}$$

As a final step, now that we have calculated all the voltages and currents, we can calculate the powers consumed by each of the four resistors.

$$P_{R1} = V_1 I_1 = 3.7656 \text{ V} \cdot 1.1411 \text{ mA} = 4.2969 \text{ mW}$$
$$P_{R2} = V_2 I_2 = 5.2344 \text{ V} \cdot 0.6887 \text{ mA} = 3.6049 \text{ mW}$$
$$P_{R3} = V_3 I_3 = 4.8407 \text{ V} \cdot 0.4524 \text{ mA} = 2.1899 \text{ mW}$$
$$P_{R4} = V_4 I_3 = 0.3936 \text{ V} \cdot 0.4524 \text{ mA} = 0.1781 \text{ mW}$$

▼ Example 11.13

Our final example uses the circuit shown in Figure 11.31. As with the previous example, our R_{EQ} value is in kilohms so we need to continue using the individual resistance values in kilohms in our calculations.

From the R_{EQ} calculation in Example 11.7 above, we have $R_{EQ} = 3.5324 \text{ k}\Omega$. Calculating I_1, we get

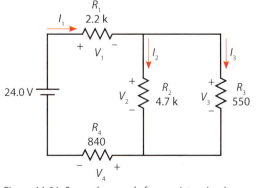

Figure 11.31. Second example four-resistor circuit.

$$I_1 = \frac{V_B}{R_{EQ}} = \frac{24.0 \text{ V}}{3.5324 \text{ k}\Omega} = 6.7942 \text{ mA}$$

Now that we have I_1, we see that we can calculate the voltage drop V_1 using Ohm's law.

$$V_1 = I_1 R_1 = 6.7942 \text{ mA} \cdot 2.2 \text{ k}\Omega = 14.9472 \text{ V}$$

Now looking to the right into the resistor network, we ask what we are able to calculate next. We do not have enough information to use the junction law at the node above R_2 because we only know one of the three currents entering and leaving that node. Likewise, we don't have enough information to use the voltage drop law because in the loop on the left containing the battery we only know two of the four voltage drops. It is beginning to look like we are out of luck. But wait! Notice that the current I_1 flows through three different devices in the left-hand branch. It obviously flows through the battery and R_1, but notice that the same current also flows through R_4! To be more specific, the current I_1 flows through R_1 and then splits into I_2 and I_3 at the upper node. These two currents flow through R_2 and R_3, respectively, and then join back together to become I_1 again at the node just to the right of R_4. This current I_1 then flows through R_4 and the battery and then is back where it started. Now that we have noticed this, we can use I_1 and R_4 to calculate V_4.

$$V_4 = I_1 R_4 = 6.7942 \text{ mA} \cdot 0.840 \text{ k}\Omega = 5.7071 \text{ V}$$

Now that we have V_4, we are able to use the voltage drop law around the left-hand loop in the circuit to determine V_2. Beginning below the battery and scanning clockwise around the loop, we have

$$24.0 \text{ V} = V_1 + V_2 + V_4$$
$$V_2 = 24.0 \text{ V} - V_1 - V_4 = 24.0 \text{ V} - 14.9472 \text{ V} - 5.7071 \text{ V} = 3.3457 \text{ V}$$

Now that we have V_2, the rest of this circuit is a snap. Notice that since R_2 and R_3 are in parallel, their voltage drops are the same and with these two voltages we can calculate I_2 and I_3. So,

$$V_2 = V_3 = 3.3457 \text{ V}$$
$$I_2 = \frac{V_2}{R_2} = \frac{3.3457 \text{ V}}{4.7 \text{ k}\Omega} = 0.7119 \text{ mA}$$
$$I_3 = \frac{V_3}{R_3} = \frac{3.3457 \text{ V}}{0.550 \text{ k}\Omega} = 6.0831 \text{ mA}$$

We have finished calculating all the voltages and currents. Now let's quickly calculate the powers consumed by each of the four resistors.

$$P_{R1} = V_1 I_1 = 14.9472 \text{ V} \cdot 6.7942 \text{ mA} = 101.5543 \text{ mW}$$
$$P_{R2} = V_2 I_2 = 3.3457 \text{ V} \cdot 0.7119 \text{ mA} = 2.3818 \text{ mW}$$
$$P_{R3} = V_3 I_3 = 3.3457 \text{ V} \cdot 6.0831 \text{ mA} = 20.3522 \text{ mW}$$
$$P_{R4} = V_4 I_1 = 5.7071 \text{ V} \cdot 6.7942 \text{ mA} = 38.7752 \text{ mW}$$

To conclude this example, let's verify that the conservation of energy is at work in this circuit. The power dissipated by each of the resistors in this circuit is the rate at which energy is being used. If energy is being conserved (which is it), then energy must be supplied to the circuit at the same rate. The power supplied to the circuit by the battery is equal to the battery voltage times the current flowing through it (I_1), or

$$P_B = V_B I_1 = 24.0 \text{ V} \cdot 6.7942 \text{ mA} = 163.0608 \text{ mW}$$

The total power being consumed by the four resistors is the sum of the four values calculated above, which is 163.0635 mW. The difference between these two values is entirely due to the rounding involved in the calculations of voltages, currents, and powers. But even with rounding errors, these two power values agree to six digits of precision, a clear demonstration of conservation of energy at work in electrical circuits.

Well, all those examples took us a while to work through. However, there is no point in studying DC circuits at all unless we are going to put in the effort necessary to actually master these calculations, and before you begin tackling the exercises, I wanted to make sure you had seen enough examples to get you prepared.

Now it's your turn. Many students find these problems difficult at first and feel like they can't get anywhere without a lot of help. But after you work a few of them, the light bulb eventually comes on in your head and you suddenly realize that you get it! After that you still need to work lots of problems to become proficient and actually reach the point of mastery. And after that, as with all kinds of problems, you need to work review problems regularly to develop long-term retention.

Enjoy the exercises, and remember that mastery is the goal. Don't give up until you are there!

Thomas Edison

George Westinghouse

Was there really a War of Currents?

In the late 19th century the *War of Currents* took place. (This sounds like a dumb science joke but that's what this industrial competition was and still is actually called.) The battle was between DC and AC electricity and the outcome determined which type of system became the eventual standard for power distribution in the United States and around the world. Needless to say, big money was at stake.

Thomas Edison, inventor of the incandescent light bulb, was the backer of the DC systems. Motors and generators that run on or produce DC electricity were easier to design, so Edison got an early lead. In 1889, the first DC distribution system in the United States was set up at Willamette Falls in Oregon City, Oregon. But the system was destroyed by a flood the next year. In 1892, Edison's company installed a DC system to light up the streets in lower Manhattan.

George Westinghouse was the big force behind the AC systems. He sought out designs from European and American engineers for motors and generators that would operate on AC, and he found them. By 1893, Westinghouse had been successful in getting AC generation systems installed at Niagara Falls, with the hope of providing for the industrial power needs of Buffalo, New York.

Who won? Well, as you know from reading this chapter, Westinghouse and AC won. The question is *why* AC won out over the DC system sponsored by the famous inventor of the light bulb. The answer has everything to do with the *transformer*, which you will learn about in the next chapter.

Chapter 11 Exercises

Introductory Circuit Calculations

For this set of exercises, use the basic single-resistor circuit shown in Figure 11.4. Use our ordinary significant digits rules in your answers for this set (not the special rule for multi-resistor circuits).

1. The current in a circuit is 13.00 A. The voltage powering the circuit is 25.00 V. Calculate the resistance in the circuit.

2. Determine the current that flows from a 24-V battery into a 250-Ω resistance. Express your answer in mA.

3. The resistance in a circuit is 12.20 kΩ. If the supply voltage is 4.500 V, calculate the current in the circuit. Express your answer in mA.

4. Calculate the supply voltage in a circuit if 0.0300 mA of current flows through 33.3 MΩ of resistance.

5. Calculate the power supplied by the power supply in problems 1, 2, 3, and 4.

6. An older-style standard light bulb is powered by a voltage of 120 V and consumes 60.00 W of power. Calculate the resistance of a standard light bulb and the current that flows in it.

7. Household electrical appliances operate at a voltage of 120 V and the maximum continuous current allowed to flow in household circuits is 12 A. Determine the maximum power that can be consumed by a household appliance (such as a hair dryer) and express your answer in kW.

8. A circuit in a pocket calculator draws 13.5 μA from a 6.0-V battery. Determine the power the calculator consumes and express your answer in μW.

9. A generator at a power station produces 155 MW of power at a voltage of 762 V. Determine the current the generator produces.

Answers

1. 1.923 Ω
2. 96 mA
3. 0.3689 mA
4. 999 V
5. a) 325.0 W; b) 2.3 W; c) 1.660 mW; d) 30.0 mW
6. $R = 240\ \Omega, I = 0.50$ A
7. 1.4 kW
8. 81 μW
9. 203,000 A

Equivalent Resistance Calculations

Determine R_{EQ} for these resistor networks. Use 4 decimal places in every calculation.

1.

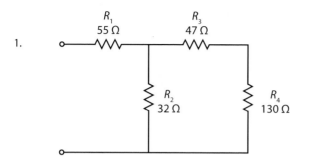

2.

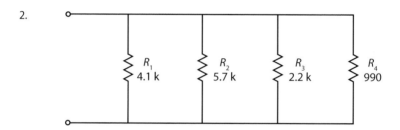

3.

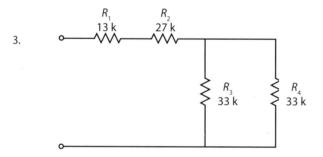

4.

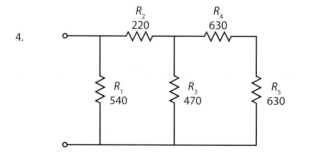

5.

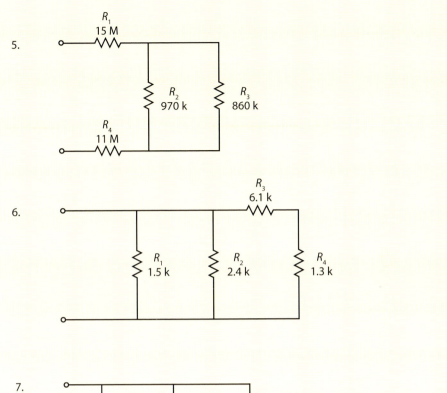

6.

7.

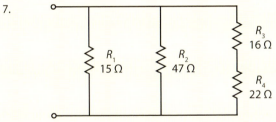

8.

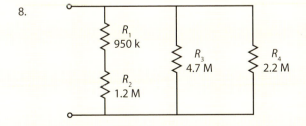

Answers

1. 82.1005 Ω
2. 0.5308 kΩ
3. 56.5000 kΩ
4. 275.4651 Ω
5. 26,455.8470 kΩ or 26.4558 MΩ
6. 0.8207 kΩ
7. 8.7520 Ω
8. 0.8831 MΩ

Multi-Resistor Circuit Calculations I

Calculate the voltage, current, or power as indicated. It is necessary to calculate R_{EQ} first for every problem except the second one. Use four decimal places.

1. Calculate the voltage across resistor R_1.

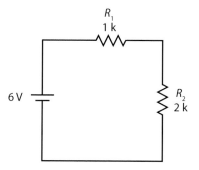

2. Calculate the current in resistor R_2.

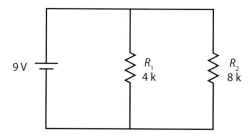

3. Calculate the power consumed by resistor R_3.

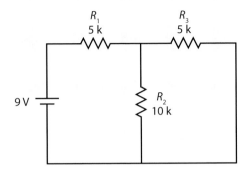

4. Calculate the voltage drop across resistor R_4.

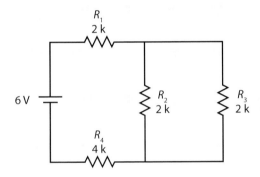

Answers

1. $V = 2.0000\ \text{V}$
2. $I = 1.1250\ \text{mA}$
3. $P = 2.5920\ \text{mW}$
4. $V = 3.4284\ \text{V}$

Multi-Resistor Circuit Calculations II

Use four decimal places in all your calculations. Your answer should match the given answer in at least the first three decimal places.

1. Compute the voltage and current for R_2.

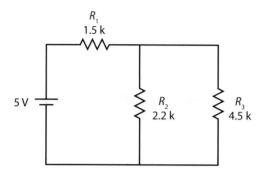

2. Compute the voltage across R_3.

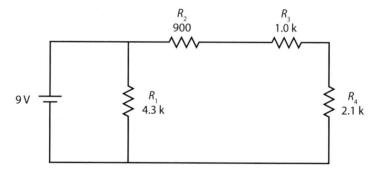

3. Compute the power consumed by R_4.

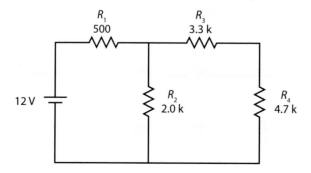

4. Compute the voltages and currents R_2 and R_3.

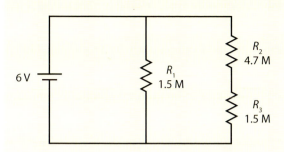

Answers

1. $R_{EQ} = 2.9776$ kΩ; For R_2, $V = 2.4812$ V and $I = 1.1278$ mA

2. $R_{EQ} = 2.0723$ kΩ; For R_3, $V = 2.2500$ V

3. $R_{EQ} = 2.1000$ kΩ; For R_4, $P = 6.1383$ mW to 6.1392 mW (depending on which power equation you use)

4. $R_{EQ} = 1.2078$ MΩ
    R_2: $V = 4.5482$ V $I = 0.9677$ µA
    R_3: $V = 1.4516$ V $I = 0.9677$ µA

Multi-Resistor Circuit Calculations III

Use four decimal places in all your calculations. Your answer should match the given answer in at least the first three decimal places.

1. Compute the voltage and current for R_3.

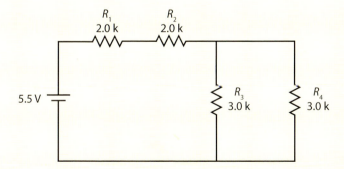

2. Compute the voltage, current and power for R_2.

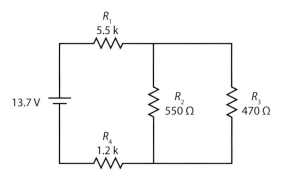

3. Compute the voltage, current, and power for R_2 and R_3.

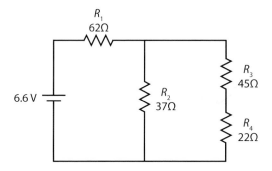

4. Compute the voltage, current, and power for R_2.

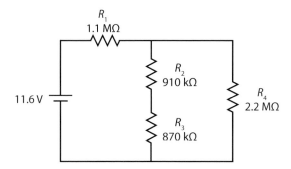

Answers

1. $R_{EQ} = 5.5$ kΩ; For R_3, $V = 1.5$ V and $I = 0.5$ mA
2. $R_{EQ} = 6.9534$ kΩ;
    For R_2, $V = 0.4989$ V, $I = 0.9071$ mA, $P = 0.4526$ mW
3. $R_{EQ} = 85.8365$ Ω
    For R_2, $V = 1.8322$ V, $I = 0.0495$ A, $P = 0.0907$ W
    For R_3, $V = 1.2330$ V, I $= 0.0274$ A, $P = 0.0338$ W
4. $R_{EQ} = 2.0839$ MΩ
    For R_2, $V = 2.8001$ V, $I = 3.0770$ µA, and $P = 8.6159$ µW

Multi-Resistor Circuit Calculations IV

Put away your solutions for Multi-Resistor Circuit Calculations II and do the entire set over again.

Multi-Resistor Circuit Calculations V

Put away your solutions for Multi-Resistor Circuit Calculations III and do the entire set over again.

Do You Know ...   What is an uninterruptible power supply?

There are many occasions when AC electrical power needs to be converted to DC or vice versa. All electronic devices run on DC, but the power we run them with comes to us as AC. So the AC we receive from the power company must be converted to DC inside the device. A device that converts AC to DC is called a *rectifier*.

There are also times when we need to convert from DC to AC. Some electrical systems are so important or sensitive that major trouble can be caused if a power outage disrupts power to the system. Examples are control systems at chemical plants, life support equipment in hospitals, and many computer systems. Sometimes back-up AC generators can be installed to provide back-up power. But a generator takes a few seconds to start up and many systems are so sensitive that they cannot tolerate a power outage of even a few seconds.

The easiest way to provide *uninterruptible* back-up power for an electronic system is with a battery. But a battery is a DC device, and our power systems are AC systems. The way they are connected together is with an *inverter*, a device that converts DC to AC. The inverter is placed in between the battery and the AC system it is backing up, and the battery is connected to the AC system through the inverter at all times. If the AC system fails, electrical power immediately flows from the battery through the inverter, and the systems using the power experience no interruption at all.

I am writing this book on a nice Mac computer, but I would not want to risk losing a bunch of data if the power to my office suddenly went out. So I have a small uninterruptible power supply (UPS) connected to my computer. If the power goes out, the UPS can run my computer for 15 minutes, giving me time to save files and shut down properly.

CHAPTER 12
Fields and Magnetism

Magnetic Levitation

A candle sitting on a magnet is levitated by the magnetic fields produced around a superconductor. The black disk is made of superconducting material, a material that has zero resistance to the flow of electricity. At present, all known superconducting materials exhibit superconductivity at −140°C or colder, which is why the black disk is shown sitting in a bath of liquid nitrogen (−196°C). The magnetic fields holding up the magnet are produced by the superconductor when the magnet is nearby.

OBJECTIVES

After studying this chapter and completing the exercises, students will be able to do each of the following tasks, using supporting terms and principles as necessary:

1. Explain what a field is.
2. Describe three major types of fields, the types of objects that cause each one, and the objects or phenomena that are affected by each one.
3. State Ampère's law.
4. State Faraday's law of magnetic induction.
5. Explain a key difference between the theories of gravitational attraction of Einstein and Newton.
6. Apply Ampère's law and Faraday's law of magnetic induction to given physical situations to determine what happens.
7. Explain the general principles behind the operation of generators and transformers.
8. Use the right-hand rule to determine the direction of the magnetic field around a wire or through a coil of wire.
9. Explain why transformers work with AC but not with DC.

12.1 Three Types of Fields

You have been hearing about fields all your life. I'm sure you know about the *gravitational field* around the earth, the *magnetic field* around a magnet, and maybe even *electric fields*. A force caused by the gravitational field around the earth pulls you toward the earth's center. The force on you is your weight. You have felt the effects of a magnetic field when holding the ends of two magnets near each other. You feel the resulting forces when the magnetic poles attract or repel each other. And you have probably seen the results of the electric field present when static electricity is around. Dry, clean hair that stands up when you brush it, synthetic clothes clinging together when removed from the dryer, a balloon that sticks to the wall after being rubbed, the leaves swinging out in the electroscope—these are all instances of the forces present due to the electric fields caused by the build up of electric charge we call static electricity.

My definition of a *field* consists of two parts. First, a field is a mathematical abstraction that describes a region in space that influences specific kinds of matter or radiation. Second, this influence—the result of a field being present—is a *force*. Different kinds of fields affect different kinds of things, but the effect is always a force on the thing.

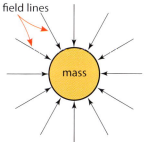
Figure 12.1. The shape of a gravitational field around a spherical mass.

The diagrams below help in visualizing what is going on when a field is present. The arrows in these diagrams are called *field lines* and represent the direction of the force on an object placed in the field, assuming that the object in question is the type of object that is affected by that type of field.

The gravitational field caused by an isolated mass is depicted in Figure 12.1. This mass could be a planet or the sun, or actually anything made of atoms. The gravitational field has a spherical shape around a spherical object. Remember that in both Newton's and Einstein's theories of gravity, everything with mass causes a

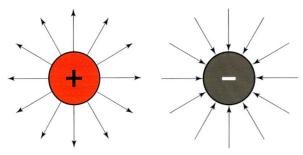

Figure 12.2. The electric fields surrounding positive and negative charges.

gravitational field, but gravity is such a weak force that the force isn't usually noticeable unless the object causing the field is very massive, like the earth, moon, or sun. So far as we know right now, there is only one kind of mass, so there is only one way to draw the gravitational field diagram. (As mentioned in the box on page 8, scientists now think there may be more than one kind of mass. If so, and we confirm it experimentally, that would be yet another fact that changed! Remember, *facts* can change; *truth* does not.)

There is an important difference between the gravitational theories of Einstein and Newton. In Newton's theory of universal gravitation, only an object with mass is affected by the gravitational field that exists around another object with mass. However, recall from Chapter 2 that Einstein's general theory of relativity treats gravity as a curvature in space-time rather than as a mysterious attraction between objects. As a result, Einstein was able to predict that starlight bends in its path as it passes near another star such as our sun, even though light does not have mass. As we saw, this prediction was confirmed by Sir Arthur Eddington's photographs during the solar eclipse of 1919 (shown on page 2). The fact that gravitational fields affect electromagnetic radiation is reflected in the summary table I present below.

Unlike mass, there are two kinds of electric charge. Electric charge is the source of electric fields, depicted in Figure 12.2. In this case, our convention is that the arrows represent the direction of the force on a positive "test charge" placed in the field. Since like charges repel, the arrows point away from the positive charge in the diagram on the left. Since opposite charges attract, the arrows point toward the negative charge on the right.

With both gravitational and electric fields, the object causing the field—a particle of mass or charge—can exist by itself. But with the magnetic field, illustrated in Figure 12.3, this is not the case (so far as we know). Magnetic fields are caused by magnets. A magnet always has two poles, which we call north and south. If you cut a magnet in half, you don't have separate north and south poles. Instead, you have two magnets, each with north and south poles. Physicists like to express this by saying, "there is no magnetic monopole." The result of this is that the field lines in a mag-

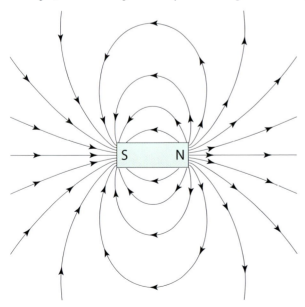

Figure 12.3. Magnetic field around a magnet.

	Gravitational Field	Electric Field	Magnetic Field
What causes this type of field?	mass	electric charge	magnets, current-carrying wires
What is affected by this type of field?	mass, electromagnetic radiation	electric charge	magnets, current-carrying wires, ferrous metals

Table 12.1. Causes of fields and what they affect.

netic field don't just go off into space as they do with gravitational and electric fields. Instead, they start on a north pole and land on a south pole of the same magnet.

When we think about these three different types of fields, it is helpful to know what causes each one and what kinds of things are affected by each one. Table 12.1 summarizes this information. This table is not exhaustive but it covers the basics. There are some other strange substances affected by magnetism (liquid oxygen, for example), but those I have listed will do for us. The term *ferrous* in the table means "made of iron." This term comes from the Latin word for iron, *ferrum*. This same Latin word gives iron its chemical symbol, Fe.

12.2   Laws of Magnetism

There are several important magnetic laws commonly studied in physics courses. We will address two of them. These two laws happen to be intimately related to the operation of the electrical devices that surround us in the modern world, so it is particularly interesting and helpful to know these two laws. They were named after their discoverers, André Ampère and Michael Faraday, whom we met a couple of chapters back. I will present the two laws in this section, and in the next section we will explore their important applications.

12.2.1 Ampère's Law

Ampère's law states:

> When current flows in a wire, a magnetic field is created around the wire and the strength of the magnetic field is directly proportional to the current. In a coil of wire, the magnetic field is proportional not only to the current but also to the number of loops, or *turns*, in the coil.

An important aspect of this law relates to what happens if a current-carrying wire is wound into a coil. As I explain a bit later, the magnetic field around such a coil of wire is magnified over what it is if the wire is not wound into a coil.

12.2.2 Faraday's Law of Magnetic Induction

Before discussing the next law we have to get a grip on the abstract concept of *magnetic flux*. As discussed in Section 12.1, we graphically show the presence of fields in space by drawing field lines indicating the forces that are present on an object placed in the field. In diagrams of this type, the closer together the field lines are, the stronger the field is. Near the magnet in Figure 12.3, the field lines are closer together, and as we all know, the magnetic field is stronger near the magnet than it is farther away.

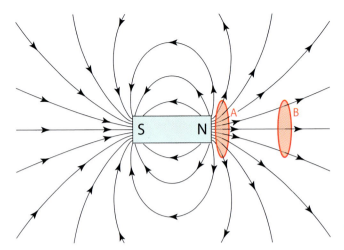

Figure 12.4. More magnetic flux is passing through loop A than through loop B.

Now consider Figure 12.4. Imagine that we place a coil of wire near the end of this magnet at position A and just hold it there. Does it seem reasonable to say that something is passing through this coil of wire? Well, whatever it is that appears to be passing through the coil, our word for it is *magnetic flux*. In fact, there is nothing physically passing through the coil and despite the suggestive sound of the word flux, nothing is flowing anywhere. What we do have passing through the coil is part of this mathematically abstract thing called the magnetic field, as indicated by the field lines. Now consider our little coil of wire being at position B instead of A. The difference, as suggested by the figure, is that there is less flux passing through the coil at position B than there is at position A. You can easily imagine that when our little coil of wire is very far away from the magnet, there is essentially no flux passing through it at all because the magnetic field out there is so weak.

Now we can state the next law, *Faraday's law of magnetic induction*, or Faraday's law, for short. Faraday's law is as follows:

If a changing magnetic flux is passing through a coil of wire, a current is induced in the coil. The strength of the induced current is proportional to the number of turns in the coil and to the rate at which the flux through the coil is increasing or decreasing.

Recall from our study of static electricity that induction comes from the word induce, which means "force to happen." Here the magnetic field interacting with this coil of wire is forcing a current "happen" in the coil. This principle is wonderful—we are able to create electrical current in wires without even touching them or connecting a power supply to them! Because of this principle, we can generate electrical power and transport it to our homes and factories. When this technology became possible in the 19th century, it utterly transformed society. Notice that as with Ampère's law, the strength of the effect depends on the number of turns in the coil.

It is critical for you to notice that current is only induced in a coil if the magnetic flux passing through the coil is *changing*, that is, increasing or decreasing. Holding the coil steady at location A in Figure 12.4 does nothing, even though a lot of flux is passing through the coil. But if you rotate the coil, for example, like a coin spinning on a table top, the flux passing through the coil will be highest when the coil is facing the magnet, and essentially zero when the coil is on edge to the magnet (because the flux is then passing by the coil without passing through it). Anything else you can imagine that causes the amount of flux passing through the coil to increase or decrease also induces current in the coil, but only as long as the change in flux is happening. These possibilities include moving the coil back and forth closer to and farther away from the magnet, moving the magnet away from the coil, or even somehow making the diameter of the coil increase or decrease.

12.2.3 The Right-Hand Rule

Our convention for determining the direction of the magnetic field lines around a current-carrying wire is called the *right-hand rule*, illustrated in Figure 12.5. According to the rule, if you grasp the wire with your right hand, with your thumb pointing in the direction of the flowing current, the direction your fingers point as they wrap around the wire is the circular direction the magnetic field lines point around the wire.

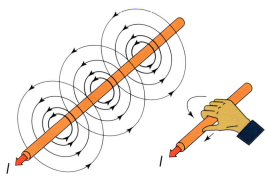

Now, an interesting thing happens if we form this wire into a coil, as illustrated in Figure 12.6. To help you visualize the way the conductor is wound, I made the front parts of the loops—those nearest to the viewer—black, and the back parts—those farthest away—charcoal gray. Applying the right-hand rule to the wire at various places, we establish the direction of the magnetic field at different places, as shown in the figure. Now

Figure 12.5. The magnetic field around a current-carrying wire. The direction of the field lines is given by the right-hand rule.

consider two different regions relative to this coil of wire, the region outside the coil (where the viewer is) and the region down through the center of the coil (where the long, straight, red field lines are). Notice, from the field lines making little closed loops, that outside the coil, both above it and below it, the field lines point to the left. Inside the coil, where the long, straight field lines are, the field lines all point to the right.

To get another view of this, imagine that we slice through the coil, cutting every through every loop simultaneously. Then imagine we throw away one half and look at the cut ends of

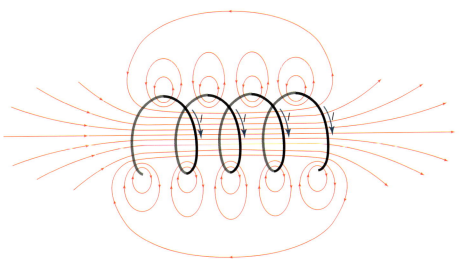

Figure 12.6. Magnetic field lines around a loosely wound coil carrying a current I.

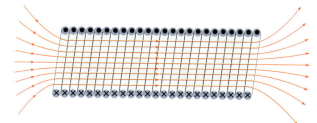

Figure 12.7. Magnetic field lines through a tightly wound coil.

the wire in the other half. In Figure 12.7, the black dots represent the cut ends of the wire where the current is coming out of the page; the black crosses represent the cut ends of the wire where the current is going into the page. Shaded in the background, you can see the remaining pieces of the turns connecting the black dots. What we have done with this coiled wire geometry is create an electric magnet, or *electromagnet*. As with a regular magnet, the electromagnet has its north pole on the end of the coil where the field lines come out (the right end) and its south pole on the end where the field lines enter the coil.

12.3   Magnetic Devices

12.3.1 Solenoids

In a coil of wire used as an electromagnet, the tighter the turns are wound the stronger the electromagnet is because no flux leaks out between the turns and circulates around the coil wires (as the flux is doing in the loosely wound coil of Figure 12.6). In a tightly wound coil, almost all the flux stays inside the coil as it passes straight down the center. (Remember, the word "passes" is a metaphor here; nothing really moves or flows anywhere.) A long, tightly wound coil of wire like this is called a *solenoid*. If a solenoid is wound onto a steel rod, called a *core*, the strength of the magnetic flux is increased considerably. So a solenoid is a powerful electromagnetic that is easily switched on and off.

By placing small moving parts made of ferrous metal near the end of the solenoid, the parts are made magnetically to move back and forth as the current to the solenoid is turned on and off. Thus, solenoids are devices that take an on-or-off electric current and use it to cause mechanical back and forth motion. It turns out that solenoids are hugely practical and hundreds of devices have been invented that use solenoids.

A common example of a solenoid consumers use every day is in the starting circuit in automobiles. When the driver turns the key to start the car, an electric motor begins spinning and turns the car engine just long enough for the ignition system to start up and begin running the car. The starter motor has a gear on it to turn the engine, but once the engine starts the starter motor with its little starter gear needs to get out

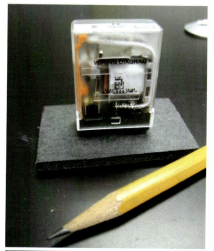

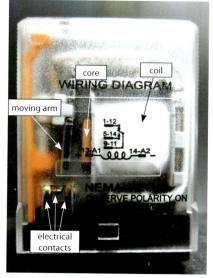

Figure 12.8. A relay (above) and a close-up showing the major internal parts (below).

of the way of the running engine or it will be torn to shreds. This is neatly managed by a solenoid that quickly moves the starter gear in one direction to engage the engine while starting the car and moves in the opposite direction when the key is released to get the starter gear out of the way once the car engine is running on its own.

Solenoids are especially useful in the world of industrial electric circuits because they function as electrically operated switches in many different applications. These switches, called *relays*, are used extensively in industrial control systems. A small control relay is shown in Figure 12.8. In the close-up of the lower photo, I have indicated the coil (covered in white tape), the solenoid core, the moving arm that is pulled back and forth as the coil is energized or de-energized, and the electrical contacts at the bottom of the moving arm that are opened and closed as the relay switches on and off.

12.3.2 Motors and Generators

While solenoids are little gadgets that a lot of people don't even know about, everyone knows about electric *motors*. Likewise, electric *generators* are massive devices that everyone knows about. Everyone knows that power stations are located all over the country generating electrical power to run the electrical devices in houses, factories, offices, and industrial plants. What you should know is that Faraday's law of magnetic induction is the simple principle that allows an electric generator to generate hundreds of megawatts of electrical power.

The operation of motors and generators is essentially the same. Electric motors are basically generators running backwards. Whereas a generator uses mechanical power from an engine to make electric current, a motor uses electric current to run a mechanical machine. Since these two devices are so similar, they are more or less the same on the inside. I will describe how a generator works, but keep in mind that an electric motor is essentially the same thing.

Consider Figure 12.9, which depicts the poles of a C-shaped magnet with a coil of wire in between them. There is a vertical axle attached to the coil of wire, which allows it to rotate. This axle is connected to an engine that rotates the coil. Because the north and south poles of the magnet are close together and facing each other on each side of the coil, there is a strong magnetic flux passing from the north pole to the south pole right through the space where the coil is mounted. In the figure, the field lines are shown pointing from the north pole of the magnet to the south pole of the magnet. Faraday's law of induction says that when the magnetic flux passing through a coil of wire is changing, a current is induced in the coil of wire. Also, as mentioned above, the amount of current induced depends on the number of turns of wire there are in the coil,

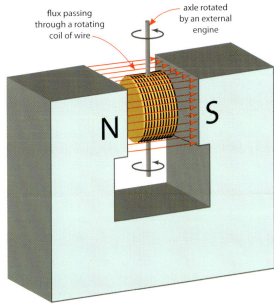

Figure 12.9. Rotating coil and magnetic poles inside a basic generator.

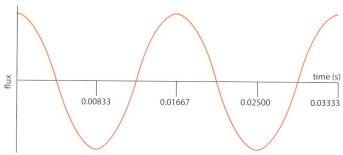

Figure 12.10. Flux through the rotating coil vs. time, at a 60 Hz rotation rate.

so the coil I have drawn in the figure has many turns. The position in which the coil is shown in the figure is the position where a maximum amount of flux passes through the coil. As the coil rotates from there, the flux passing through it decreases from this maximum value. The flux falls to zero, rises to maximum again but in the reverse direction (because the coil has rotated half way around), falls to zero again, and so on.

I have sketched what this rising and falling flux looks like in Figure 12.10. This curve has the same sinusoidal shape that we saw in Chapter 9 for graphically modeling waves. The sinusoidal variation of the flux through the coil produces a sinusoidally shaped current in the coil, which is exactly the shape of the voltage and current curves for the AC power distribution system. The curves look exactly like the wave curves we explored in Chapter 9 and they can be examined with the same mathematics. In America, the frequency of the AC voltage and current is 60 Hz. To produce this frequency of oscillating flux, the engine or turbine turns the coil so that it completes 60 rotations per second. In Europe, the frequency is 50 Hz. Since the frequency in America is 60 Hz (which I will write as 60.00 Hz), this means the period is

$$\tau = \frac{1}{f} = \frac{1}{60.00 \text{ Hz}} = 0.01667 \text{ s}$$

I have shown the time values in Figure 12.10 for the changing flux to complete one half cycle, one cycle, one and one half cycles, and two cycles.

In the sketch in Figure 12.9, I drew the magnet as if it were an ordinary permanent magnet. In practice, this works for a tiny demonstration generator, but using a permanent magnet large enough for an industrial generator is impractical. Instead, large coils of wire operating according to Ampère's law are used to create the magnetic field for the generator. These coils are called *field coils*. Part of the generator's own output is used to power these coils. Figure 12.11 is a photograph of a device that is designed in just this way. The photo is actually of a small motor, not of a generator, but as I said earlier, a motor is just a generator running backwards. Both devices depend on coils of wire to produce the stationary magnetic flux, and have rotating coils on a shaft inside the magnetic field. The

Figure 12.11. Inside look at a small motor showing the stator and rotor coils.

entire winding system that is fixed in place (the field coils) is called the *stator*. All the rotating coils and the frames they are wound on is called the *rotor*. In the photograph, you see that on the rotor there is more than just a single coil of wire. In fact, the motor shown has six coils, which is pretty common for a small motor like this one.

Before we move on, aren't you curious how the electric current gets in or out of the rotor while the rotor is spinning? I assumed you would be, so I took one more photo to show you a common way this is done. Figure 12.12 shows a close-up of a round device on the end of the rotor shaft called a *commutator*. Pressing against the commutator is

Figure 12.12. The brushes and commutator allow the current to get in (motor) or out (generator) of the spinning rotor coils.

a chunk of carbon called the brush, or, since there are usually two or more of them, the *brushes*. The wiring from outside the motor attaches to the brushes and a spring inside the brush holder keeps the brushes pressed against the commutator. The copper segments on the commutator are connected to the different coils on the rotor, so as the rotor turns different coils are connected one after the other through the brushes to the outside and are interacting each in turn with the magnetic field. Of course, the carbon bushes wear down because they are constantly pressing against the spinning commutator, so the brushes must eventually be replaced. If you own a car long enough, you may eventually have to replace the brushes inside the starter motor, as I once did.

12.3.3 Transformers

There is one more type of device, ubiquitous in our power distribution system, that you should know about and understand. These devices are called *transformers*, and they have nothing whatsoever to do with the movies popular a few years ago. Before I tell you about transformers, I want to describe the background that makes them necessary. The box at the end of Chapter 11 describes the "War of Currents" between the DC power distribution system envisioned by Thomas Edison and the AC system supported by George Westinghouse. Westinghouse's AC system won the war, and the reason is because of the advantage of being able to use transformers to reduce power losses in the wires of the distribution system.

In the circuits we studied in Chapter 11, we neglected the voltage drops in the wires. But in a large electrical distribution system, the long wires have some significant resistance, and thus there is a voltage drop in the wires. This voltage drop means that power is lost from the system due to heating in the wires caused by the wires' own resistance.

A typical large generator at a power station puts out 100 MW or more of power at a voltage of around 700 or 800 volts. The power station might be located 10 miles from a city and the electricity has to flow through wires to get there. The resistance of such wire is around 0.05 Ω per 1,000 ft. Well, I won't get tedious with the math, but the fact is that even with resistance in the wiring this low, we lose over 2/3 of our total power just heating up the wires as the current flows in the wires to get to the city! Obviously, losing 70,000,000 W of power after paying for the energy necessary to generate it is not a practical way to transport electricity.

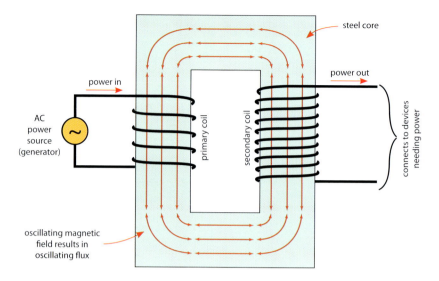

Figure 12.13. Primary and secondary coils in a transformer.

The problem is the current in the wires. Power losses are proportional to the *square* of the current and if we transport 100 MW at a voltage of 700 V, we have high currents. ($P = VI$, so $I = P/V$ = 100,000,000 W/700 V = 143,000 A!) Now, since $P = VI$, for a fixed amount of power, increasing the voltage decreases the current by the same factor. If, for example, we increase the voltage by a factor of 300, the current decreases by a factor of 300 and the power loss in the wires drops from 70% of the total power to less than 1%.

This is where transformers come in, which operate on the principles of Ampère's and Faraday's laws. A transformer is a simple device (no moving parts) that uses these laws to raise or lower an AC voltage while simultaneously lowering or raising the current. In Figure 12.13, we have a square ring, called the *core*, made of steel plates. Turns of wire are wound around the core on two sides. The coil where the electric power enters the transformer is called the *primary coil*. Electric power exits the transformer at the *secondary coil*. Here is the beautiful part: the ratio of the primary voltage to the secondary voltage is the same as the "turns ratio" of the transformer. The turns ratio is simply the ratio of the number of turns in the primary coil to the number of turns in the secondary coil. Whatever the ratio is of the number of turns in the primary to the number of turns in the secondary, the voltage is changed by this same ratio. If the voltage goes up, the current goes down, and vice versa. So we use transformers to boost the voltage way up at the power station, which drops the current way down, since the product stays the same. This drastically reduces power losses in the wires. Then when the wires get to the city where the electricity is to be used, we use transformers again to drop the voltage down to practical levels for people to use in their homes, offices, and factories.

Now let's take a look at the magnetic principles that make transformers work. Referring again to Figure 12.13, we apply an AC voltage to the primary coil. An AC voltage is sinusoidal, just like the sinusoidal flux depicted in Figure 12.10. A sinusoidal voltage like this causes a sinusoidal AC current to flow in the primary coil. According to Ampère's law, a magnetic flux is created inside the coil, just as with the solenoids we studied before. Here, however, the flux is not steady. Since the current in the primary coil is sinusoidal, the flux in the transformer core is also sinusoidal, reaching a peak, decreasing to zero, reversing direction to a negative peak, back to zero, over and over. The flux in the core is oscillating at the same frequency as the current in the primary coil.

Now, magnetic flux "flows" very easily inside of iron or steel, like our transformer core. So the flux created by the primary coil goes around the core, passing through the secondary coil. In the figure, I have drawn lines in the transformer core representing the magnetic field in the core. I put arrows on the field lines to indicate that the field is oscillating back and forth, clockwise then counter-clockwise, changing direction 120 times per second. So inside the transformer core the flux is continuously changing. This means that in the secondary coil, Faraday's law of magnetic induction kicks in. Since a changing magnetic flux is passing through the secondary coil, a current is induced in the secondary coil. This secondary current, which is also sinusoidal, flows out of the transformer to whatever electrical devices it is connected to.

Notice what happens if we connect a battery (DC voltage) instead of an AC voltage source to the transformer primary. Ampère's law still operates in the primary coil, magnetic flux is still created in the primary coil, and the flux still travels around the transformer core and passes through the secondary coil. But the flux is static or steady, that is, not changing. Thus, no current is induced in the secondary coil at all because Faraday's law requires the flux through a coil to be changing to induce current in the coil. Transformers do not work with DC currents and voltages.

Edison's DC distribution system had no way of reducing the heavy power losses that occur when power is transported in wires at useful voltages. Edison was generating DC power at around 100 V. This was viewed back then, as now, as a safe voltage to have around where people live and work. One of Edison's arguments against the AC system/transformer idea was that the high voltages used to transport the electrical power would be unsafe. Westinghouse argued that the high voltages could be made safe, and the fact that transformers could be used to reduce power losses down to manageable levels was unquestionable. The advantage of being able to transport power efficiently won the day and we have had AC power distribution systems everywhere in the world ever since. The transformer made it all possible.

You probably have no idea how many transformers you use and depend on every day. There is a transformer at your house or apartment that lowers the power distribution voltage from 10 or 15 kV down to the 240 V that goes into your house or apartment building. In neighborhoods where the power lines are underground, house transformers are often inside green metal boxes in the yards of the houses. If the power lines are above ground on poles, the house transformers are on the power poles in gray canisters. There are also huge transformers at the power stations to raise the voltage from around 700 V up to several hundred thousand volts for transport, and more huge transformers in substations at the destination communities to drop the voltage back down.

Nearly every piece of electronic equipment you use has a transformer in it. For portable devices such as mobile phones and laptop computers, the devices run on DC batteries. The transformers are in the chargers that recharge the batteries.

Figure 12.14 shows a couple of small transformers of the type used in electronic devices such as a stereo system or computer. In the close-up on the right, if you look carefully at the slots in the two white plastic pieces you see the windings of the two coils in the transformer. Wires are attached to the coils in various places and connected to the silver-colored terminals, and from there wires are connected to the rest of the electronics.

We are surrounded all the time by motors, generators, transformers, and relays, so I went into a fair amount of detail in this chapter describing how these electrical devices

Figure 12.14. Small transformers.

Do You Know ...

Where were the first transformers made?

As we have seen in this chapter and the last, the transformer was the device that enabled AC to win the War of Currents (see page 264). In the picture at left are shown two of the first transformers, designed in Budapest in 1885.

Below is a large contemporary power transformer such as one sees in electrical substations. The particular transformer shown increases the power transmission voltage from 110 kV to 380 kV for a transmission line running from Germany to the Czech Republic.

work. One of the things I enjoyed about getting my undergraduate degree in electrical engineering was that I then understood *how* they all work. If this kind of stuff stimulates your curiosity, then maybe you should consider getting an engineering degree, too. It was when I was 15 years old in 9th grade, perhaps just like you, that I decided to get an engineering degree myself!

Chapter 12 Exercises

Fields Study Questions

1. What kinds of things cause a gravitational field?
    a. Does the Holy Spirit have a gravitational field around Him?
    b. Does the archangel Gabriel have a gravitational field around him?
    c. Does an atom of nitrogen create a gravitational field?
    d. Does a neutron in an atom of nitrogen create a gravitational field?

2. What kinds of things are affected by a gravitational field?
    a. Is heat affected by a gravitational field?
    b. Is your soul affected by gravity?
    c. Is the broadcast signal from your favorite radio station affected by gravity?
    d. Is Beethoven's 9th Symphony affected by gravity?
    e. When you are listening to Beethoven's 9th Symphony, is the sound wave affected by gravity?

3. Describe circumstances under which your body is and is not affected by an electric field.

4. Which of these creates an electric field?
    a. An isolated electron.
    b. An ordinary atom.
    c. A Van de Graaff generator.
    d. A neutron.
    e. A basketball.
    f. Beethoven's 9th Symphony.
    g. A plasma. (A plasma is a gas made of ions, which are atoms that do not have equal numbers of protons and electrons.)

5. Which of these is affected by a magnetic field?
    a. A copper wire with no current flowing in it.
    b. A copper wire with current flowing in it.
    c. A block of cast iron.
    d. An aluminum window frame.
    e. A banana.
    f. The stainless steel screws used to repair a person's bones after a bad injury.
    g. A magnetic compass.
    h. A magnetic compass in a vacuum.
    i. A magnetic compass under water.

j. The magnetic compass in Jonah's pocket after he was swallowed by the fish.
k. The steel girders in a large office building.

Magnetism Study Questions

1. Does the current in a solenoid have to be changing in order for a magnetic field to be produced? Why or why not?

2. The figure at the right shows a solenoid with a current flowing in it due to the battery. Use the right-hand rule to determine the direction of the magnetic field inside of the solenoid. Using this information, identify which end of the solenoid (the left or right) is the magnetic north pole of this electromagnet.

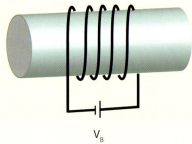

3. The figure below depicts a magnetic field produced by a solenoid and a coil of wire placed in the magnetic field. Using Faraday's law, briefly but clearly describe what occurs in the coil in each of the following circumstances:

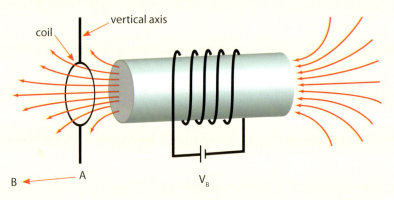

a. The coil is moving rapidly to the left from point A to point B.
b. The coil is stationary as shown but its diameter is rapidly getting smaller and smaller.
c. The coil and the solenoid are both stationary but the current in the solenoid is increasing.
d. The coil is rotating in place about a vertical axis, like a coin spinning on a table top.

4. In the preceding problem, what difference does it make if the coil is made with three turns of wire instead of one?

5. Use the laws of magnetism to write a few sentences explaining how a generator works.

6. Use Ampère's and Faraday's laws to write a few sentences explaining how a pow-

er transformer works.

7. Explain why a transformer works with AC current in the primary coil but not with DC current.

Do You Know ...   What is Nikola Tesla's claim to fame?

Because of the transformer, AC power systems are vastly more efficient than DC systems for long-distance power distribution. But without motors and generators that work with AC, AC would have been dead in the water. Enter Serbian-American Nikola Tesla.

Tesla was an incredibly brilliant inventor, the king of high-powered electrical experiments, sometimes even referred to as the proverbial mad scientist. You may have heard of Tesla from the movie *The Prestige* (2006), in which Tesla's research facility in Colorado Springs plays a role. Tesla did operate such a facility, for less than a year, in 1899. But that was years after his invention of the AC induction motor allowed George Westinghouse to capture the market for electrical power distribution.

Ironically, when Tesla first immigrated to the U.S. in 1884, he went to work for Thomas Edison! (See page 264 for why this is ironic.) But he soon set up shop for himself and developed the AC induction motor (below) and the AC generator (right). Tesla made a fortune selling the patents for these inventions to Westinghouse, but he spent it all on more experiments. He died penniless and in debt, in a New York hotel, where he had been living on a small stipend paid to him by Westinghouse.

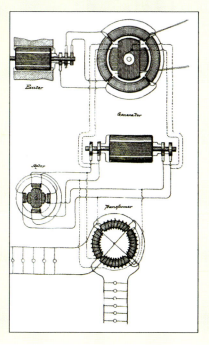

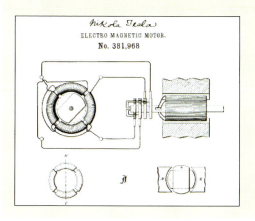

CHAPTER 13
Geometric Optics

Laser Optics

Laser weapons are bound to show up on the battlefield sooner or later. Soldiers will need to be equipped with protective eyewear that blocks laser light at all wavelengths while still allowing the wearer to see, a technology that does not exist today. Scientists like the one shown above use systems of lenses, mirrors, and lasers in a project to develop new materials that can do the job. (By the way, one cannot see laser beams in the air as this image suggests, unless there is fog or dust to reflect the light to our eyes, or to the camera. The light shining on the test components is realistic, but the yellow and red beams may have been Photoshopped in.)

OBJECTIVES

Memorize and learn how to use these equations:

$$\frac{1}{d_o} + \frac{1}{d_i} = \frac{1}{f} \qquad M = \frac{y_i}{y_o} = -\frac{d_i}{d_o}$$

After studying this chapter and completing the exercises, students will be able to do each of the following tasks, using supporting terms and principles as necessary:

1. Explain the difference between a real image and a virtual image.
2. Define the *focal length* of a lens.
3. Explain why a real image formed by a biconvex lens is inverted.
4. Describe, in general, the images formed by convex and concave mirrors and convex and concave lenses.
5. Use the lens equation to compute the locations of objects and images and to determine required focal lengths for spherical mirrors and lenses.
6. Use the magnification equation to determine magnification in an optical system and to determine object or image distances.
7. Describe the general operation of these five single-lens systems:
   a. eye
   b. camera
   c. photocopier
   d. projector
   e. magnifying glass
8. Describe the general operation of these three two-lens systems:
   a. corrective lenses (glasses)
   b. refracting telescope
   c. laser beam expander and collimator
9. Compute the new diameter of a laser beam after expansion and collimation.

In this last chapter, we look into the interesting subject of *geometric optics*. Back in your elementary school days, you may have studied a little about convex and concave lenses and you may have learned about how Galileo's telescope worked. We will cover those subjects and several others in this chapter. But the study of geometric optics gets conceptually difficult and mathematically crazy very fast. For this reason, we have to be cautious about how far we go into the subject.

Here's my plan for this chapter. First, we address the basic laws for convex and concave mirrors and lenses, including the math. We consider a few very common applications of mirrors and single-lens optical systems, including the human eye, the camera, the photocopier, and the magnifying glass. Then we take a brief look at two-lens systems, touching as lightly as possible on the operation of telescopes and one or two other applications.

13.1   Ray Optics

13.1.1 Light As Rays

In Chapter 9, we looked at light as waves of electromagnetic radiation. But many familiar phenomena lend themselves to the simpler analysis of treating light as rays. The study of geometric optics involves developing the properties of images formed by lenses and mirrors by treating light as thin rays, similar to the laser beam from a laser pointer. A *ray* is simply a straight arrow pointing in the direction the light is propagating. This type of analysis is called *ray optics*, or *geometric optics*. The goal is not in understanding the nature of light itself, but in discovering general principles pertaining to the behavior of light as it reflects and refracts in optical systems such as telescopes, microscopes, cameras, and other systems. We refer to this study as geometric optics because of the geometry used to construct the ray diagrams.

Recall from Chapter 9 that when light reflects it obeys the law of reflection. This law states that the angle of incidence equals the angle of reflection. The angles are measured between the light ray (incident or reflected) and the *normal line*, an imaginary line perpendicular to the reflecting surface.

Recall also from Chapter 9 that when a ray of light strikes the surface of a transparent material at an angle, the ray that enters the material refracts, that is, it slows down and changes direction at the surface of the new material.

13.1.2 Human Image Perception

Figure 13.1. Diverging rays are perceived to originate at the point where they appear to converge.

We are able to see objects because ambient light reflects off the objects and into our eyes. Rays of light reflecting off an irregularly-shaped object always diverge (spread out), and our visual perception is such that we subconsciously calculate where the rays appear to be diverging *from*. If we trace these diverging rays backward to their apparent source, that's where they come together, or converge. Wherever that is (where the light rays appear to originate), that is where our brain places the image we are perceiving, as indicated in Figure 13.1. This is the case regardless of whether the object forming the image is actually at that location. If rays are converging instead of diverging when they reach our eyes, then the object that is the source of the rays appears out of focus.

The conventions and methods of ray optics make use of the fact that we perceive images to be located where diverging rays of light appear to converge, when following them backwards. In our analysis, we refer to objects separately from the images of the objects. And to locate where the image of an object will be, we just follow the rays.

13.1.3 Flat Mirrors and Ray Diagrams

Figure 13.2 depicts an object (*O*) and the image (*I*) it forms in a flat mirror. Rays of light generally emanate from the object in all directions, but only three rays are shown in the figure to suggest the light rays coming from the object toward the mirror. At the location of the observer, the rays are diverging. But if we trace the rays back, as shown by the dashed lines in the figure, the place where they appear to converge is the location of the image we see. The distance of the object from the mirror or lens is the object distance, d_o, and the distance of the image from the mirror is the image distance, d_i.

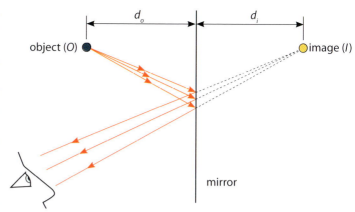

object (O)

image (I)

mirror

Figure 13.2. Image formation in a flat mirror.

The object shown in Figure 13.2 is small, and the light rays are all emerging from this small point. In geometric optics, we usually perform our analysis by considering the paths of rays from a single convenient point on an object. But the same principles apply to every other point on the object, so one of the ways to think about rays optics is that the image of an object could be constructed point by point, using ray optics on every point of the object to determine where the image of that point will be. This is a convenient way to think about what we are doing, but fortunately we don't have to actually track rays from every point on an object. Usually, it is sufficient to track two or three rays from the top of an object, or possibly the top and bottom, to determine where the image forms in an optical system.

Figure 13.3 shows the *ray diagram* representing the situation of the mirror in the previous illustration. In a ray diagram, it is customary to depict the object and image as vertical arrows located on the center line of the optical component (the mirror or lens). This center line is called the *optical axis*. It is also customary in ray diagrams of lenses to place the object on the left, with light rays directed from left to right as they pass through the lenses in the optical system. Ray diagrams allow us to see the relative sizes of the object and image, which indicates what, if any, magnification is involved. We can also see if the image is upright or inverted relative to the object.

There are a few more features of ray diagrams that we use all the time in geometric optics. First, when we look at an object through an optical system such as a telescope or a pair of glasses, we obviously center the object in the field of view. In a ray diagram, this means the object straddles the optical axis. But since lenses are symmetrical about the optical axis, we keep the analysis simple by considering only what happens to rays originating from the top of an object above the optical axis. For this reason, the object is typically represented by an arrow sitting atop the optical axis.

Second, notice that two rays from the object are shown in Figure 13.3. One ray is parallel to the optical axis; the other points toward the center of the optical component, where the optical axis passes through. A third ray is typically shown as well, aligned with the *focal point* of the optical component, which we address

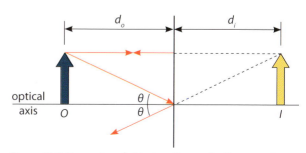

optical axis

O

θ

θ

I

Figure 13.3. Geometry of object and image for the case of a flat mirror.

below. These three rays are usually sufficient to figure out anything we need to know about how the optical system operates.

Finally, look again at Figure 13.3 as we identify a few things about the image formed. The ray directed toward the center hits the mirror and obeys the law of reflection, so the angles this ray and its reflection make with the optical axis are both labeled θ. Tracing the reflected ray back to where it intersects the horizontal reflection of the horizontal ray locates the image. From the geometry of the triangles in this diagram, we can say all of the following:

- The image is the same size as the object. Mathematically, this means the magnitude of the *magnification*, M, is $M = 1$. This is always the case with a flat mirror. We address magnification in more detail when we get to the math for spherical lenses and mirrors.

- The image is upright (not inverted). Mathematically, this means the sign of the magnification, M, is positive.

- The image is located on the other side of the mirror, the same distance from the mirror as the object. Mathematically, this means that $d_o = d_i$.

- The image is *virtual*, a concept we address next.

13.1.4 Real and Virtual Images

One of two possibilities arises for any given optical image. The first is when the light rays forming the image actually pass through the location where the image is. If this is the case, the image is said to be a *real image*. Real images can be seen and can be projected onto a screen.

The other possibility is that the light rays forming an image do not actually pass through the image location. In this case, we have a *virtual image*. Virtual images can be seen (by eye and camera) but they cannot be projected onto a screen. This is for the simple reason that a screen must be placed at the image location, but the light rays don't actually go there. Referring again to Figure 13.3, you see that since the light rays from the object reflect off the mirror, they do not pass through the image location. Thus, the image in a flat mirror is a virtual image.

An important clarifying point to make here is that to see an image your eye does not have to be at the image location. (In fact, it can't be—see page 310.) Your eye (or a camera lens) only has to be in the path of the light rays. If the light rays hit your eye, the image you see appears to be of a certain size and at a particular location, namely, those of the image. The size and location of the image are determined from the ray diagram.

13.2   Optics and Curved Mirrors

13.2.1 Concave and Convex Optics

Many common optical devices such as lenses and mirrors have surfaces that are either flat or spherically shaped. Flat surfaces are called *planar* surfaces. A spherical surface curving toward the object or image in front of it (depending on which way the component is facing) is a *concave* surface. A spherical surface curving away from the object or image is a *convex* surface. Lenses that are convex or concave on both sides (a common

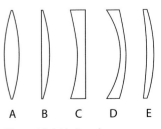

Figure 13.4. Various lenses: A) biconvex, B) plano-convex, C) plano-concave, D) negative meniscus, E) positive meniscus.

type) are called *biconvex* and *biconcave* lenses, respectively. Examples of several lens configurations are depicted in Figure 13.4.

When light rays traveling parallel to the optical axis of a spherical mirror or lens strike the surface of the optical component, they reflect or refract toward or away from a point on the optical axis called the *focal point*. The location of the focal point in the case of concave and convex mirrors is shown in Figures 13.5 and 13.6. The case of the biconvex lens is shown in Figure 13.7 The distance from the focal point to the surface of a mirror or to the center of a lens is called the *focal length*, denoted as f in the figures.

Now just as parallel rays striking a spherical surface are directed toward the focal point, rays emanating from a point source of light at the focal point reflect or refract to form parallel lines after hitting a spherical mirror or lens. For a concave mirror, simply look again at Figure 13.5 and imagine a tiny light source at the focal point. The light rays emerge in parallel lines from the mirror, just as in the figure except with the arrows pointing the opposite direction. The same thing works for lenses, similar to the ray diagram in Figure 13.7 but with the arrows pointing in the opposite direction.

The focal length is the most important parameter used in the selection of optical components, and we will use it repeatedly in describing the behavior of mirrors and lenses in the balance of this chapter.

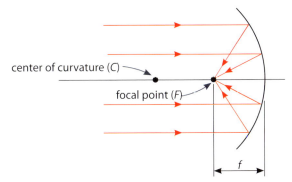

Figure 13.5. Focal point for a concave mirror.

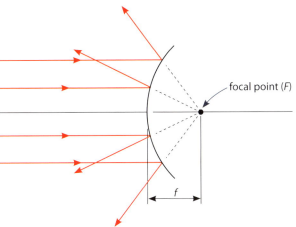

Figure 13.6. Focal point for a convex mirror.

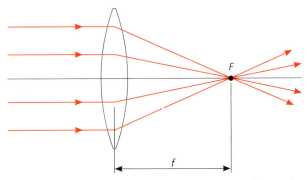

Figure 13.7. Parallel rays passing through a biconvex lens and converging to the focal point.

13.2.2 Approximations In Geometric Optics

There are several important approximations commonly used in an introductory study of geometric optics. First, the statement above about parallel rays being reflected toward the focal point is only approximately correct for spherical mirrors and lenses. It is exactly correct for a parabolic mirror, but light refracting through different points on a spheri-

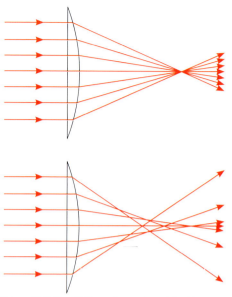

cal lens arrives at slightly different points along the optical axis, as illustrated by the exaggerated depiction in Figure 13.8. The deficiency in lens or mirror performance due to this inherent inaccuracy in spherical components is called *spherical aberration*. There are other lens designs that reduce spherical aberration, but these introduce trade-offs such as chromatic aberration, astigmatism, or the inability to focus properly except at the center of the image. In introductory optics courses and labs, we stick with spherical lenses and ignore spherical aberration.

For the second approximation, refer again to Figure 13.7 and notice that the rays are shown refracting in the center of the lens. They don't actually do this, of course. Refraction occurs at both lens surfaces: once as the rays pass from air into the lens, and again as the rays pass from the lens back into the air, as depicted in Figure 13.9. The second approximation customary in geometric optics is to use the so-called *thin-lens approximation* in which all the refraction is assumed to occur at a single place in the middle of the lens. For lenses that are thin relative to their diameter, the approximation works well and simplifies the analysis considerably.

Figure 13.8. Parallel rays through an ideal lens converge to a point (top). Spherical aberration is illustrated in the lower diagram (exaggerated).

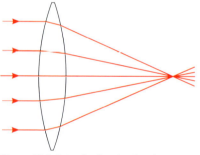

Figure 13.9. Actual refraction in a lens occurs at both lens surfaces.

And while we are on the subject of approximations, we might as well mention one more. The equations we present in the following sections for lenses and mirrors depend on the assumption that the angles the light rays make with the optical axis are small angles. With small diameter, long focal length lenses, this approximation works fairly well, and use of this approximation is customary in geometric optics (even though the angles may not look that small in the ray diagrams).

We are now ready to consider the relationships between objects and their images for concave and convex mirrors and lenses.

13.2.3 Spherical Mirrors

We first consider the image formed by a concave mirror, illustrated by the ray diagrams in Figures 13.10 and 13.11. Looking first at Figure 13.10, three rays are shown originating at the top of the object. As alluded to previously, light rays emanate from all over the object, in all directions. But the three rays shown allow us to locate the top of the image, and thus the location and orientation of the image.

The ray heading toward the center of the mirror's curvature, *C*, reflects straight back because the mirror is spherical. This ray is basically traversing the diameter of a circle, and bouncing right back through the center again. The ray heading toward the focal point reflects parallel to the optical axis, as we discussed previously. The ray initially parallel to the optical axis reflects toward the focal point. All three of these rays intersect at the location of the top of the image. With the image located, we see that it is smaller than the object and inverted. If you stand in front of this mirror and look at your reflection (that is, you are the object), this is what you actually see—a smaller (minified), inverted image of yourself. The image is also a real image because the rays forming the image pass through the image location.

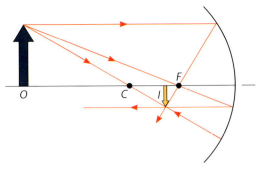

Figure 13.10. Ray diagram for a concave mirror, with the object relatively far away.

Figure 13.11 shows the same mirror, but with the object close to the mirror, to the right of the focal point. The same three rays are drawn: one in-line with the center that reflects off the mirror and goes through the center, one in-line with the focal point that reflects parallel to the optical axis, and one parallel to the axis that reflects toward the focal point. Here the intersection point, shown by the dashed lines, lies to the right of the mirror. Since none of the rays actually pass through this point, this is a virtual image. The image is also upright. And since it is larger than the object, the image is magnified.

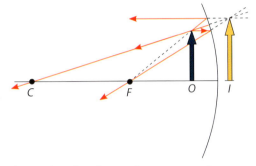

Figure 13.11. Ray diagram for a concave mirror, with the object close to the mirror.

This arrangement is used in those magnifying mirrors people use for applying cosmetics. The mirror has only a very slight curvature, so the center of the mirror is several feet away. The person looking in the mirror is quite close, much closer than the focal point. The result is an image that is larger than life and (fortunately) upright.

The convex mirror is shown in Figure 13.12. Again, the same three rays are shown originating at the top of the object: one in-line

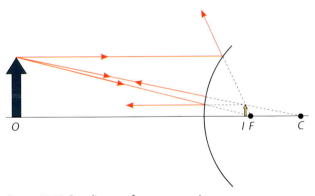

Figure 13.12. Ray diagram for a convex mirror.

with the center that reflects off the mirror and bounces straight back, one in-line with the focal point that reflects parallel to the optical axis, and one parallel to the axis that reflects in-line with the focal point. Extending these rays with dashed lines beyond the reflecting surface we find their intersection point, which is the top of the image. The image is upright, smaller than the object (minified), and virtual.

Convex mirrors are frequently seen in parking facilities and retail shops. Images in convex mirrors look small, but the mirror affords a wide-angle view useful for helping to see what's around the corner. Convex mirrors are also used for the side mirrors on the passenger side of cars, which explains why the mirrors have a warning label printed near the bottom of the mirror stating, "OBJECTS IN MIRROR ARE CLOSER THAN THEY APPEAR."

13.2.4 The Mirror Equation

In performing calculations with spherical mirrors, we use the distances from the object and image to the mirror surface, d_o and d_i, respectively. We also use the focal length, f, which is related to the radius of the mirror's curvature.

Using the approximation referred to previously—that the ray angles relative to the optical axis are small—it can be shown that the following *mirror equation* describes the location of the image in a spherical mirror:

$$\frac{1}{d_o} + \frac{1}{d_i} = \frac{1}{f}$$

This equation requires a convention for the algebraic signs used on the values for d_i and f. For real images, d_i is positive; for virtual images d_i is negative. The value of f is positive when the focal point and center are in front of the mirror (concave). The value of f is negative when the focal point and center are behind the mirror (convex).

The magnification of a mirror, M, is the ratio of the image height to the object height. If we designate the object and image heights to be y_o and y_i, respectively, M is calculated as

$$M = \frac{y_i}{y_o} = -\frac{d_i}{d_o}$$

This equation says that the ratio of image and object *distances* is the same as the ratio of image and object *sizes*. The sign convention for M is that if M is positive the image is upright. The sign conventions for spherical mirrors are summarized in Table 13.1.

Variable	Positive	Negative
d_i	real image	virtual image
f	concave mirror	convex mirror
M	upright image	inverted image

Table 13.1. Sign conventions for spherical mirrors.

▼ Example 13.1

A convex mirror with a focal length of 12 inches is used to monitor the aisles in a grocery store. If a person 65 inches tall stands 10 feet (120 inches) from this mirror, what kind of image is produced?

There is a long way and a short way to solve problems like this. The long way is to solve for the unknown variable first, before inserting any values into any equations. There is no doubt that this is always the best way to solve physics problems, and it is what I nearly always require my students to do. We will solve this first example this way. In the next example, we will use the short method.

We must first solve the mirror equation to find the image distance, d_i. This involves a good bit of algebra. We begin by multiplying both sides by $d_o d_i$ so that we can eliminate the denominators on the left.

$$\frac{1}{d_o} + \frac{1}{d_i} = \frac{1}{f}$$

$$\frac{d_o d_i}{d_o} + \frac{d_o d_i}{d_i} = \frac{d_o d_i}{f}$$

$$d_i + d_o = d_i \cdot \frac{d_o}{f}$$

Next, we get the two terms that have d_i in them together on the left side so we can factor out d_i.

$$d_i - d_i \cdot \frac{d_o}{f} = -d_o$$

$$d_i \left(1 - \frac{d_o}{f} \right) = -d_o$$

Finally, we divide both sides by the term in parentheses to isolate d_i.

$$d_i = \frac{-d_o}{\left(1 - \frac{d_o}{f} \right)} = \frac{d_o}{\left(\frac{d_o}{f} - 1 \right)}$$

Multiplying the numerator and denominator by f simplifies this expression.

$$d_i = \frac{d_o}{\left(\frac{d_o}{f} - 1 \right)} \cdot \frac{f}{f} = \frac{d_o f}{d_o - f}$$

Inserting values (with a negative focal length for a convex mirror), we find the image distance:

$$f = -12 \text{ in}$$

$$d_o = 120 \text{ in}$$

$$d_i = \frac{d_o f}{d_o - f} = \frac{120 \text{ in} \cdot (-12 \text{ in})}{120 \text{ in} - (-12 \text{ in})} = -\frac{120 \text{ in} \cdot 12 \text{ in}}{120 \text{ in} + 12 \text{ in}} = -10.9$$

This negative distance indicates the image is virtual. The magnification is

$$M = -\frac{d_i}{d_o} = -\frac{(-10.9)}{120} = 0.091$$

The positive magnification indicates the image is upright. The size of the image is just under 1/10th the size of the object. The size of the image is obtained by multiplying the object height by the magnification, or

$$M = \frac{y_i}{y_o}$$

$$y_i = y_o M = 65 \text{ in} \cdot 0.091 = 5.9 \text{ in}$$

13.3   Lenses

13.3.1 Light Through a Lens

For the rest of this chapter, we address optical systems using lenses. To me, this is much more interesting than mirrors because there is no comparison between a low-tech application like parking lot mirrors and the high-tech lens applications that are all around us today. (There are high-tech mirror applications, too, they just aren't as commonly known.)

We begin by looking at the biconvex lens (Figure 13.7). Because parallel rays converge at the focal point after passing through a biconvex lens, this lens is called a *converging lens*. The lens in a basic magnifying glass is a simple biconvex lens, as are the objective lenses in microscopes and telescopes.

Figure 13.13 depicts a biconvex lens with an object in front of it. Since light passes through the lens, we identify the focal point on both sides of the lens. Again, for rays parallel to the optical axis, the focal point is the place where they converge after passing through the lens. Three rays originating from the top of the object are drawn in the figure to help us locate the image. One ray is parallel to the optical axis. As with the reflections in spherical mirrors, this ray refracts toward the focal point of the lens. A second ray is drawn heading for the center of the lens and passing straight through. The third ray heads toward the focal point on the object side of the lens. When this ray hits the lens it refracts

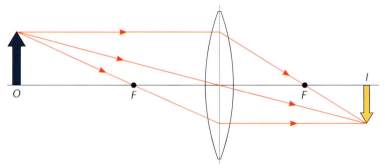

Figure 13.13. Ray diagram of a biconvex (converging) lens.

and continues parallel to the optical axis.

Now let's consider this question: how does this ray diagram of the biconvex lens help us understand how lenses work? In considering this question, consider the two observer viewing positions shown in Figure 13.14. In the diagram on the left, the observer position is between the lens and the image position. Here the rays approaching the observer's eye are not diverging from a common point as they are in the right-hand part of the figure. But these three rays are all originating from the same point, namely, the top of the object of the other side of the lens. This situation makes the object appear *out of focus*. When rays

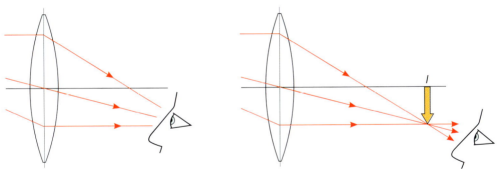

Figure 13.14. Seeing a focused image requires being farther away from the lens than the image distance.

that should be arriving at the eye from the same point appear to be coming from different points, the object appears out of focus.

In the right-hand part of the figure, the observer is farther away from the lens than the image location. Notice that now it doesn't matter how far the observer is away from the lens. At any position to the right of the image, the rays arriving at the eye are all emerging from the same point, so the observer at any location sees a focused image.

The image formed is real and inverted. The size and location of the image depend on the focal length of the lens and the object distance. We address the mathematics of this in the next section.

Try holding up a magnifying glass at arm's length and looking at an object that is about arm's length on the other side of the lens. The image you see is in focus and inverted. (Note that words on the spine of a book are still readable, but from the reverse direction.) Now draw the lens toward your eye. The image distances and object distances are both changing, but initially you continue to see a focused, inverted image. When the lens gets close enough to your eye so that your eye is about at the focal point, the image goes out of focus and becomes upright as your eye passes through the focal point. (You can probably amuse yourself for quite a while with this experiment if you actually try it.)

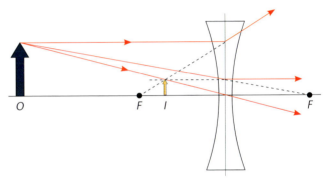

Figure 13.15. Ray diagram of a biconcave (diverging) lens.

Next consider the ray diagram of a biconcave lens, depicted in Figure 13.15. This lens is called a *diverging lens*, because rays parallel to the optical axis are refracted away from the axis and diverge. This is the case with the upper ray in the figure. Again, to help us locate the image,

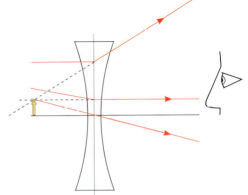

Figure 13.16. Any viewing position results in a focused image because the rays always appear to originate at the top of the image.

301

we follow three specific rays from the top of the object. The first ray is parallel to the optical axis and refracts to be in-line with the focal point. The second ray is directed at the focal point on the other side of the lens and refracts to be parallel to the optical axis. The third ray goes straight through the center of the lens and is unaffected. As with all the optics we have seen so far, these three rays intersect at the top of the image. The image is virtual and upright.

Once again let's consider what this diagram tells us about how the lens behaves. As shown in Figure 13.16, no matter where the observer is the rays coming through the lens all appear to be originating from the top of the image. This means the image is in focus regardless of where the observer is.

13.3.2 Single-Lens Applications

We now consider the ray diagrams for several common single-lens applications. We begin with a brief look at the eye. The following four applications describe how a single biconvex lens works with an object in four distinct locations, from far to near.

The Eye   Refraction of light rays entering the eye occurs both at the cornea and in the lens. Together, they act like a biconvex lens with focal lengths of 15.6 mm in front of the eye and 24.3 mm inside the eye. As shown

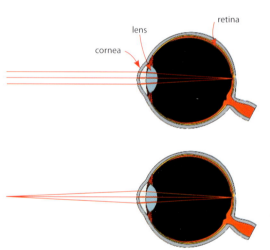

in the upper part of Figure 13.17, the rays from objects viewed at a distance enter the eye essentially parallel to the optical axis and are focused on the retina at the back of the eye. When viewing distant objects like this, the lens is relaxed. As you might expect from our discussion of the biconvex lens, the image on the retina is inverted, but happily our wonderful brains compensate so that we perceive the image to be upright.

To view near objects, as in the lower part of the figure, the muscles at the edges of the lens cause the lens to change shape in order to direct the light rays to a point at the retina. This action of the lens is called *accommodation* and it changes the focal length of the eye's lens. The farthest point on which a person's eye can focus is called the *far point* and the nearest point on which the eye can focus is called the *near point*. With

Figure 13.17. The eye focusing light on the retina from a distant spot (top) and a near spot (bottom).

age, the lens becomes less flexible and cannot provide the accommodation necessary to focus on objects outside of a narrower range of distances. This is where corrective lenses (glasses or contacts) come in. A lens placed in front of the eye forms a multiple-lens system, which we will consider later.

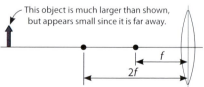

The Camera   Objects far away from a biconvex lens, as depicted in Figure 13.18, produce a minified,

Figure 13.18. The camera records small images of large objects, requiring the object to be much farther away than 2f from the lens.

inverted image. This is the arrangement used by a basic camera, and the ray diagram is the one shown in Figure 13.13. The image location is at the film or CCD (charge-coupled device) sensor where the image is to be recorded. To focus the camera so that the image is at the film or CCD sensor, the lens is moved in or out by adjusting the camera focus. This changes both the object and image distances until the image is focused at the proper location.

The Photocopier The job of a photocopier is to make a life-sized copy. The original placed on the machine is the object and the paper the copy is being made on is where the image falls. With a life-sized copy, the object and image are the same height. This occurs when both the object and image are a distance 2f from the lens, as illustrated in Figure 13.19. This is the symmetry point for a biconvex lens, where the object and image are the same size. The ray diagram is similar to Figure 13.13, but with the object and image each at a distance 2f from the lens.

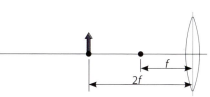

Figure 13.19. The photocopier records life-sized images of objects, requiring the object to be a distance 2f from the lens.

The Projector If the object distance is in between 2f and f as in Figure 13.20, the inverted image is larger than the object. This is how a projector works. The ray diagram for this arrangement is shown in Figure 13.21. The object is the film or slide, or with digital projectors, a transparent LCD (liquid crystal display) device that acts as a pixelated color filter to form an image generated by a computer. As with the previous applications, the image is inverted. This is accommodated by making the film, slide, or LCD device

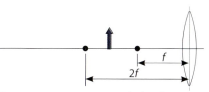

Figure 13.20. A projector works by placing the object in between the 2f and f positions to create a large image.

upside down in the projector so that the projected image comes out right side up. The light source itself is placed to the left of the object and the light shines through the object and the lens to create the projected image. The display screen is placed at the image location. As with a camera, the lens position is moved back and forth to adjust the focus.

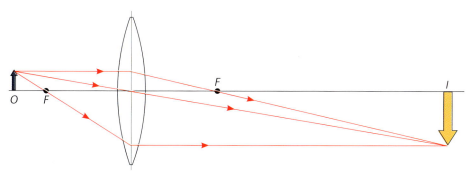

Figure 13.21. In a projector, the object is in between the 2f and f points on the optical axis, producing a magnified image.

The Magnifying Glass To use a biconvex lens as a magnifying glass, the object distance is less than the focal length of the lens, as shown in Figure 13.22. The purpose of a magnifying glass is to magnify small objects and produce an image that is upright. (Inverted images are fine for cameras and photocopiers, but it's hard for us to think straight about what we are looking at if the image is inverted. Scientists working with micro-scopes

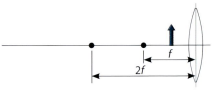

Figure 13.22. An upright, magnified image occurs when the object distance is less than f.

and telescopes must get used to this.) The ray diagram illustrating how this is accomplished is shown in Figure 13.23. When the small object is placed in between the focal point and the lens, we get a magnified, upright image, as the dashed lines indicate. Notice that on the right side of the lens (where the viewer is), the three rays from the top of the object all appear to converge at the top of the image. The image is virtual, which means you can take a picture of what you see in a magnifying glass but you cannot project it onto a screen because the light rays do not actually go to where the image is.

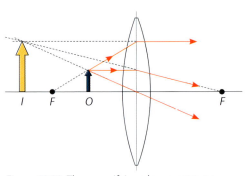

Figure 13.23. The magnifying glass creates an upright, magnified, virtual image when the object is between the focal point and the lens.

13.3.3 The Lens Equation

As we did with mirrors, we now address the math for these lenses. As before, we use d_o, d_i, and f to represent the object and image distances and the focal length. Here, however, distances are measured to the center of the lens, rather than to the surface. It may amaze you to learn that *the lens equation is exactly the same as the mirror equation*,

$$\frac{1}{d_o} + \frac{1}{d_i} = \frac{1}{f}$$

You may also be amazed to learn that the magnification equation for lenses is also the same as for mirrors,

$$M = \frac{y_i}{y_o} = -\frac{d_i}{d_o}$$

The sign convention used for these equations is that d_i and f are both negative for the case when a diverging lens produces a virtual image. The magnification, M, is positive when the image is upright and negative when the image is inverted. These conventions are summarized in Table 13.2. Now we will run through several examples to show how the lens and magnification equations are applied.

Variable	Positive	Negative
d_o	real object (in front of lens)	virtual object (behind lens)
d_i	real image (behind lens)	virtual image (in front of lens)
f	converging lens	diverging lens
M	upright image	inverted image

Table 13.2. Sign conventions for spherical lenses.

▼ Example 13.2

A simple digital camera design uses a biconvex lens to focus the light from objects onto the CCD sensor in the camera for recording. The camera is designed to capture an image of a person 7.0 feet away and focus the image on the sensor in the camera. The sensor inside the camera is 1.75 cm from the lens. Determine the required focal length for the lens and the magnification provided by this optical system. State the focal length in centimeters.

As promised, we will use the short method to solve this problem, even though as a teacher I think the long way is better for all of us. But I know sometimes students think our procedures are unbearably tedious, so this time we will take the short cut.

In this problem we have some unit conversions to perform, so we begin by writing down the given distances and converting the units. The MKS units for distance are meters, of course, but since the answer is required in centimeters, we might as well do this problem in centimeters from the start. The lens equation always works fine as long as d_o, d_i, and f have the same units.

$$d_o = 7.0 \text{ ft} \cdot \frac{0.3048 \text{ m}}{\text{ft}} \cdot \frac{100 \text{ cm}}{\text{m}} = 213 \text{ cm}$$
$$d_i = 1.75 \text{ cm}$$
$$f = ?$$
$$M = ?$$

Now we write down the lens equation, insert the values of d_o and d_i right away, and let our calculators do some work for us. In order to assure that this calculation is mathematically legitimate, I show the units as separated out from the fractions. We can only add terms together if they have the same units of measure and separating the units out shows this to be the case.

$$\frac{1}{d_o} + \frac{1}{d_i} = \frac{1}{f}$$

$$\frac{1}{213 \text{ cm}} + \frac{1}{1.75 \text{ cm}} = \frac{1}{f}$$

$$\frac{1}{213} \cdot \frac{1}{\text{cm}} + \frac{1}{1.75} \cdot \frac{1}{\text{cm}} = \frac{1}{f}$$

$$\left(\frac{1}{213} + \frac{1}{1.75} \right) \cdot \frac{1}{\text{cm}} = \frac{1}{f}$$

$$0.576 \cdot \frac{1}{\text{cm}} = \frac{1}{f}$$

Now we just have to take the reciprocal of both sides to get the focal length in centimeters.

$$0.576 \cdot \frac{1}{\text{cm}} = \frac{1}{f}$$

$$f = \frac{1}{0.576} \text{ cm} = 1.7 \text{ cm}$$

To get the magnification, we use the magnification equation.

$$M = -\frac{d_i}{d_o} = -\frac{1.75 \text{ cm}}{213 \text{ cm}} = -0.0082$$

The negative sign indicates that the image is inverted, which we already knew since the system uses a converging lens.

 Example 13.3

The camera in the previous example might have a CCD sensor in it that is 1.0 cm wide. If the camera is used to take a picture of a child 3.0 ft tall, determine if the image fits on the CCD sensor.

The magnitude of M indicates that the image is 0.0082 times the size of the object. For an object 3.0 ft high, the height of the image, in centimeters, is

$$3.0 \text{ ft} \cdot 0.0082 = 0.0246 \text{ ft} \cdot \frac{0.3048 \text{ m}}{\text{ft}} \cdot \frac{100 \text{ cm}}{\text{m}} = 0.75 \text{ cm}$$

This is smaller than the width of the sensor, so the image fits.

In the next example, we look at how the lens equation and the magnification equation are used together in a two-part problem to determine the type of lens that must be purchased for a particular application.

▼ Example 13.4

An entertainment company wants to use a digital projector to project live images of the band during concerts. The LCD device, which serves as the object for the optical system, is 2.00 inches tall and mounted 7.5 inches from the projector lens. The show producers want the image on the screen to be 15 feet tall. Determine the focal length required for the projector lens.

We note first that we have both the object height and the image height, so we can calculate the magnification. We also know that the magnification is negative since from Figure 13.21 and Table 13.2 the image is inverted in a projection system. We will use inches for the computations in this problem. Designating the object height as y_o and the image height as y_i, we have

$$y_o = 2.00 \text{ in}$$

$$y_i = 15 \text{ ft} \cdot \frac{12 \text{ in}}{\text{ft}} = 180 \text{ in}$$

Thus, the magnification is

$$M = -\frac{y_i}{y_o} = -\frac{180 \text{ in}}{2.0 \text{ in}} = -90.0$$

Knowing M, we now calculate the image distance because $M = -\dfrac{d_i}{d_o}$. So,

$$M = -\frac{d_i}{d_o}$$

$$d_i = -d_o \cdot M = -(7.5 \text{ in}) \cdot (-90.0) = 675 \text{ in}$$

This value for d_i is equivalent to 56.25 ft, so that is how far the projector lens must be from the screen. To calculate f, we use the lens equation (with short-cut math to make this easy):

$$\frac{1}{f} = \frac{1}{d_o} + \frac{1}{d_i}$$

$$f = \frac{1}{\dfrac{1}{d_o} + \dfrac{1}{d_i}} = \frac{1}{\dfrac{1}{7.5 \text{ in}} + \dfrac{1}{675 \text{ in}}} = 7.4 \text{ in}$$

Notice that our answer makes sense in light of what we know about where the lens must be placed in a projector system. As shown in Figure 13.20, the object (the LCD device) must be placed at a distance between f and $2f$ from the lens. Our object distance is, in fact, just a bit more than the focal length.

▲

13.3.4 Multiple-Lens Systems

Systems making use of more than one lens are used in many different types of optical systems today. You may know that Galileo's telescope made use of two lenses. We will look at his arrangement shortly. A camera with a zoom lens is another example of a multiple-lens system.

In this section we examine a few applications of two-lens systems. The basic idea for two-lens systems is that *the image produced by the first lens becomes the object for the second lens*. Moreover, the total magnification of the system, M_T, is the product of the magnifications of the individual lens magnifications M_1 and M_2, or

$$M_T = M_1 \cdot M_2$$

Multiple-lens calculations is quite complicated, so we will not get much into it. Our goal in this last section is limited to reviewing the ray diagrams for a few well-known applications just so you can have a feel for how the optical systems work.

Corrective Lenses When people suffer from nearsightedness, the image produced by the eye falls in front of the retina, as shown in the upper part of Figure 13.24. The lenses in glasses use meniscus lenses, as shown in Figure 13.4 (D) and (E). To correct nearsightedness, a lens is used that is thicker at the edges than in the center (a diverging lens). As shown in the lower part of Figure 13.24, when the lens is placed in front of the eye a virtual image is formed that is close enough to the eye so that the eye can focus on it.

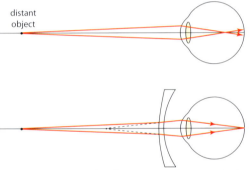

distant
object

Figure 13.24. Nearsightedness causes the image in the eye to be focused in front of the retina. Correction of the image distance is enabled by placing a negative meniscus lens in front of the eye.

To correct farsightedness requires a meniscus lens that is thicker in the middle than at the edges, as shown in Figure 13.4 (E). Farsightedness causes the image to be placed behind the retina (outside the eyeball). The corrective lens again produces a virtual image that is far enough away from the eye for the eye to focus on it.

The Refracting Telescope Figure 13.25 shows the ray diagram for a refracting telescope using two biconvex lenses. This arrangement is equivalent to placing a camera and a magnifying glass one after the other. The two lenses in the telescope are called the objective and the eyepiece. The objective works like a camera lens. It produces an intermediate image that is minified and inverted. The intermediate image is located in between the focal point for the eyepiece and the eyepiece itself, which means the eyepiece works as a magnifying glass, or magnifier. As we saw before, the magnifier produces an image that is upright (meaning, in this case, that the final image remains inverted, since it was already inverted when the magnifier took over), virtual, and magnified. The viewer focuses the image by moving the eyepiece. To make the telescope work effectively, the objective must have a long focal length, and the eyepiece must have a short focal length.

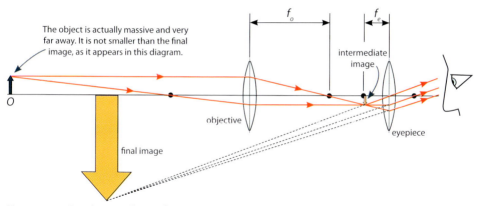

The object is actually massive and very far away. It is not smaller than the final image, as it appears in this diagram.

Figure 13.25. Ray diagram for a refracting telescope.

One could say that the role of the objective is simply to create an image that can be viewed up close with the magnifier (eyepiece). This intermediate image must fit inside the telescope tube, which makes it obvious that the image must be greatly minified. (An image of the moon or a star that is not greatly minified does not fit in there.)

This lens arrangement is called a *Keplerian telescope*, since Johannes Kepler used two biconvex lenses this way in his own telescope. The *Galilean telescope*, so-named for the obvious reason that Galileo used one, uses a concave lens in place of one of the convex lenses in an arrangement similar to the laser beam expander discussed next.

A Laser Beam Expander and Collimator   Our final application is a two-lens system for expanding and collimating a laser beam. We need to begin by explaining *collimation*.

The light in a laser is very close to being collimated, which means that the diameter of the beam remains constant as the beam propagates. In fact, a perfectly collimated beam of light is impossible; all beams of light spread out, or diverge, as they propagate. But with a laser, the angle of divergence, shown as θ in Figure 13.26, is very small—about 0.05° is typical. But even though all lasers do have an angle of divergence, for lab work the beam diameter is essentially constant over a short distance, so we speak of laser beams as collimated.

Figure 13.26. The angle of divergence in a laser beam.

In many applications, a laser beam needs to be expanded to a larger diameter, and then re-collimated after the expansion. This can be done with a Keplerian telescope, allowing the light to move through the telescope in the reverse direction, and placing the focal points of the two lenses at the same place. Another way to do it is with a Galilean telescope. This arrangement is shown in Figure 13.27. In the figure, the light moves through the lenses in the direction opposite that used for a telescope.

In a beam expander, the laser beam enters through a concave lens, L_1 (or, usually, a plano-concave lens, as shown). Since the incoming light rays are collimated, they are parallel, and thus the concave lens refracts all the rays so they diverge in a direction in line with the L_1 focal point. This point becomes the ob-

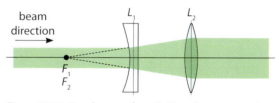

Figure 13.27. Ray diagram for a Galilean beam expander.

ject for the biconvex lens, L_2, which has its focal point in the same location as the concave lens focal point. Since all the rays are diverging from the same point, and this point is at the focal point, the biconvex lens collimates them, refracting them so they are parallel to the optical axis.

For this arrangement to work, the two lenses must obviously have different focal lengths since the focal points are in the same place. As long as the focal lengths are long so that we are dealing with small angles, the math of this arrangement works out to give a magnification for the expander equal to the ratio of the focal lengths. Designating the focal lengths for lenses L_1 and L_2 as f_1 and f_2, respectively, the magnification is

$$M = \frac{f_2}{f_1}$$

This same equation works for both Keplerian and Galilean beam expanders.

As mentioned before, the lens arrangement in Figure 13.27 was used by Galileo in his telescope. As a telescope, the lens arrangement shown is reversed so that the light from a distant object enters through the biconvex lens. Just as with the Keplerian telescope shown in Figure 13.25, this produces a magnified image of a distant object.

▼ Example 13.5

A laser beam with a diameter of 0.84 mm is expanded with a Keplerian beam expander. Determine the new diameter of a laser beam if the lens focal lengths are 100.0 mm and 1000.0 mm.

Using the magnification equation above,

$$M = \frac{1000.0 \text{ mm}}{100.0 \text{ mm}} = 10.0$$

Multiplying the factor by the original beam diameter gives the new diameter.

$$0.84 \text{ mm} \cdot 10.0 = 8.4 \text{ mm}$$

13.3.5 Imaging with The Eye

Early in the chapter, I mentioned that to see an image your eye needs to be in the path of the rays but that it cannot be at the location of the image formed by a lens. The reason is that the eye itself is a lens. As with all of the multiple-lens systems we have reviewed, when two lenses are used together, the image of the first lens becomes the object for the second lens. So for the eye to focus on the image from a lens and produce its own image on the retina, the image from the first lens needs to be between the eye's near point and far point (see page 302). The distance from this image to the eye becomes the object distance for the cornea/lens lens system in the front of the eye.

Do You Know ...   *How are rainbows formed?*

Rainbows form when water droplets in the air act as spherical lenses. The effect depends on reflection, refraction, and lens geometry. It also depends on a phenomenon called *dispersion*. A close look at what happens to a refracting ray of light reveals that different wavelengths of light—meaning different colors—are refracted by slightly different amounts. This means that as white light enters a prism, the different colors spread apart. This causes the familiar rainbow produced by a prism or piece of cut glass. The same thing happens when sunlight enters a droplet of water.

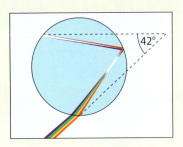

During or after a rain storm, the atmosphere is chock full of tiny, spherical water droplets. As depicted in the first illustration, when a beam of white light enters the water droplet it begins splitting into separate colors. Some of the light inside the droplet reflects off the back side of the droplet and heads back to the front as the different wavelengths of light continue to spread apart. (Some of the light passes through the rear of the droplet and back out into the air. This light is not shown in the diagram.) To see a rainbow, the sun must be overhead and behind you. Large numbers of water droplets in the air refract the sunlight in the same way, and the combined effect produces the rainbow.

The angle produced by the incoming and outgoing rays of light is 42°, so a person can only see the rainbow when standing in an area covered by light refracting in the droplets and returning at this angle. This is why you can't drive to where the end of the rainbow is. As you drive out of the coverage zone of the refracted, returning light, the rainbow disappears.

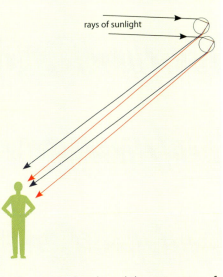

rays of sunlight

As the colors emerge from the droplets, the red angles downward at a steeper angle than the blue. As you can see from the lower diagram, when the rays arrive at your eyes, the steeper angle of the red light makes it appear to be coming from higher up in the sky than the blue you see, so red is on the outside (top) of the rainbow. (By the way, on the evening of the very day I worked out this sketch and the geometry of the rainbow, a friend just happened to give me a book showing the exact same sketch drawn in about 1310 by Theodoric of Freiberg. How about that!)

Chapter 13 Exercises

Mirror and Lens Calculations

1. A convex mirror with a focal length of 8.00 inches is used as a wide-angle mirror on the side of a truck. If the driver looks in the mirror at a vehicle 4.50 ft tall and 25.0 ft away from the mirror, what does the driver see?

2. A fun house at an amusement park has a large concave mirror with a focal length of 6.0 m. If you are 5 feet 6 inches tall and you stand 2.0 m front of this mirror, describe the image you see.

3. A biconvex lens has a focal length of 250.0 mm. If an object in front of this lens is at a distance 3.88 m, determine the distance from the lens to where the image forms.

4. If the object in the previous problem is a chair 4.10 feet high and 2.75 feet wide, determine the dimensions of the image of the chair and state your result in inches.

5. Consider a photographer using an early 20th-century camera like the one shown to take a portrait of a person sitting 8.75 feet away. Inside the camera, the lens position is 15.5 inches from the film. Determine the focal length required for this camera lens.

6. In a modern DSLR camera, one of the standard lens focal lengths is 28.00 mm. If this lens is focused so the lens is 28.20 mm from the CCD sensor, how far away from the lens must the object be for a focused image to form? State your answer in feet.

7. A magnifying glass has a focal length of 12.00 inches. If the lens is held 3.00 inches from the typed characters on a piece of paper, determine where the image forms and what the magnification is.

8. A movie projector uses 35-mm film (that is, the film is 35.0 mm wide) to project an image on a screen 82.5 feet away. The film path is 7.50 inches from the lens. Determine the required focal length of the lens and the width of the image on the screen.

9. A student uses a Keplerian beam expander to expand and collimate a laser beam. The laser beam is 0.88 mm in diameter. The two lens focal lengths are 125 mm and 650 mm. Determine the diameter of the expanded beam.

10. A certain digital projector is designed with a lens focal length of 125.0 mm. If the screen is 27.0 feet away, how far must the LCD device in the projector be from the lens? If the height of the screen is 8.00 ft, how tall must the LCD device be? State your answers in mm.

Answers

1. He sees an upright image, 1.48 inches high.
2. You see an upright image of yourself, 8 ft 3 in tall.
3. $d_i = 267$ mm
4. 3.39 in H × 2.27 in W
5. $f = 13.5$ in
6. $d_o = 12.95$ ft
7. $d_i = -4.00$ in; $M = 1.33$
8. $f = 7.44$ in; image width = 4.62 m
9. $D = 4.6$ mm
10. $d_o = 127$ mm; 37.6 mm

Optics Study Questions

1. Describe the conditions under which a biconvex lens produces an inverted image or an upright image.

2. Both concave and convex mirrors produce minified images when the object is far away. Why are the mirrors used for safety and traffic monitoring always convex?

3. If an extremely small light source is placed at the focal point of a biconvex lens, what do the light rays do as they pass through the lens?

4. To see an upright, magnified image in a concave mirror, where does the object need to be?

5. Imagine that you have a diverging lens with a focal length of 100 mm. Describe what you see if you hold this lens up in front of you and look through it at a tree 30 feet away. (That is, does the tree appear large or small, and is the image inverted or upright?)

6. With a Galilean beam expander and collimator system, why must the lenses be positioned so that their focal points coincide?

7. The text states that a Keplerian telescope works like a camera and a magnifier placed one after the other. Explain this comment.

8. Why do flat mirrors always produce virtual images?

9. Explain why it is that when using a magnifying glass as a magnifier to enlarge some tiny print, the image is in focus no matter where your eyes are located.

10. What is a meniscus lens?

11. Why is the lens in Figure 13.4 (D) identified as a negative lens?

Glossary

A

absolute pressure	A pressure measurement based on vacuum as the zero reference.
absolute zero	The lowest temperature theoretically possible; the temperature at which all molecular motion ceases.
AC	Alternating current.
acceleration	Changing velocity; the rate at which an object's velocity is changing.
accommodation	The action of the lens in the eye in which the muscles at the edges of the lens cause the lens to change shape in order to direct light rays to a point at the retina in order to view near objects.
accuracy	The degree to which a measurement is free of error.
acoustic energy	Energy transported by sound waves.
alloy	A solution of metals, formed by melting the metals, mixing them thoroughly, and allowing the solution to harden.
alpha particle	A particle composed of two protons and two neutrons, released during beta decay.
alpha radiation	The emission of alpha particles during the nuclear process of alpha decay.
alternating current	Oscillating electric current; an electric circuit in which the voltages and currents are sinusoidal, and thus continuously increase, decrease, and switch direction. In the US, the frequency of AC electrical systems is 60 Hz; in Europe, 50 Hz.

ampere	The SI System unit for electric current, representing a flow of charge of one coulomb per second; one of the seven SI System base units.
amplitude	The height of a wave, measured from the centerline of the wave to the top of a peak.
anechoic chamber	A room designed for audio testing with absorbing material on all surfaces so that sound is completely absorbed and does not reflect back into the room.
anode	The positive pole of a battery.
aqueous solution	A solution in which water is the solvent.
arc	The passage of electric current through the air by means of ionization of the air; in the process the ionized particles in the air gain energy and emit an intense violet light; the violent process of ionization produces shock waves in the air that we hear as a snapping, crackling, or zapping sound.
atmosphere	The layer of gas surrounding the earth, composed of 78% nitrogen, 21% oxygen, and small amounts of other gases.
atom	The smallest part of an element, composed of a nucleus of protons and neutrons surrounded by a cloud of electrons located in orbitals according to their energies.
atomic mass	The average number of nucleons (protons and neutrons) in the nuclei of the atoms of a particular element.
atomic number	The number of protons in the nuclei of the atoms of a particular element; the elements are listed in the Periodic Table of the Elements according to atomic number.

B

barometer	A device for measuring atmospheric pressure.
barometric pressure	The local value of atmospheric pressure.
base unit	One of the seven fundamental units of measure in the SI System of units. The base units are meter (length), kilogram (mass), second (time), ampere (electric current), candela (luminous intensity), kelvin (temperature), and mole (amount of substance).
beta decay	When a neutron in the nucleus of an atom spontaneously converts into a proton, emitting an electron and a neutrino in the process (see nuclear decay).
Big Bang	The theoretical origin of the universe as an immense explosion of energy from a single point; occurred 13.77 billion years ago according to widely accepted theory.
boiling point	The temperature at which a substance enters a phase transition from liquid to gas (or vapor), or vice versa.

brittleness	A physical property that describes a substance's tendency to shatter.
brushes	Blocks of carbon that press against the commutator of a motor, forming the electrical connection between the external power source and the coils in the rotor.
buoyancy	The tendency of objects to float.
buoyant force	An upward force acting on an object submerged or partially submerged in a fluid medium; equal to the weight of the displaced fluid, according to Archimedes' principle.
buret	A piece of laboratory apparatus used for measuring and dispensing liquid.

C

cathode	The negative pole of a battery.
cathode ray tube	A vacuum tube in which a high-voltage power supply causes a beam of electrons to travel from the cathode to the anode.
charge	An electrical property of all protons and electrons. Protons have positive charge and electrons have an identical amount of negative charge; charge gives rise to an electric field.
chemical formula	A sequence of chemical symbols and subscripts used to denote a particular compound.
chemical potential energy	The energy stored in the bonds between atoms in compounds.
chemical symbol	The symbol for one of the 118 elements listed in the Periodic Table of the Elements.
circular wave	A wave on water formed by the wind; referred to as circular because a floating object moves in a circular pathway as water waves pass beneath it.
coherence	The property, inherent in laser light, that all the individual photons in the laser beam are in phase with each other.
coil	Turns or loops of electrically conductive wire, usually wrapped around a ferromagnetic core.
collimation	The shaping of a beam of light so that the beam does not diverge. In other words, the diameter of the beam remains relatively constant over distances typically found in laboratories. A property of laser light.
colloidal dispersion	A heterogeneous mixture formed by substances in two different phases; also called a colloid.
commutator	The segmented copper ring on the shaft of a motor where the electric current enters the coils in the rotor.
compound	Two or more different elements chemically bonded together.

compression	In a sound wave, a region in the wave where the air pressure is higher than normal atmospheric pressure.
compressive strength	A physical property that describes the ability of a substance to withstand crushing forces.
concave	A surface of a mirror or lens that curves toward a viewer in front of the mirror or lens.
condensation	The phase transition from vapor to liquid.
conduction	In electricity: When electric charge moves in an electrically conductive substance, such as a metal. In heat transfer: The process of heat transfer by the propagation of vibrations of atoms in a solid.
conduction electron	Electrons in a metal that are free to move around in the metallic crystal lattice.
conductor	A substance that conducts electricity, such as a metal.
conservation of energy	The principle that energy is never created or destroyed, only changed from one form to another.
conservation of momentum	The principle that momentum is conserved in any interaction.
convection	The process of heat transfer by molecular motion in fluids. As fast molecules mingle with slower molecules, they collide and kinetic energy is transferred from the faster molecules to the slower molecules.
converging lens	A lens that causes parallel rays passing through the lens to converge to the focal point on the image side of the lens. Converging lenses produce real images. Also called a positive lens since converging lenses have positive focal lengths.
convex	A surface of a mirror or lens that curves away from a viewer in front of the mirror or lens.
core	A ferrous bar or rectangle on which are wound the turns of wire in a solenoid or transformer.
covalent bond	The chemical bond that forms between nonmetal elements by means of sharing electrons. Hydrogen also forms covalent bonds.
crystal	The atomic structure of a substance held together by ionic bonds, characterized by the orderly arrangement of atoms in a rigid lattice structure.
crystal lattice	The structural framework in which the atoms are arranged in a crystal.
Cycle of Scientific Enterprise	The process of forming a theory to account for a body of facts, then formulating testable hypotheses from the theory, then testing the hypotheses experimentally and analyzing the

results. If a hypothesis is confirmed, the theory it came from is strengthened; otherwise the theory is weakened.

D

DC	Direct current.
density	The ratio of mass to volume for a substance; a physical property that describes the amount of mass in a given volume.
derived unit	A unit of measure in the SI System based on a combination of various of the seven base units.
diatomic gas	A gas that exists in the form of molecules composed of two atoms of an element chemically bonded together. The common diatomic gases are hydrogen, oxygen, nitrogen, fluorine, and chlorine.
diffraction	When waves encounter corners or obstructions, bending and spreading out as a result.
direct current	Electric current that flows steadily in one direction; an electric circuit in which the voltages are constant and the currents flow at steady rates in one direction.
dissolve	When a solute mixes with a solvent so that the particles of solute and solvent are mixed together uniformly all the way down to the molecular level.
dispersion	The effect produced by the fact that the angle by which light is refracted as it passes from one medium to another is slightly dependent on wavelength. As a result of dispersion, white light refracting through a prism is separated into individual colors.
distance	The length traveled by an object.
diverging lens	A lens that causes parallel rays passing through the lens to diverge away from the focal point on the object side of the lens. Diverging lenses produce virtual images. Also called a negative lens since diverging lenses have negative focal lengths.

E

efficiency	In any system or process, the ratio of usable energy out to energy in.
elastic potential energy	Energy stored in a stretched or compressed object or substance.
electrical energy	Energy flowing in electrical conductors.
elastic collision	A collision in which kinetic energy is conserved.
electric current	Moving electric charge.
electric field	A field around electric charges; produces a force on other charges.

electrical conductivity	A physical property that describes the ability of a substance to conduct electricity. Metals typically have high electrical conductivity.
electricity	When the electrical effects of charged particles are made evident through electric current or static electricity.
electrolyte	An electrically conductive liquid such as an aqueous salt solution.
electromagnet	A magnetic formed by winding a wire into a coil and passing a current through it.
electromagnetic force	A force caused by an electric or magnetic field.
electromagnetic radiation	Waves (or photons) of pure energy at any wavelength in the electromagnetic spectrum.
electromagnetic spectrum	The entire range of wavelengths of electromagnetic radiation from long wavelength radio waves to short wavelength X-rays and gamma rays; includes microwaves, infrared radiation, visible light, and ultraviolet radiation.
electron	One of the three fundamental particles inside atoms; located in orbitals by energy around the nucleus; possesses negative charge, and has a mass of 1/1,836 the mass of a proton.
electron sea	The conduction electrons in a metal.
electroscope	A metal rod with two loose metal foil leaves attached at the bottom, used to demonstrate static electricity.
electrostatic force	A force between stationary objects that are charged with static electricity.
element	A substance characterized by atoms all possessing the same number of protons in the nucleus.
Empyrean	In the Ptolemaic model of the heavens, the region beyond the spheres where God or the gods dwell.
energy	One of the basic constituents of the universe, the others being matter and intelligence.
epicycle	A mathematical idea in obsolete planetary models in which a planet moves in an orbit (the epicycle) around a point which in turn circles another heavenly body.
evaporation	When a substance undergoes the phase transition from liquid to vapor without first being heated to the boiling point.
experiment	A scientific test of a hypothesis.
explanatory variable	A variable manipulated by researchers in an experiment that produces variation in the response variable.

F

faith-science debate	A conflict in contemporary Western culture fueled on one hand by those who incorrectly claim that science shows there is no God and on the other hand by Christians whose interpretations of the Bible lead them to reject widely accepted scientific theories.
far point	The farthest point on which a person's eye can focus.
ferrous metal	A metal containing iron.
field	A geometrical description of the way one object exerts a force on another without contact.
field coils	The coils in the stator of a motor or generator used to produce the magnetic field surrounding the rotor.
field lines	Arrows in a graphical representation of a field that indicate the direction of the force present on an object placed in the field, assuming that the object is the type affected by the type of field in question.
Firmament	The sphere containing the fixed stars in the Ptolemaic model of the heavens.
fission	A nuclear reaction in which a large nucleus splits to form two or three smaller nuclei (fission fragments) along with a few individual neutrons; energy is released in the process.
fluid	A substance that flows; a liquid or a gas.
flux	The nonmaterial stuff filling a region in which a field is present.
focal length	The distance from the focal point to the center of a spherical lens or surface of a spherical mirror.
focal point	For a spherical lens or mirror, the point toward which parallel rays propagating along the optical axis converge (or from which the rays diverge) after encountering the lens or mirror.
force	A push or pull.
frequency	In an oscillating system, the number of cycles completed per second.
friction	A force that converts the kinetic energy of motion into heat.
fundamental	The lowest frequency present when a series of harmonics is produced during resonance.
fusion	A nuclear reaction in which two hydrogen nuclei (protons) fuse together to form a helium nucleus, releasing energy in the process.

G

galaxy	A large spiral structure in outer space consisting of some 100 billion stars. There are about 100 billion galaxies.

gas	A phase of matter in which the internal energy of the atoms is high enough that they are completely independent of each other and free to move around.
gauge pressure	A pressure measurement based on atmospheric pressure as a reference so that atmospheric pressure equals zero, and the measured pressure is the net pressure above atmospheric pressure.
general theory of relativity	Albert Einstein's theory (1915) of gravity; an explanation of gravity in terms of acceleration, and of the curvature of the geometry of spacetime around a massive object.
generator	A machine that uses an engine to rotate a coil of wire in a magnetic field, thus producing electric current through the principle of Faraday's law of magnetic induction.
geocentric	A solar system model in which the sun, moon, and planets all orbit the earth.
gold foil experiment	Ernest Rutherford's experiment in which alpha particles were fired at a thin gold foil, leading to the discovery of the atomic nucleus.
graduated cylinder	A piece of laboratory apparatus used for measuring liquid volume.
gravitational field	A field around an object possessing mass; produces a force on other massive objects; also affects electromagnetic radiation.
gravitational force	The weakest of the four fundamental forces; the force of attraction between two objects due to their gravitational fields. The force is so weak that it is typically only noticeable in the case of very large masses.
gravitational potential energy	Stored energy due to an object's location in a gravitational field.
ground	An electrical connection to the earth so that the earth may be used as a zero-voltage reference for electrical systems.

H

harmonic	During resonance, a series of resonating frequencies called harmonics are produced. The lowest frequency harmonic is the fundamental. Higher frequency harmonics have frequencies that are multiples of the frequency of the fundamental.
harmonic spectrum	A graph of the harmonic frequencies resonating in a system, showing their relative intensities.
heat	Energy in transit from a warm body to a cooler one by means of infrared radiation, thermal conduction, or convection.
heat of fusion	The energy required to melt one gram of a solid substance while keeping its temperature the same.

heat of vaporization	The energy required to vaporize (boil) one gram of a liquid substance while keeping its temperature the same.
heat transfer	When energy is transferred from a warm object to a cooler one by means of infrared radiation, thermal conduction, or convection.
heliocentric	A solar system model in which the earth and other planets all orbit the sun and the moon orbits the earth.
heterogeneous mixture	A mixture in which the presence of at least two different substances is visible to the eye or with a microscope.
homogeneous mixture	A mixture with a composition that is uniform all the way down to the molecular level; a solution.
hypothesis	A testable, informed prediction, based on a theory, of what will happen in certain circumstances.

I

incident	A term that describes a ray or wave as it approaches the boundary of a new medium, where it reflects and/or refracts.
induction	In electricity: When the presence of nearby charges causes movable charges in a metal to accumulate. In magnetism: The spontaneous generation of a current in a coil of wire due to a changing magnetic flux passing through the coil.
inelastic collision	A collision in which some of the initial kinetic energy is converted into other forms of energy, and thus kinetic energy is not conserved.
inertia	A property of all matter that causes matter to prefer its present state of motion; quantified by the variable *mass*.
infrared radiation	Electromagnetic radiation with wavelengths slightly longer than the wavelengths of visible (red) light.
infrasonic	A sound wave with a frequency lower than about 20 Hz, the lower limit of human hearing.
insulator	In electricity: A substance that does not conduct electricity. In heat: A substance that resists the flow of heat.
intelligence	The wisdom of God embedded everywhere in the order and structure in nature; one of the three basic things the universe is made of, the other two being energy and matter.
interaction	An alternative way of understanding forces in terms of fields and subatomic particle exchanges.
interference	A wave interaction produced when two waves arrive at the same place at the same time. If the waves arrive in phase (constructive interference), the wave amplitudes add to produce a wave of greater amplitude. If the waves arrive out of phase (destructive interference), the waves cancel out.

internal energy	The sum of all the kinetic energies possessed by all the particles in a substance.
International System of Units	The system of units of measure administered in France and accepted by most of the countries in the world; the SI or metric system.
inverse-square law	An inverse proportion in which one quantity decreases as a second quantity increases; the first quantity decreases in proportion to the square of the second quantity; generically represented as $y = k/x^2$.
inverter	A device that converts DC electricity into AC electricity.
ion	An atom that has gained or lost one or more electrons, thus acquiring a net electric charge.
isotope	Any variety of the atoms of an element, differing in the number of neutrons the atoms have in their nuclei.

K

kilogram	One of the seven base units in the SI System; the unit for mass.
kinetic energy	Energy an object possesses by virtue of its motion.
kinetic theory of gases	The theory that the kinetic energy of gas particles is a function of the internal energy of the gas, and thus the gas temperature. The kinetic theory of gases explains gas pressure in terms of the collisions of gas particles with the walls of a container.

L

law of universal gravitation	Isaac Newton's theory (1687) that all objects with mass attract all others, with a force that is directly proportional to the product of two masses and inversely proportional to the square of the distance between them.
laws of motion	Isaac Newton's three laws (1687) summarizing the behavior of objects with mass. When no net force is present on such an object, the object retains its present state of motion (first law); when a net force is present, the object accelerates in proportion to the force, and in inverse proportion to its mass (second law); and any force always produces an equal and opposite reaction force (third law).
laws of nature	The orderly and very mathematical system of rules that objects in nature obey; a manifestation of the intelligence in nature that points to the creator.
Leyden jar	A glass jar with an internal metal lining that can store electric charge.
liquid	A phase of matter in which the internal energy of the atoms is high enough that they are free enough from each other to move around, and yet low enough that they are still loosely at-

tracted to each other so that particles cling together enough to form drops and fill containers without lids.

liter — A unit of volume equal to 1,000 cubic centimeters; not an official SI System unit, but commonly used in scientific study.

longitudinal wave — A wave with the characteristic that the oscillation producing the wave is moving in a direction parallel to the direction in which the wave is propagating.

lurking variable — A variable in an experiment that causes variation in the response variable without the researchers being aware that it is doing so.

M

magnetic field — A field around a magnet or electromagnet; causes a force on other magnets, electromagnets, current-carrying wires, and magnetic materials.

magnification — In an optical system, the ratio of the image size to the object size.

mass — The variable used to quantify the amount of inertia in an object.

mass-energy equivalence — Einstein's theory (1905) that mass and energy are different forms of the same thing and that amounts of mass and energy may each be expressed in terms of the other; the theory can be used to calculate the energy released when a quantity of matter is converted into energy in a nuclear reaction.

matter — Anything that has mass and volume; one of the three basic things the universe is composed of, the other two being energy and intelligence.

mechanical equivalent of heat — The theory that heat and kinetic energy are different forms of the same thing—energy, and may be converted from one form to the other.

mechanical wave — Any wave requiring a medium in which to propagate; all waves other than electromagnetic waves.

medium — Matter of some kind through which the energy of mechanical waves can propagate.

melting point — The temperature at which a substance undergoes the phase transition from solid to liquid or vice versa.

meniscus — The bowl-shaped curve on the top of a liquid in a narrow cylinder.

meniscus lens — A lens for which the curvature on the two sides of the lens tends in the same direction, producing a bowl shape.

mental model — A theory.

metalloid	A group of seven elements in between the metals and the non-metals in the Periodic Table of the Elements, and possessing some of the properties of both.
meter	One of the seven base units in the SI System; the SI System unit for length.
metric prefix	Prefixes added to units of measure in the SI System to indicate multiples or fractional multiples. For example, the prefix kilo– indicates a multiple of 1,000, so a kilometer is 1,000 meters. The prefix micro– indicates a fractional multiple of 1/1,000,000, so a micrometer is 1/1,000,000 of a meter.
metric system	The common term in the U.S. for the International System of Units.
Milky Way	The name for our galaxy.
MKS system	A subset of the units used in the metric system using only base units such as the meter, kilogram, and second, and units derived from these.
mixture	A combination of substances that occurs without any chemical reaction. Substances in the mixture retain their own properties and may be separated from one another by physical (non-chemical) means.
molecule	A chemically bonded cluster of atoms; the smallest particle in compounds formed by covalent bonding between nonmetals (including hydrogen).
momentum	The product of an object's mass and velocity.
monopole	A hypothetical single pole of a magnet (north or south), which does not exist.
motion	A state in which an object is not at rest.
motor	A machine that uses electric current to produce rotation of mechanical systems or machines through the principle of Ampère's law.
multi-resistor network	A network of resistors in an electric circuit consisting of more than one resistor.

N

near point	The nearest point on which a person's eye can focus.
neutron	One of the three basic particles inside atoms; located in the nucleus; possesses no electric charge and has a mass very close to the mass of a proton (but slightly greater).
net force	A force or combination of forces on an object that is not balanced by any other force, thus causing acceleration according to Newton's second law of motion.
network	A system of two or more resistors connected together.

normal line	A line perpendicular to the boundary between two media, used as a reference for measuring angles involved in wave reflection and refraction.
nuclear decay	A naturally occurring nuclear reaction in which an atom of a radioactive element spontaneously changes the composition of its nucleus by the emission of particles and/or electromagnetic radiation.
nuclear energy	The use of controlled nuclear fission to produce steam which then powers a turbine and generates electricity; energy released from a nuclear reaction.
nuclear radiation	Particles and/or electromagnetic radiation released during nuclear decay.
nucleon	A proton or neutron.
nucleus	The center of an atom where the protons and neutrons are.
nuclide	Any isotope of any element.

O

ohm	The SI System unit for resistance.
oil drop experiment	Robert Millikan's experiment by which the charge on a single electron was discovered.
orbital	An energy region in an atom where up to two electrons may reside. Electrons are arranged in orbitals according to the energy they possess.
oscillation	Continuous, periodic movement back and forth or up and down.

P

parallel resistance	Two or more resistors connected together at both ends.
period	The length of time required for an oscillating system to complete one full cycle of its oscillation.
phase diagram	A diagram showing the temperature, pressure, and/or energy associated with phases and phase transitions for a particular substance.
phase transition	The process of a substance changing phase from solid to liquid, liquid to gas, gas to liquid, liquid to solid, solid to gas, etc.
photon	A single quantum of electromagnetic energy.
physical property	Any characteristic of a substance other than those describing its chemical behavior.
pink noise	Sound energy possessing equal energy in each octave of the audible spectrum (20–20,000 Hz).

plane wave	A wave is space in which the phases of all the waves are aligned in a flat geometric plane in space.
plasma	An ionized gas, recognized as one of the four phases of matter, the other three being solid, liquid, and gas (or vapor).
plum pudding model	The atomic model of J.J. Thomson in which atoms are envisioned as tiny clouds of massless, positive charge sprinkled with large numbers of negatively charged electrons.
polar	The circumstance that arises when the electrons in a molecule are not evenly distributed, so one end or region of the molecule is more positively charged and another region is more negatively charged.
polarization	A condition of light in which the oscillating electric fields in the light waves are all aligned in the same direction.
pole	One end of a magnet, identified as either north or south.
potential	Voltage.
Potential difference	Voltage drop; voltage difference.
potential energy	Energy that is stored and can be released by conversion into a different form of energy; potential energy is stored in springs, in objects elevated in gravitational fields, in chemical bonds between atoms, etc.
power	The rate at which energy is produced or utilized.
precision	The degree of resolution or fine-ness in a measurement.
pressure	The amount of force present per unit area on a surface.
Primum Mobile	Latin for "first mover." In the Ptolemaic model of the heavens, this is the 9th sphere that drives the other spheres, causing them all to rotate around the earth once each day.
product	The compounds formed by a chemical reaction.
proton	One of the three basic particles inside atoms; located in the nucleus; possesses positive charge and has a mass very close to the mass of a neutron (slightly less).
pure substance	An element or a compound.

Q

quantized	Divided into discrete lumps or packets; non-continuous.
quantum	Albert Einstein theorized in 1905 that all energy is divided into discrete packets called *quanta*. A single packet of energy is called a *quantum* of energy.
quantum model	The contemporary atomic model in which electrons are confined to specific orbitals according to their energies inside atoms.

R

radiation	Electromagnetic waves. In heat transfer, refers specifically to electromagnetic waves in the infrared region of the electromagnetic spectrum. Not to be confused with nuclear radiation, which can involve massive particles such as protons, electrons, and neutrons.
radioactive	A term for substances that spontaneously undergo nuclear decay, producing nuclear radiation in the process.
rarefaction	In a sound wave, a region in the wave where the air pressure is lower than normal atmospheric pressure.
ray optics	The study of optical systems in which light is treated as thin rays similar to laser beams.
real image	An image produced by an optical system where the light rays pass through the image location.
rectifier	A device that converts AC electricity into DC electricity.
reflection	When a wave bounces off the boundary between two media, obeying the law of reflection in the process.
refraction	When a wave passes through the boundary separating two media, changing velocity and direction in the process.
relay	A device that uses a solenoid to actuate electrical contacts thus forming an electrically operated switch.
resistance	Electrical resistance to the flow of electric current; a property of resistors.
resistor	A device in electric circuits used to regulate the flow of electric current.
response variable	A variable monitored by researchers in an experiment that varies in response to changes in another variable being manipulated (the explanatory variable).
resonance	A wave interaction produced when the dimensions of the medium correspond to an integral multiple of half the wavelength of waves propagating in the medium. When this occurs, standing waves in the medium combine to produce a wave with much greater amplitude that the amplitude of the original wave causing the effect.
retrograde motion	When a planet appears to reverse the direction of its nightly progress against the fixed background of stars.
retro-rocket	A rocket pointing in the same direction that a space vehicle is traveling, and used to slow the vehicle down.
rotor	The rotating coils, along with the supporting mechanical parts, in a motor or generator.

S

scalar	A physical quantity that may be expressed in terms of a magnitude only.
schematic diagram	An abstract representation of the devices in an electric circuit and how they are connected.
science	The process of using experiment, observation, and reasoning to develop mental models (theories) of the natural world.
scientific fact	A scientific statement supported by a great deal of evidence that is correct as far as we know.
second	The base unit of time in the SI System; one of the seven base units.
series resistance	Two or more resistors joined together one after the other in a single electrical pathway.
sibilance	The hissing sound produced between the teeth when saying a word like "Susan."
significant digits	The digits in a measurement that represent its precision.
sine wave	A wave with a specific mathematical shape characterized by the trigonometric function $y = \sin \theta$.
SI System	The International System of Units administered in France and accepted by most of the countries in the world; commonly known in the U.S. as the metric system.
solar system	The sun and the planets, along with the planets' moons and other smaller bodies.
solar wind	A stream of high-energy charged particles (protons and electrons) continuously emitted by the sun.
solenoid	A long, current-carrying coil of wire that produces a magnetic field down its center.
solid	A phase of matter in which the internal energy of the atoms is so low that they are rigidly bound together and not free to move around but able to vibrate in place.
solute	A solid, liquid, or gas that is dissolved into a fluid.
solution	A homogeneous mixture; formed when one substance (a solute) dissolves into another substance (the solvent).
solvent	A fluid into which a solute is dissolved.
sonic boom	The loud boom produced by a shockwave in the air caused by the flight of an object such as a jet traveling at supersonic speed.
Sound Pressure Level	A scale for measuring the loudness of sounds, abbreviated SPL.

specific heat capacity	The amount of heat that must be added to or removed from one gram of a substance to change its temperature by one Celsius degree.
speed	The rate at which an object is moving.
spherical aberration	The property inherent in spherical lenses that parallel rays passing through the lens are not refracted to exactly the same point on the optical axis.
standing wave	A waved formed in a medium when waves and their reflections align in a fixed relationship. Under the right conditions, standing waves produce resonance.
state of motion	The state of an object characterized by its possessing a particular velocity; that is, possessing a particular speed in a particular direction; includes the state of being at rest (a speed of 0 m/s).
static discharge	The release of a static accumulation of charge produced when an electrostatic voltage is high enough to cause the surrounding air to ionize, forming a conductive channel in which charge flows; produces an arc (see arc).
static electricity	An accumulation of charge.
stator	The stationary coils (field coils) in a motor or generator.
stellar parallax	The change in the relative positions of the stars due to the change in earth's location as it orbits the sun.
steady-state universe	The discarded theory that the universe is neither expanding nor contracting, and has no beginning, but has always been here.
strong nuclear force	One of the four fundamental forces; the strongest of all forces; binds the protons together in the nucleus of atoms.
subatomic particle	Mainly protons, electrons, and neutrons, although there are approximately 60 lesser-known particles.
sublimation	A phase transition in which a substance transitions from a solid to a gas without going through the liquid phase.
subsonic	A velocity lower than the speed of sound in air.
substance	Anything that contains matter.
supersonic	A velocity higher than the speed of sound in air.

T

telos	A Greek word meaning purpose, goal, or end.
theory	A model or explanation that seeks to account for the related facts, and provide means for producing new hypotheses.
thermal conductivity	A physical property that indicates the ability of a substance to conduct heat.
thermal energy	Energy an object possesses due to being heated.

thermal equilibrium	The state that exists when two objects (or an object and its environment) are at the same temperature and there is no heat flow between them.
thermal properties	Properties that describe how a substance behaves as it is heated or cooled.
threshold of human hearing	A loudness of 0 dB(SPL), which corresponds to the quietest sound humans can hear.
threshold of pain	A loudness of approximately 130 dB(SPL), which corresponds to the onset of physical pain.
time interval	A specific period of time.
transformer	A magnetic device that uses the principles of Ampère's law and Faraday's law to raise or lower an AC voltage while simultaneously lowering or raising the associated AC current in inverse proportion.
transverse wave	A wave with the characteristic that the oscillation producing the wave is moving in a direction perpendicular to the direction in which the wave is propagating.
triple point	The combination of temperature and pressure at which all three phases of a substance coexist in equilibrium.
truth	The way things really are; revealed to us by God through his word and through nature, or known to us by the direct and immediate testimony of our senses, or known by valid reasoning from true premises; not a part of the language of scientific claims.
turbine	A system of blades mounted on a rotating shaft such that when steam, water, or air flows past the blades the shaft turns.
turn	A single loop of wire in magnetic devices such as solenoids, motors, generators, and transformers.

U

ultrasonic	A sound wave with a frequency higher than about 20 kHz, the upper limit of human hearing.
uniform acceleration	A situation caused by the presence of a constant net force, in which the velocity of an object increases or decreases by the same amount each second.
unit conversion factor	An expression with a value of one, written by placing equivalent quantities expressed in different units of measure in the numerator and denominator of a fraction.

V

Van de Graaff generator	A machine that produces a large accumulation of static electric charge on a metal dome by means of a motor-driven rubber belt.

vapor	The gaseous phase of matter; a term applied to describe the gaseous state of substances that are solids or liquids at room temperature
vaporization	The process of transitioning to the gas or vapor phase; boiling.
vector	A physical quantity that must be expressed in terms of both magnitude and direction.
velocity	A quantity that describes how fast an object is moving and in which direction.
virtual image	An image produced by an optical system where the light rays do not pass through the image location.
visible spectrum	The portion of the electromagnetic spectrum that we can sense with our eyes by perception of the colors red, orange, yellow, green, blue, and violet.
volt	The SI System unit of measure for voltage.
voltage	The electrical "pressure" that forces charge to flow.
voltaic pile	An invention that generated electricity, composed of copper and zinc plates, which was the predecessor to the modern battery.
volume	The variable used to quantify the amount of space an object takes up.

W

wave	A disturbance in space and time that carries energy from one place to another.
wavelength	The spatial separation between the peaks of a wave.
wave packet	A synonym for photon.
wave train	A continuous sequence of waves.
weak nuclear force	One of the four fundamental forces; the agent involved in beta decay.
weight	The force acting on an object due to gravity.
work	A mechanical process whereby energy is transferred from one object to another by a force pushing through a certain distance.

Z

zodiac	A belt of twelve constellations around the earth. The term derives from the Latin and Greek terms meaning "circle of animals," and is so named because many of the constellations in the zodiac represent animals.

APPENDIX A
Reference Data

In the following four tables, the values in uncolored cells are exact and thus do not affect the significant digit count associated with a computation. Values in yellow cells are rounded, and you must include the precision of the value in your assessment of the significant digits associated with a computation.

You should commit the conversion factors, prefixes, and physical constants listed in Tables A.2, A.3, and A.4 to memory. You will use these values throughout the course in computations on quizzes and assignments.

Table A.1 Complete list of the 20 SI prefixes.											
Multiples	Prefix	deca–	hecto–	kilo–	mega–	giga–	tera–	peta–	exa–	zetta–	yotta–
	Symbol	da	h	k	M	G	T	P	E	Z	Y
	Factor	10	10^2	10^3	10^6	10^9	10^{12}	10^{15}	10^{18}	10^{21}	10^{24}
Fractions	Prefix	deci–	centi–	milli–	micro–	nano–	pico–	femto–	atto–	zetto–	yocto–
	Symbol	d	c	m	μ	n	p	f	a	z	y
	Factor	1/10	$1/10^2$	$1/10^3$	$1/10^6$	$1/10^9$	$1/10^{12}$	$1/10^{15}$	$1/10^{18}$	$1/10^{21}$	$1/10^{24}$

Table A.2 Metric prefixes for memory.

Fractions	Multiples
centi– (c) = 1/100 = 10^{-2} (Thus, there are 100 cm in 1 m.)	
milli– (m) = 1/1000 = 10^{-3} (Thus, there are 1000 mm in 1 m.)	kilo– (k) = 1000 = 10^{3} (Thus, there are 1000 m in 1 km.)
micro– (μ) = 1/1,000,000 = 10^{-6} (Thus, there are 1,000,000 μm in 1 m.)	mega– (M) = 1,000,000 = 10^{6} (Thus, there are 1,000,000 m in 1 Mm.)
nano– (n) = 1/1,000,000,000 = 10^{-9} (Thus, there are 1,000,000,000 nm in 1 m.)	giga– (G) = 1,000,000,000 = 10^{9} (Thus, there are 1,000,000,000 m in 1 Gm.)

Table A.3 Conversion factors for memory.

5280 ft = 1 mi	2.54 cm = 1 in
3 ft = 1 yd	$1000 \ cm^3 = 1 \ L$
365 days = 1 year	$1000 \ L = 1 \ m^3$
3600 s = 1 hr	$1 \ mL = 1 \ cm^3$

Table A.4 Physical constants for memory.

$c = 3.00 \times 10^8$ m/s (speed of light in a vacuum or in air)	ρ_{water} = 998 kg/m³ = 0.998 g/cm³ (density of water at room temperature)
$g = 9.80$ m/s² (acceleration due to gravity)	

Table A.5 Additional conversion factors, constants, and equations.
You do not need to memorize these. They are written here for your convenience.

0.3048 m = 1 ft			
1609 m = 1 mi	4.45 N = 1 lb	3.786 L = 1 gal	101,325 Pa = 14.7 psi (atmospheric pressure at sea level)
$T_C = \dfrac{5}{9}\left(T_F - 32°\right)$	$T_F = \dfrac{9}{5}T_C + 32°$	$T_K = T_C + 273.2$	$T_C = T_K - 273.2$

APPENDIX B
Chapter Equations and Objectives Lists

To assist your study and review, all the equations and Objectives Lists are listed here together.

B.1   Chapter Equations

$d = vt$

Relation between distance, velocity and time, valid for objects moving at constant velocity.

$a = \dfrac{v_f - v_i}{t}$

Relation for acceleration, based on change in velocity and time.

$a = \dfrac{F}{m}$

Newton's Second Law of Motion.

$F_w = mg$

Weight equation.

$E_G = mgh$

Equation for gravitational potential energy.

$E_K = \frac{1}{2}mv^2$

Equation for kinetic energy.

$v = \sqrt{\dfrac{2E_K}{m}}$

Equation for velocity, as a function of kinetic energy.

$W = Fd$

Work equation.

$p = mv$

Momentum equation.

$\rho = \dfrac{m}{V}$

Density equation.

$P = \rho g h$ Equation for pressure under a liquid.

$P = \dfrac{F}{A}$ Equation relating pressure to force applied over an area.

$P_{abs} = P_{gauge} + P_{atm}$ Equation for converting gauge pressure to absolute pressure.

$v = \lambda f$ Wave equation. (Memorization not required.)

$\tau = \dfrac{1}{f}$ Period equation. (Memorization not required.)

$V = IR$ Ohm's Law.

$P = VI$ Electrical power equation.

$\dfrac{1}{d_o} + \dfrac{1}{d_i} = \dfrac{1}{f}$ Mirror and lens equation.

$M = \dfrac{y_i}{y_o} = -\dfrac{d_i}{d_o}$ Magnification equation.

B.2   Chapter Objectives Lists

CHAPTER 1 THE NATURE OF SCIENTIFIC KNOWLEDGE

1. Define science, theory, hypothesis, and scientific fact.
2. Explain the difference between truth and scientific facts and describe how we obtain knowledge of each.
3. Describe the difference between General Revelation and Special Revelation and relate these to our definition of truth.
4. Describe the "Cycle of Scientific Enterprise," including the relationships between facts, theories, hypotheses, and experiments.
5. Explain what a theory is and describe the two main characteristics of a theory.
6. Explain what is meant by the statement, "a theory is a model."
7. Explain the role and importance of theories in scientific research.
8. State and describe the steps of the "scientific method."
9. Define explanatory, response, and lurking variables in the context of an experiment.
10. Explain why experiments are designed to test only one explanatory variable at a time. Use the procedures the class followed in the Pendulum Experiment as a case in point.
11. Explain the purpose of the control group in an experiment.
12. Describe the possible implications of a negative experimental result. In other words, if the hypothesis is not confirmed, explain what this might imply about the experiment, the hypothesis, or the theory itself.

CHAPTER 2 MOTION

1. Define and distinguish between velocity and acceleration.
2. Use scientific notation correctly with a scientific calculator.

3. Calculate distance, velocity, and acceleration using the correct equations, MKS and USCS units, unit conversions, and units of measure.
4. Use from memory the conversion factors, metric prefixes, and physical constants listed in Appendix A.
5. Explain the difference between accuracy and precision and apply these terms to questions about measurement.
6. Demonstrate correct understanding of precision by using the correct number of significant digits in calculations and rounding.
7. Describe the key features of the Ptolemaic model of the heavens, including all of the spheres and regions in the model.
8. State several additional features of the medieval model of the heavens and relate them to the theological views of the Christian authorities opposing Copernicanism.
9. Briefly describe the roles and major scientific models or discoveries of Copernicus, Tycho, Kepler, and Galileo in the Copernican Revolution. Also, describe the significant later contributions of Isaac Newton and Albert Einstein to our theories of motion and gravity.
10. Describe the theoretical shift that occurred in the Copernican Revolution and how Christian officials (both supporters and opponents) were involved.
11. State Kepler's first law of planetary motion.
12. Describe how the gravitational theories of Kepler, Newton, and Einstein illustrate the way the Cycle of Scientific Enterprise works.

CHAPTER 3 NEWTON'S LAWS OF MOTION

1. Define and distinguish between matter, inertia, mass, force, and weight.
2. State Newton's laws of motion.
3. Calculate the weight of an object given its mass and vice versa.
4. Perform calculations using Newton's second law of motion.
5. Give several examples of applications of the laws of motion that illustrate their meaning.
6. Explain why Newton's first law of motion is called the law of inertia.
7. Use Newton's laws of motion to explain how a rocket works.
8. Apply Newton's laws of motion to application questions, explaining the motion of an object in terms of Newton's laws.

CHAPTER 4 ENERGY

1. State the law of conservation of energy.
2. Describe how energy can be changed from one form to another, including:
   a. different forms of mechanical energy (kinetic, gravitational potential, elastic potential)
   b. chemical potential energy
   c. electrical energy
   d. elastic potential energy
   e. thermal energy
   f. electromagnetic radiation
   g. nuclear energy
   h. acoustic energy
3. Briefly define each of the types of energy listed above.

4. Describe two processes by which energy can be transferred from one object to another (work and heat), and the conditions that must be present for the energy transfer to occur.
5. Describe in detail how energy from the sun is converted through various forms to end up as energy in our bodies, as energy used to run appliances in our homes, or as energy used to power machines in industry.
6. Explain why the efficiency of any energy conversion process is less than 100%.
7. Calculate kinetic energy, gravitational potential energy, work, heights, velocities, and masses from given information using correct units of measure.
8. Define friction.
9. Using the pendulum as a case in point, explain the behavior of ideal and actual systems in terms of mechanical energy.
10. Explain how friction affects the total energy present in a mechanical system.

CHAPTER 5 MOMENTUM

1. Use the momentum equation to calculate mass, velocity, and momentum using correct units of measure.
2. Define *interaction*.
3. State the law of conservation of momentum.
4. Distinguish between elastic and inelastic collisions.
5. Use the principle of conservation of momentum to solve problems involving linear interactions between objects with equal or unequal masses in elastic collisions, with one object initially at rest.
6. Describe how the law of conservation of momentum relates to Newton's second and third laws of motion.

CHAPTER 6 ATOMS, MATTER AND SUBSTANCES

1. Define and describe *atom* and *molecule*.
2. State the five points of John Dalton's atomic model and the year it was published.
3. Write a brief description of each of the following three major experiments:

   • J.J. Thomson's cathode ray tube experiment
   • Robert Millikan's oil drop experiment
   • Ernest Rutherford's gold foil experiment

   For each written description, include the following:

   • A brief description of the experimental setup (one or two sentences)
   • The name of the scientist and the year the experiment was performed
   • The major discoveries or new atomic model that resulted

4. Write brief descriptions of the Bohr and quantum models of the atom.
5. Use the density equation to calculate the density, volume, or mass of a substance.
6. Calculate the volume of right rectangular solids and use the volume in density calculations.
7. Draw the "family tree" of substances, including in it the following terms: substance, pure substance, alloy, mixture, compound, heterogeneous mixture, homogeneous mixture, colloid, element, and solution.
8. Name two examples for each substance listed in the family tree.
9. Define each of the terms in the substances family tree.

10. Distinguish between compounds and mixtures.
11. Describe the two basic types of structures atoms can form when bonding together to form compounds.
12. Recognize and state the chemical symbols for 24 common element names and vice versa (see Table 6.2).
13. Use the concept of internal energy to define the three common phases of matter—solids, liquids, and gases—and distinguish between them at the molecular level.
14. Explain how evaporation and sublimation occur.
15. Define *heat of fusion* and *heat of vaporization*.

CHAPTER 7 HEAT AND TEMPERATURE

1. Define and distinguish between *heat, internal energy, thermal energy, thermal equilibrium, specific heat capacity*, and *thermal conductivity*.
2. State the freezing/melting and the boiling/condensing temperatures for water in °C, °F, and K.
3. Convert temperature values between °C, °F, and K using given formulas.
4. Describe and explain the three processes of heat transfer: conduction, convection, and radiation.
5. Describe how temperature relates to the internal energy of a substance and to the kinetic energy of its molecules.
6. Explain the kinetic theory of gases and use it to explain why the pressure of a gas inside a container is higher when the gas is hotter and lower when the gas is cooler.
7. Apply the concepts of specific heat capacity and thermal conductivity to explain how common materials such as metals, water, and thermal insulators behave.

CHAPTER 8 PRESSURE AND BUOYANCY

1. Explain the cause of pressure on objects submerged in a fluid.
2. Explain the cause of air pressure.
3. Distinguish between gauge pressure and absolute pressure.
4. Convert absolute pressure to gauge pressure and vice versa.
5. Calculate the gauge pressure and the absolute pressure at a given depth below the surface of a liquid.
6. Relate Blaise Pascal's explanation of how Torricelli's barometer works, including Pascal's bold new theory about the vacuum.
7. Determine the pressure created by a given force applied to a particular surface area.
8. Explain Archimedes' principle of buoyancy and use it to explain why things float in liquids.
9. Calculate the weight that can be supported by a regular, buoyant solid when the solid is fully or partially submerged in water.

CHAPTER 9 WAVES, SOUND AND LIGHT

1. Explain what a wave is.
2. On a graphical representation of a wave, identify the wave parameters and parts: crest, trough, amplitude, wavelength, and period.
3. Define the *frequency* and *period* of a wave.
4. Describe the following five wave interactions, giving examples of each:
   a. Reflection

b. Refraction
c. Diffraction
d. Resonance
e. Interference
   i. Constructive interference
   ii. Destructive interference
5. Give examples of longitudinal, transverse, and circular waves and the media in which they propagate.
6. Define *infrasonic* and *ultrasonic* and give examples of these types of sounds.
7. Define *infrared* and *ultraviolet* and give examples of these types of radiation.
8. Calculate the velocity, frequency, period, and wavelength of waves from given information.
9. Given the frequency of a wave, determine the period, and vice versa.
10. List at least five separate regions in the electromagnetic spectrum, in order from low frequency to high frequency.
11. State the frequency range of human hearing in hertz (Hz) and the wavelength range of visible light in nanometers (nm).
12. State the six main colors in the visible light spectrum in order from lowest frequency to highest.
13. Identify the relations: frequency and pitch; amplitude and volume.
14. Explain how waves of different frequencies (harmonics) contribute to the timbre of musical instruments.

CHAPTER 10 INTRODUCTION TO ELECTRICITY

1. Describe the roles of Alessandro Volta and James Clerk Maxwell in the development of our knowledge of electricity.
2. Define and distinguish between static electricity and electric current.
3. Explain what static electricity and static discharges are.
4. Describe three ways static electricity forms and apply them to explain the operation of the Van de Graaff generator and the electroscope.
5. Describe the general way the atoms are arranged in metals.
6. Use the idea of the "electron sea" to explain why electric current flows so easily in metals.

CHAPTER 11 DC CIRCUITS

1. Using the analogy of water being pumped through a filter, give definitions by analogy for *voltage*, *current*, *resistance*, and *potential difference*.
2. Explain what electric current is and what causes it.
3. State Kirchhoff's two circuit laws.
4. Calculate the equivalent resistance of resistors connected in series, in parallel, or in combination.
5. Use Ohm's law and Kirchhoff's laws to calculate voltages, currents, and powers in DC circuits with up to four resistors.

CHAPTER 12 FIELDS AND MAGNETISM

1. Explain what a field is.

2. Describe three major types of fields, the types of objects that cause each one, and the objects or phenomena that are affected by each one.
3. State Ampère's law.
4. State Faraday's law of magnetic induction.
5. Explain a key difference between the theories of gravitational attraction of Einstein and Newton.
6. Apply Ampère's law and Faraday's law of magnetic induction to given physical situations to determine what happens.
7. Explain the general principles behind the operation of generators and transformers.
8. Use the right-hand rule to determine the direction of the magnetic field around a wire or through a coil of wire.
9. Explain why transformers work with AC but not with DC.

CHAPTER 13 GEOMETRIC OPTICS

1. Explain the difference between a real image and a virtual image.
2. Define the *focal length* of a lens.
3. Explain why a real image formed by a biconvex lens is inverted.
4. Describe, in general, the images formed by convex and concave mirrors and convex and concave lenses.
5. Use the lens equation to compute the locations of objects and images and to determine required focal lengths for spherical mirrors and lenses.
6. Use the magnification equation to determine magnification in an optical system and to determine object or image distances.
7. Describe the general operation of these five single-lens systems:
   a. eye
   b. camera
   c. photocopier
   d. projector
   e. magnifying glass
8. Describe the general operation of these three two-lens systems:
   a. corrective lenses (glasses)
   b. refracting telescope
   c. laser beam expander and collimator
9. Compute the new diameter of a laser beam after expansion and collimation.

APPENDIX C
Laboratory Experiments

C.1   Important Notes

Important Note: Please refer to pages xii–xiii in the Preface for Teachers for important information pertaining to the terms "experimental error" and "percent difference" as used in this text.

The following pages contain your guidelines for the five laboratory experiments you will conduct in *Introductory Physics* during the year. For each of these experiments, you will submit an individual written report. It is your responsibility to study *The Student Lab Report Handbook* thoroughly so that you can meet the expectations for lab reports in this course.

The instructions written here are given to help you complete your experiment successfully. However, your report must be written in your own words. This applies to all sections of the report. Do not copy the descriptions in this appendix into your report in place of writing your report for yourself in your own words.

C.2   Lab Journals

You must maintain a proper lab journal throughout the year. Your lab journal will contribute to your lab grade along with your lab reports. Chapter 1 of *The Student Lab Report Handbook* contains a detailed description of the kind of information you should carefully include in your lab journal entries. The following are highlights from that description.

A good lab journal includes the following features:

1. The pages in the journal are quadrille ruled (graph paper) and the journal entries are in pencil.

2. The journal is neatly maintained and free of sloppy marks, doodling, and messiness.

3. Each entry includes the date and the names of the team members working together on the experiment.

4. Every experiment and every demonstration that involves taking data or making observations is documented in the journal.

5. Entries for each experiment or demonstration include:

    •   the date

- the team members' names

- the team's hypothesis

- an accurate list of materials and equipment, including make and model of any electronic equipment or test equipment used

- tables documenting all the data taken during the experiment, including the units of measure and identifying labels for all data

- all support calculations used during the experiment or in preparation of the lab report

- special notes documenting any unusual events or circumstances, such as bad data that require doing any part of the experiment over, unexpected occurrences or failures, or changes to your experimental approach

- little details about the experiment that need to be written in the report that you may forget about later

- important observations or discoveries made during the experiment.

C.3   Experiments

Experiment 1 The Pendulum Experiment

Variables and Experimental Methods   We are going to conduct an investigation involving a simple pendulum. This experiment is an opportunity for you to learn about conducting an effective experiment. In this investigation, you will learn about controlling variables, collecting careful data, and organizing data in tables in your lab journal.

To make your pendulum, bend a large paper clip into a hook. Then connect the hook to a string, and connect the string to the end of a meter stick. Then lay the meter stick on a table with the pendulum hanging over the edge and tape the meter stick down. Finally, hang one or more large metal washers on the hook for the weight.

Your goal in this experiment is to identify the explanatory variables that influence the period of a simple pendulum. A pendulum is an example of a mechanical system that is *oscillating*, that is, repeatedly "going back and forth" in some regular fashion. In the study of any oscillating system, an important parameter is the *period* of the oscillation. The period is the length of time (in seconds) required for the system to complete one full cycle of its oscillation. In this experiment, the period of the pendulum is the response variable you monitor. (Actually, for convenience you will monitor a slightly different variable, closely related to the period. This is explained on the next page.) After thinking about the possibilities and forming your team hypothesis, construct your own simple pendulum from string and some weights and conduct tests on it to determine which variables actually do affect its period and which ones do not.

In class, explore the possibilities for variables that may affect the pendulum's period. Within the pendulum system itself there are three candidates, and your instructor will lead the discussion until the class has identified them. (We ignore factors such as air friction and the earth's rotation in this experiment. Just stick to the obvious variables that clearly apply to the problem at hand.)

Then, as a team, continue the work by discussing the problem for a few minutes with your teammates. In this team discussion, form your own team hypothesis stating which variables you think affect the period. To form this hypothesis, you need not actually do any new research or tests. Just use what you know from your own experience to make your best guess.

The central challenge for this experiment is to devise an experimental method that tests only one explanatory variable at a time. Your instructor will help you work this out, but the basic idea is to set up the pendulum so that two variables are held constant while you test the system with large and small values of the third variable to see if this change affects the period. You must test all combinations of holding two variables constant while manipulating the third one. All experimental results must be entered in tables in your lab journal. Recording the data for the different trials requires several separate tables. For each experimental setup, time the pendulum during three separate trials and record the results in your lab journal. Repeating the trials this way enables you to verify that you have valid, consistent data. To make sure you can tell definitively that a given variable is affecting the period, *make the large value of the variable at least three times the small value in your trials.*

Here is bit of advice about how to measure the period of your pendulum. The period of your pendulum is likely to be quite short, only one or two seconds, so measuring it directly with accuracy is difficult. Here is an easy solution: assign one team member to hold the pendulum and release it on a signal. Assign another team member to count the number of swings the pendulum completes, and another member as a timer to watch the second hand on a clock. When the timer announces "GO" the person holding the pendulum releases it, and the swing counter starts counting. After exactly 10.0 seconds, the timer announces "STOP" and the swing counter states the number of swings completed by the pendulum during the trial. Record this value in a table in your lab journal. If you have four team members, the fourth person can be responsible for recording the data during the experiment. After the experiment, the data recorder reads off the data to the other team members as they enter the data in their journals.

This method of counting the number of swings in 10 seconds does not give a direct measurement of the period, but you can see that your swing count works just as well for solving the problem posed by this experiment, and is a lot easier to measure than the period itself. (The actual period is equal to 10 seconds divided by the number of swings that occur in 10 seconds.)

One more thing on measuring your swing count: your swing counter should state the number of swings completed to the nearest 1/4 swing. When the pendulum is straight down, it has either completed 1/4 swing or 3/4 swing. When it stops to reverse course on the side opposite from where it was released, it has completed 1/2 swing.

When you have finished taking data, review the data together as a team. If you did the experiment carefully, your data should clearly indicate which potential explanatory variables affect the period of the pendulum and which ones do not. If your swing counts for different trials of the same setup are not consistent, then something is wrong with your method. Your team must repeat the experiment with greater care so that your swing counts for each different experimental setup are consistent.

Discuss your results with your team members and reach a consensus about the meaning of your data. Expect to spend at least four hours writing, editing, and formatting your report. Lab reports count a significant percentage of your science course grades throughout high school, so you should invest the time now to learn how to prepare a quality report.

Your goal for this report is to begin learning how to write lab reports that meet all the requirements outlined in *The Student Lab Report Handbook*. One of our major goals for this year is to learn what these requirements are and become proficient at generating solid reports. Nearly all scientific reports involve reporting data, and a key part of this first report is your data tables, which should all be properly labeled and titled.

After completing the experiment, all the information you need to write the report should be in your lab journal. If you properly journaled the lab exercise, you will have all of the data, your hypothesis, the materials list, your team members' names, the procedural details, and everything else you need to write the report. Your report must be typed and will probably be around two or three pages long. You should format the report as shown in the examples in *The Student Lab Report Handbook*, including major section headings and section content.

Here are a few guidelines to help you get started with your report:

1. There is only a small bit of theory to cover in the Background section, namely, to describe what a pendulum and its period are. You should also explain why we are using the number of swings completed in 10 seconds in our work in place of the actual period. As stated in *The Student Lab Report Handbook*, the Background section must include a brief overview of your experimental method and your team's hypothesis.

2. Begin your Discussion section by describing your data and considering how they relate to your hypothesis. In this experiment, we are not making quantitative predictions, so there are no calculations to perform for the discussion. We are simply seeking to discover which variables affect the period of a pendulum and which do not. Your goal in the Discussion section is to identify what your data say and relate that to your reader.

    a. What variables did you manipulate to determine whether they had any effect on the period of the pendulum?

    b. What did you find? Which ones did affect the period? How do the data show this? Refer to specific data tables to explain specifically how the data support your conclusion.

    c. Were you surprised by what you found?

Experiment 2 The Soul of Motion Experiment

Newton's Second Law of Motion   Note: The report for this experiment requires you to set up a graph showing predicted and experimental curves on the same set of axes. Procedures for creating such a graph on a PC or Mac are described in detail in *The Student Lab Report Handbook*.

You will have a great time with this experiment. You meet out in the parking lot as a class. The idea is to push a vehicle from the rear, using scales that measure the force the pushers are applying to the vehicle. You time the vehicle as it accelerates from rest through a ten-meter timing zone and use the time data to calculate the experimental values of the vehicle's acceleration. Using the mass of the vehicle and Newton's second law, you calculate a predicted acceleration for each amount of pushing force used. Your goal is to compare your predicted accelerations to the experimental values for four different force values. You then graph the results and calculate the percent difference to help you see how they compare.

This experiment is an excellent example of how experiments in physics actually work. The scientists have a theory that enables them to predict, in quantitative terms, the outcome of an experiment. Then the scientists carefully design the experiment to measure the values of these variables and compare them to the predictions, seeking to account for all factors that affect the results. If the theory is sound and the experiment is well done, the results should agree well with the theoretical predictions and the percent difference should be low.

In our case, when a force is applied to a vehicle at rest, we expect the vehicle to accelerate in accordance with Newton's second law of motion, $a = \dfrac{F}{m}$, which predicts that the acceleration depends on the force applied. So Newton's second law is our theoretical model for the motion of an accelerating object. Now, we know that a motor vehicle has a fair amount of friction in the brakes and wheel bearings, which means that not all the force applied by the pushers serves to accelerate the car. Some of it simply overcomes the friction. Also, if the ground is not be perfectly level, this affects the acceleration as well. So to make the model as useful as possible, you must use the actual *net* force on the vehicle in your predictions so to make them as accurate as possible. Details are discussed below.

For the data collection, you need a way to measure the actual vehicle's acceleration so that you can compare it to your predictions. You already know an equation that gives the acceleration based on velocities and time. However, you have no convenient way of measuring the vehicle's velocity. (The vehicle moves too slowly for the speedometer to be of any use.) Fortunately, there is another equation you can use if you time the vehicle with a stop watch as it starts from rest and moves through a known distance. If you know the distance and the time, and the acceleration is uniform, you can calculate the vehicle's acceleration with the equation $a = \dfrac{2d}{t^2}$. Use this equation to determine the experimental acceleration value for each force, using the average time for each set of trials.

Here are some crucial details to help make this experiment as successful as possible:

1. Always have two students pushing on the vehicle. Thus, for each force value the pushers use, the total applied force is twice that amount. (You will use four different force values in the experiment.)

2. Measure the friction on the vehicle so you can subtract it from the force the pushers are applying to get the net force applied for your predictions. To measure the friction, use

one pusher and estimate the absolute minimum amount of force needed to keep the vehicle barely moving at a constant speed. As you know from our studies of the laws of motion, vehicles move at a constant speed when there is no net force. So if the vehicle is moving at a constant speed, it means that the friction and the applied force are exactly balanced. This allows you to infer what the friction force is.

3. Use four different values of pushing force. For each force value, time the vehicle over the ten-meter timing zone at least three times. The forces the pushers apply to the vehicle always vary quite a bit, so if you get three valid trials at each force you have three reliable data points for the time. You then calculate the average of these times and use it to calculate the experimental value of the acceleration of the vehicle for that force.

4. The major factor introducing error into this experiment is the forces applied by the pushers. Pushing at a constant force while the vehicle is accelerating is basically impossible. (The dial on the force scale jumps all over the place.) But if the pushers are careful, they can push with an *average* force that is pretty consistent. You need a standard to judge whether you have had a successful run with consistent pushing. Here is the criterion to use: when you obtain three trials that have time measurements that are all within a range of one second from highest to lowest, accept those values as valid. If your times are not this close together, assume that the pushing forces are not consistent enough and keep running new trials until you get more consistent data.

5. The instructor will take the vehicle, with a full gas tank, to get it weighed and report this weight to the class. Make sure to measure the weight of the driver and the weight of the scale support rack (if there is one). Add these weights to the weight of the vehicle and determine the mass for this total weight. (Of course, the instructor must also make sure the gas tank is full on the day of the experiment, since the fuel in the tank typically amounts to 1–2% of the vehicle weight.)

Considerations for Your Report

In the Background section of your report, be sure to give adequate treatment to the theory you are using for this experiment. In the Newton's second law equation, acceleration is directly proportional to force, so a graph of *acceleration* vs. *force* should be linear. In the Background, you should use this concept to explain why you expect your experimental acceleration values to vary in direct proportion to the force. Explain the equations you are using to get the predicted and experimental acceleration values. Since you are using two different equations, your Background section should include explanations for both of them and why they are needed. The force you are using to make your predictions takes friction into account. You need to explain how friction is taken into account, why you are doing so, and how this relates to the equations.

In the Procedure section, don't forget the important details, such as how you measured the friction force, weighed the driver, and judged the validity of your time data.

In the Results section, present all time data in a single table, along with the average times for the trials at each force value applied by the pushers. Present all the predicted values, experimental values, and percent differences (see Preface, pages xii–xiii) in another table or two. Do not forget to state all the other values used in the experiment, such as the vehicle weight, the weights of the driver and support rack, the distance, the total mass you calculate, and the friction force you measure. (As *The Student Lab Report Handbook* describes, in any report, all the data collected must be presented, and they all must be placed either in a table or in complete sentences.)

Variable	Equation	Comments
force	net force = (2 × force for each pusher) – friction force estimate	There are four values of net force, one for each set of trials.
predicted acceleration	predicted accel = (net force)/(total mass)	Net force is as calculated above. Mass is determined from the total weight. There is a predicted acceleration for each value of net force.
experimental acceleration	experimental accel = (2 × distance)/(avg time)^2	Distance is the length of the timing zone. Average time is the average of the three valid times for a given trial. There is an experimental acceleration for each value of net force.

Table C.1. Summary of equations for the calculations.

In the Discussion, the main feature is a graph of *acceleration* vs. *force*, showing both the predicted and experimental values on the same graph for all four force values. Carefully study Chapter 7 on graphs in *The Student Lab Report Handbook* and make sure your graph meets all of the requirements listed.

For your predicted values of acceleration, use the total mass of the vehicle, driver, and support rack. The instructor will tell you the weight of the vehicle, which you record in your lab journal. Also record the weights of the driver and support rack determined during the experiment. Convert the total weight from pounds to newtons, then determine the mass in kilograms by using the weight equation, $F_w = mg$.

For the force values in your predictions, use the nominal amount of force applied (the two pushers' forces combined) less the amount of force necessary to overcome the friction (which is determined during the experiment).

Table C.1 summarizes the calculations you need to perform for each set of trials.

The heart of your discussion is a comparison of the two curves representing acceleration vs. force (displayed on the same graph), and a discussion of how well the actual values of acceleration match up with the predicted values. In addition to this graphical comparison, compare the four predictions to the four experimental acceleration values by calculating the percent difference for each one, presenting these values in a table and discussing them.

To compare the curves, think about the questions below. Do not write your discussion section by simply going down this list and answering each question. (Please spare your instructor the pain of reading such a report!) Instead, use the questions as a guide to the kinds of things you should discuss and then write your own discussion section in your own language.

Thought Questions and Considerations for Discussion

1. Are both of the curves linear? What does that mean?

2. Do they both look like direct proportions? What does that imply?

3. Do the curves have similar slopes? What does that imply?

4. How successful are the results? A percent difference of less than 5% for an experiment as crude as this is a definite success. If the difference is greater than 5%, identify and discuss the factors that may have contributed to the difference between prediction and result. In this experiment, there are several such factors, including wind that may have been blowing on the vehicle.

5. Do not make the mistake of merely assuming that the fluctuation in the pushers' forces explains everything without taking into account the precautions you took to eliminate this factor from being a problem (the time data validity requirement).

6. Also do not make the mistake of assuming that friction explains the difference between prediction and result. Friction can only affect the data one way (slowing the vehicle down). So if friction is a factor, the data have to make sense in light of how friction affects the data. But further, since measuring friction and taking it into account in your predictions is part of your procedure, a generic appeal to friction will not do.

7. Finally, do not make the mistake of asserting that errors in the timing or the timing zone distance measurement explain the difference between prediction and result. Consider just how large the percentage error could realistically be in these measurements, and whether that kind of percentage helps at all in explaining the difference you have between prediction and result. For example, the timing zone is 10 m long. If it is carefully laid out on the pavement, it is unlikely that the distance measurement is in error by more than a centimeter or so. Even including the slight misalignments of the vehicle that crop up, the distance could probably not be off by more than, say, 10 or 20 cm. But this is only 1–2% of 10 m, and if you are trying to explain a percent difference of 5 or 10% or more this won't do it. Similar considerations apply to the time values. Given the slow speed the vehicle moves, how far off can the timing be? What kind of percentage error would this produce?

Alternate Experimental Method

If your class is using digital devices such as the PASCO Xplorer GLX to read forces, you can use a slightly different experimental method that improves results and lowers the difference between prediction and result. One of the major sources of error in this experiment is the difficulty the pushers have in accurately applying the correct amount of force to the vehicle. If you use bathroom scales to measure the force, there is nothing that can be done about this problem and the pushers simply have to do the best they can.

However, with the digital devices you can eliminate the problem of force accuracy by using the actual average values of the forces applied by the two pushers to calculate the predicted values. The Xplorer GLX can record a data file of the applied force during a given trial, and when reviewing the data file back at your computer you can view the mean value of the force during the trial. You can use this mean value to calculate the predicted acceleration from Newton's second law. Using this method to form your predictions eliminates much of the uncertainty surrounding the forces applied to the car.

Here are a few details to consider if you use this alternative approach to collecting data:

1. You do not need to select four different force values in advance and push the vehicle repeatedly at each force value. Instead, only a single trial is needed for each force.

2. Select 10 or 12 different target force values and run a single trial with each. The force targets should range from low values that barely get the vehicle to accelerate, all the

way up to the highest values the pushers can deliver. For each trial, tell the pushers the target force and tell them to do their best to stay on it during the trial. But it doesn't matter nearly as much how accurate the pushers are because you are using the average of the actual data from the digital file to make the predictions, rather than relying on the pushers to maintain the target force accurately.

3. The method for determining values of net force for the predictions is similar to that shown in Table C.1. The difference is that instead of doubling the target force for each pusher, you add together the actual mean forces obtained from the data files for each pusher and subtract out the friction force.

4. Use the time of each trial to determine the experimental value of the acceleration for that trial.

5. Calculate the percent difference for each trial and report these values in the report. Also calculate the average of the percent difference values and use this figure in your discussion of the results.

Experiment 3 The Hot Wheels Experiment

Conservation of Energy   Perform this experiment together as a class. Use the principles of conservation of energy to predict the velocity a toy car has when it rolls to the bottom of a hill, starting from rest, and compare this prediction to your experimental result. The concept is simple. You have a Hot Wheels car on a short ramp of Hot Wheels track. You let the car roll down the ramp and use a stopwatch or a digital timing system with a pair of infrared photogates to time the car as it passes through a short timing zone on a horizontal stretch of track at the bottom of the hill. Use the time data and the distance between the photogates to determine how fast the car is moving at the bottom of the hill, and compare this experimental measurement of velocity to the velocity you predict from performing a conservation of energy calculation based on the height of the hill.

Data collection involves running the car down the hill several times, recording the time for each trial. You use the average time from these trials to determine the final velocity. You also measure the initial height, final height, and mass of the Hot Wheels car. You also measure the distance between the photogates. Each of these measurements should be taken by four or five different students in the class to improve the accuracy. Let everyone write their measurements in a table on the marker board to make sharing all the data easy. Average the measurements for each variable and use these averages in your calculations. If any single measurement seems to be quite different from the others, it may mean the different measurement is inaccurate. A different student should repeat the measurement and the inaccurate measurement discarded.

You do not take friction into account in this experiment. There are a couple of reasons for this. First, the Hot Wheels cars have such low weight that the friction force is very low and difficult to measure. Second, since the friction is so low, you can neglect it and still obtain reasonably good results with a moderate percent difference. With a digital timing system to aid in determining the car's final velocity, the percent difference is lower.

Before taking measurements, review Appendix E on Making Accurate Measurements. Use correct procedures when measuring heights and the mass of the car with the triple-beam balance.

This experiment is simple enough that not much needs to be said about your report preparation. The report does not require any graphs. There are a couple dozen measurements and their averages, plus the times and their average, that you must present in a table or two. There is only one percent difference value to report. If it is low, focus your discussion section on why it is so low rather than on explaining the percent difference. If the difference is greater than you expect, then of course you must analyze your data and results to find reasonable possibilities for the source(s) of the difference between prediction and result.

In your Background section, make sure you explain the theory behind this experiment thoroughly, including the law of conservation of energy, how the law applies to your setup, and the equations you use for the two forms of energy involved. The final velocity prediction is calculated using the conservation of energy principles discussed in class. The experimental velocity measurement uses the simple equation relating distance, time, and velocity you studied in Chapter 2.

Experiment 4 Density

Accurately Determining Density with Correct Lab Technique   Note: The report for this experiment requires the student to create scatter plots for two data sets. Procedures for creating such a graph on a PC or Mac are described in detail in *The Student Lab Report Handbook*.

The purpose of this experiment is to make accurate measurements that enable you to determine the densities of some standard commercial materials and to compare these experimentally determined density values to published reference values. This experiment is an exercise in careful measurements, the goal being to have the lowest possible error.

As you know, density is defined as $\rho = m/V$. Common metric units for density are kg/m³ (MKS) and g/cm³. The g/cm³ units are common for laboratory work, which typically deals with small quantities. These units are appropriate for this activity.

We can rewrite the density equation as $m = \rho V$. In this linear equation, mass (the dependent variable) varies directly with volume (the independent variable), with the density acting as the constant of proportionality (the slope of the line). If we measure the masses and volumes of several samples of the same material, large volumes have large masses, small volumes have small masses, and in a graph of *mass* vs. *volume* the points all fall on the same line. The more accurately the masses and volumes are measured, the more perfectly the values line up in the graph.

This experiment involves working with two materials, aluminum and PVC plastic. You have four samples of each material, each of a different size. You use a triple-beam balance to determine each mass and a graduated cylinder to determine each volume using tap water and the displacement method. After measuring each of the masses and volumes, you calculate the density of each piece. The mean of these four values is your experimental density value for the material.

For each of the two materials, create a plot of all the masses versus their corresponding volumes. The graph provides a visual indication of how accurate your data are.

Safety Issues

Be careful with the glassware in this activity! There are three ways to break glass—carelessness, silliness, and improper procedures. These are all inappropriate in a laboratory.

Using the Apparatus Properly

1. Review Appendix E on Making Accurate Measurements. When placed in a graduated cylinder, most liquids form a bowl-shaped curve on the top of the liquid. This curved shaped is called a *meniscus*. For water, the proper place to read the liquid level in the graduated cylinder is at the bottom of the meniscus. To see the lower rim of the meniscus clearly, it may help to place a dark colored background behind the cylinder while you read it.

2. Avoid *parallax error* in your measurements by positioning your head and eyes directly in line with the marks on the instrument you are trying to read. If your head is slightly too high or too low, your reading of measurements on a graduated cylinder will suffer from parallax error and will be inaccurate. The same idea applies to readings made on the scale of a triple-beam balance.

3. Read your measurement with the appropriate number of significant digits. This means recording all the digits known with certainty, plus one digit you estimate between the marks inscribed on the instrument.

4. Triple-beam balances have an adjustment knob to calibrate the balance so that it reads properly. Always check the calibration before you begin making measurements and adjust it if necessary. When no masses are on the pan and the weights are all set to zero, the alignment marks on the balance should line up perfectly.

Standard Values

The two materials in this experiment are aluminum and polyvinyl chloride (PVC). Determine your percent differences by using the standard reference density value as the predicted value and your average density as the experimental value. This gives you one percent difference figure for each material. The standard density values found in references for aluminum and PVC are:

aluminum   2.70 g/cm³
polyvinyl chloride   1.4 g/cm³

Note: The aluminum value is more precise (three significant digits) than the PVC value (two significant digits). This is because in the manufacturing of PVC, its density can vary from 1.35 to 1.45 g/cm³. If the measurements your class makes are very consistent, you might be able to determine more precisely the density of the PVC samples your class is using (to three significant digits). Then you can use that more precise value for your percent difference calculations. After everyone has their density values, your instructor may analyze them to see if this is possible. This will depend on how consistent the different values are from the student groups.

Important Additional Notes

1. If you are using large, 250-mL glass graduated cylinders for your volume measurements, avoid breaking one of these costly pieces of apparatus by using the proper technique to slide your samples down into the cylinder and to extract them. Your instructor will show you this proper technique.

2. Make sure every single measurement you make is documented in your report, as well as all your calculated density values and your percent difference figures.

3. In your report, include two graphs, one showing mass vs. volume for aluminum, another for PVC.

4. Use correct precision and significant digits throughout your measurements, calculations, and report.

5. Demonstrate in your report that you used correct procedures regarding calibration of the triple-beam balance, the way volumes are determined from the liquid meniscus, avoidance of parallax error, and so on.

6. Measure masses first, while the parts are still dry, so that stray water drops do not affect your mass measurements.

Experiment 5 DC Circuits

DC Circuits and the Use of Electronic Test Equipment There are two main goals for this laboratory exercise. The first is to use the DC circuit theory you have learned (Ohm's law and Kirchhoff's laws) to make predictions and compare these predictions to experimental results. In this case, we want to use the circuit calculation techniques you have learned to determine the expected voltages and currents in a circuit (your predictions), and then compare these predictions to actual measurements. The second goal is for you to fool around with the test equipment and learn how to use it. An important learning objective for science classes is to gain experience using unfamiliar equipment. This is a valuable skill for anyone entering a technical field of study, as many of you will. However, for the purpose of writing your report, consider yourself a competent researcher and focus on the first of these two goals.

A word is in order here regarding the many technical details for this experiment. The main thing for you to focus on is predicting and measuring the voltages and currents in your circuit. There are many other details presented below, such as the resistor color code and the details of how electronic breadboards work. I hope you will enjoy learning about these things and playing around with the equipment. But the point is for you to have an enjoyable experience with the experiment, not to worry about all the technical details. If you are not really interested in learning about the resistor color code, it's no big deal. The information is included because some people are interested.

Begin this experiment by designing a DC circuit powered by an electronic DC power supply (or battery) and the following four resistor values:

$$1.1 \text{ k}\Omega \qquad 1.5 \text{ k}\Omega \qquad 2.0 \text{ k}\Omega \qquad 4.7 \text{ k}\Omega$$

In your circuit, include both parallel and series resistance combinations; however, the particular arrangement of the four resistors is up to you. Next, you select a voltage to use to power your circuit. An electronic DC power supply or battery pack ia used as your voltage source. Calculate the equivalent resistance for your circuit in advance of the experiment. But before calculating the voltages and currents for your circuit, you must measure the voltage that your power supply produces when it is actually connected to your resistor network. Use four decimal places in all of your calculations. Compare your results with those of your team members to ensure that you agree on the numbers. These voltage and current values are your predicted values. Since there are six or eight measured values to compare to your predictions, you have six or eight values of the percent difference to present and discuss in your report.

On the day of the experiment, hook up your circuit and measure the experimental voltage and current values for each resistor. You may also wish to measure the equivalent resistance of your resistor network and the current going from the power supply into the network. These values may help in the discussion of your results. You may wish to compute the percent difference for the equivalent resistance as an aid to explaining the performance of your circuit.

Technical Notes

1. Resistors come in different "tolerances." Most common resistors have a 10% tolerance, which means that the resistor value can vary by +/−10% from the value indicated by the colored bands on the resistor. In this experiment, you use "precision resistors" that

have a 1% tolerance. You should make note of this so that you can take it into consideration when you write the discussion section of your report and calculate the percent difference values.

2. To determine which of your resistors is which, you can measure the resistance of each one using your digital multimeter (DMM). You can also learn how to read the resistor code in the colored bands on the resistors. This is not hard and some of you will find it interesting, so I will explain it here. Let's begin with the way the code works for ordinary 10% tolerance resistors. (This is one of those things that *all* science geeks know! Most of them learned it in ninth grade and have known it ever since.) For 10% resistors, the resistance values are in colored-coded scientific notation, with 2 significant digits. Since the value is in scientific notation, there is an implied decimal between the first and second digits. The first colored band corresponds to the first significant digit, the second band to the second digit, the third band to the power of 10, and the fourth band (if there is one) to the resistor tolerance. No fourth band usually indicates a 20% tolerance resistor. Tolerances of 10%, 5% and 1% are represented by silver, gold and brown bands, respectively.

Table C.2 shows the resistor color code for the digits (first two bands) and the power of 10 (third band).

	black	brown	red	orange	yellow	green	blue	violet	gray	white
sig digs	0	1	2	3	4	5	6	7	8	9
power of 10	$\times 10^0$	$\times 10^1$	$\times 10^2$	$\times 10^3$	$\times 10^4$	$\times 10^5$	$\times 10^6$	$\times 10^7$	$\times 10^8$	$\times 10^9$

Table C.2. Resistor color code.

Here is an example of how the color code is used for standard resistors. (Note, however, that we are using precision resistors, not standard resistors, in our experiment.) If the first three bands are yellow, violet, and orange, the resistor value is determined as follows:

yellow = 4 violet = 7

These two bands give you the value 4.7.

orange = $\times 10^3$

Putting these together gives 4.7×10^3, or 4.7 kΩ.

For precision resistors with 1% tolerance, like those used in this experiment, the code works slightly differently. These resistors have five colored bands. The first three bands are for significant digits with no implied decimal. The fourth band indicates the power of ten needed to multiply the first three digits by to bring the total to the correct value. The fifth band is the resistance tolerance, which is brown for 1% resistors. Thus, if the first four bands are yellow, violet, green, and brown, the resistor value is

yellow = 4 violet = 7 green = 5 brown = $\times 10^1$

$475 \times 10^1 = 4750$, or 4.75 kΩ.

3. Your DMM measures voltages, currents, and resistances. To make a measurement, you must attend to these three things: a) connect the test leads to the correct terminals on the DMM, b) set the DMM selector switch to the appropriate setting, and c) connect the DMM properly to the circuit you are testing. Depending on the type of meter you have, you may also have to set a range switch. The range switch is set to a low range for low measurement values (such as voltages in the mV range or currents in the mA range), or to a higher range for higher values (such as voltages close to 1 V or currents close to 1 A). However, many DMMs these days automatically set the range so that the meter can measure any value without fooling with a range switch.

    Most DMMs have a "common" or "negative" terminal, colored black, where the black test lead connects. The red test lead connects to one of two different red terminals, depending on what measurement you are making. (Some DMMs combine these into a single red terminal.) The two red terminals on the DMM are labeled to indicate which one is to be used for voltage and resistance measurements and which one is to be used for current measurements.

    a. Voltage and resistance measurements are made by connecting the DMM *across* a resistor or power source. In other words, the DMM is connected in *parallel* to the resistor or power supply. To connect the DMM this way for a voltage measurement, you simply touch the red test lead to the higher voltage side of the device and touch the black test lead to the lower voltage side of the device. If you are measuring a resistance, be sure the resistance you are measuring is completely disconnected from everything else, and simply connect the two test leads to the ends of the resistor or resistor network. It does not matter which lead connects to which end of the resistor.

    > *Caution: Never place the DMM selector switch in the resistance setting when the DMM is connected to devices with the power supply on.* The DMM does not want to see current flowing through it when the selector switch is in the *resistance* position. If you connect it like this, you will either blow a fuse in the DMM or burn up the DMM, depending on what the voltage is at the time. Whenever your DMM is switched to the *resistance* position, the power supply to the circuit must be off.

    b. To make a current measurement, you must break the circuit and insert the DMM *into* the circuit so the current in the circuit flows through the DMM. In other words, the DMM is connected in *series*, like a resistor. The current should flow from the circuit into the red test lead and out of the black test lead back into the circuit. If you are measuring the current flowing through a resistor, it does not matter whether the current goes through the resistor first and then through the DMM, or vice versa.

4. To make it easy for you to connect your resistors together, you will use a small mounting board called a *breadboard*. The breadboard has small rows of holes in which you insert the resistors or connecting wires. Certain rows of holes in the breadboard are connected together inside the breadboard to make it easy to connect different devices together. The hole patterns in the breadboard and the ways they are internally connected are illustrated in Figure C.1.

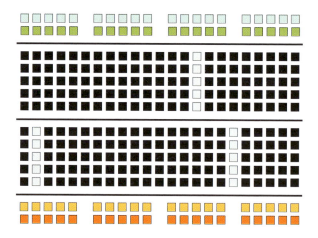

Figure C.1. Hole arrangements in an electronic breadboard.

In Figure C.1, the horizontal rows of holes shown that are the same color (other than black) are all connected together inside the breadboard. Thus, each of the holes in the four long rows of holes along the two long edges of the breadboard are connected together within a given row. As an example, every single one of the orange holes in the figure is connected together inside the breadboard. Two of these long rows of holes are typically used to connect the breadboard to the power supply. This is done by connecting the positive power supply terminal to one of the rows (such as the one shown in green) and the negative power supply terminal to a different one (such as the blue row). This way, anything on the breadboard needing power can get it by just connecting somewhere along those long rows. The short columns of five holes each in the center section of the breadboard (black or gray in Figure C.1) are connected together,

The five-way binding posts on this breadboard are spaced to accommodate a dual banana plug. In this photo, there is a dual banana plug inserted into the binding posts, and connecting wires have been installed from the binding posts to the holes on the breadboard.

The binding posts on this breadboard won't accept a dual banana plug (they are too far apart), so single banana plugs must be used. If the test leads that come with your DMM have alligator clips, single banana plugs can be inserted into the binding posts to give the alligator clips something to clamp to, as shown in this photo. Connecting wires are also shown from the binding posts to the holes on the breadboard.

Figure C.2. Connecting the power supply to the breadboard.

Figure C.3. Example resistor connections installed on the breadboard: two parallel resistors (left) and two series resistors (right).

but only within each individual column. I picked three of these columns at random and colored them gray to illustrate how this works. Each of the separate columns of five holes shown in gray are internally connected together within the column and are not connected to anything else. The same thing holds true for each of the columns of black holes shown in the figure on both sides of the horizontal center line. (The spacings of the holes are designed so that many different types of integrated circuits, computer chips, and other devices can be plugged straight into the breadboard. Obviously, we aren't using any of those devices in our experiment.)

The breadboard also has a set of terminals you can use to connect the circuit to the power supply. These terminals are called *five-way binding posts*. The five-way binding posts have a hole in the top that is sized to accept a type of plug called a *banana plug*. Sometimes these terminals or binding posts are spaced apart to accommodate a dual banana plug connector. These different connectors are shown in Figure C.2. To attach the connecting wire for the breadboard circuit to the binding posts, unscrew the red or black knob on the binding post, insert the wire into the hole through the metal post, and tighten the knob back down.

Using a test lead with the appropriate connectors on both ends, connect the power supply to the five-way binding posts on the breadboard. Then use short lengths of connecting wire to connect the binding posts to the rows that you are using for your (+/−) power supply points.

To illustrate all this, Figure C.3 shows two simple circuits, one with two resistors in parallel and one with two resistors in series. (Your circuit has four resistors.) As you see from these photos, you don't need much connecting wire to put a circuit together. The resistors themselves can be the means of connecting one device to another. However, note that if your resistors have been used before by other students and the legs are bent, they may not insert into the breadboard sockets properly. So take a moment to straighten the resistor legs out so that they insert completely and securely down into the sockets. Figure C.4 shows students making measurements with a complete circuit connected to a power supply.

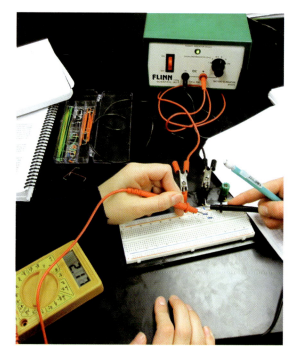

Figure C.4. Making measurements on a circuit.

Procedural Highlights for this Experiment

This is a complex experiment with a lot of details to consider. But now that you are familiar with the background, let's summarize the main elements of the procedure:

1. Design your circuit before the lab session and calculate the R_{EQ} value in advance to save time. Don't calculate anything else yet. Record your design and R_{EQ} value in your lab journal.

2. On the day of the experiment, your first step is to identify which resistor is which. If you have time, you may enjoy doing this by working through the color code. But if time is a factor, it will be much faster simply to measure the resistances with your DMM.

3. Build your circuit on your breadboard and get your instructor to check that it is connected correctly.

4. Connect the power supply to the breadboard and turn it on. Measure the power supply voltage at the terminals on the breadboard. Measure this voltage by touching the DMM leads to the power supply terminals on the breadboard, and record this reading in your lab journal.

5. Using the actual power supply voltage from the measurement you made in the previous step, calculate all the predicted values for the currents and voltages in your circuit. Enter these all in your lab journal in a table. In this table, you can list your variable names in the first column. In the second column enter the calculated values for each variable (your predictions). Make a third column in your table for the measured values.

6. Now measure each of the voltages and currents. Remember, you measure a voltage by simply touching the DMM leads to each side of the resistor. To measure a current, you

must remove one of the resistor legs from the breadboard and connect the DMM in series with the resistor so that the current flows in series through both the resistor and the DMM and then back into the rest of the circuit. Depending on the DMM you have, you may also need to connect the test leads to the DMM one way for current measurements and another way for voltage or resistance measurements.

7. Compare your measurements to your calculated values as you go. They should match closely. If they don't, you've got a problem somewhere that you need to figure out before you continue. The problem could be in your circuit, your calculations, your measurement techniques, or your equipment. If you don't solve the problem, your measurements may be a waste of time and may have to be done all over again.

Additional Notes

1. Be sure to use an accurately measured value of the power supply voltage for all your calculations of the predicted values of voltage and current. Unless you are using an expensive voltage source with an accurate output voltage display, simply depending on the labels or indicators of the power supply will almost certainly result in large errors.

2. Your report must include a schematic diagram of your circuit. The diagram should have all the resistances, voltages, and currents labeled on it. These labels should match the labels in your data tables, which must also be in the report in the Results section. It is best to place your schematic in the Procedure section of your report so you can refer to it there.

3. Your schematic diagram can be prepared in a computer application or may be drawn by hand. However, if it is drawn by hand you must prepare it very neatly using a straight-edge, and you must scan it so it can be digitally placed in your report.

4. Your report must include percent difference figures for all the voltages and currents in the circuit.

APPENDIX D
Scientists to Know About

In *Introductory Physics*, we examine several historical episodes in the history of physics. In each of them, there are well-known scientists we encounter. Many of these contributors are of such key importance in the history of science that they are worth knowing about, so I have selected some of them for you to remember. To help you organize your study of these important scientists, this table specifies what you need to know.

Tycho Brahe Johannes Kepler Galileo Isaac Newton (10 to 20 point quiz items)	• His name and nationality • Key dates related to his scientific work • Where he worked • His major areas of research and discovery • Summary of his major contribution, including the year (For Kepler, this includes stating the first law of planetary motion; for Newton, stating the three laws of motion; for Dalton, stating the five principles of his atomic model.)
Nicolaus Copernicus J. J. Thomson Robert Millikan Ernest Rutherford (10 point quiz items)	• His name and nationality • Summary of his major contribution, including the year (For Thomson, Millikan, and Rutherford this includes describing their famous experiments and the significant discoveries and new theories resulting from each one.)
Albert Einstein Democritus Alessandro Volta James Clerk Maxwell (5 point quiz items)	• His name and nationality • A single statement describing his major contribution, including the year (Einstein is famous for many things, but in this course, our interest is in his general theory of relativity.)

APPENDIX E
Making Accurate Measurements

Making accurate measurements in experiments requires care. It also requires learning practices that help you minimize error. In this appendix we review some of these practices.

Two common measurement issues involving special technical terms are avoiding *parallax error* and working correctly with the *meniscus* on liquids. These terms both have to do with using analog instruments with measurement scales that must be correctly aligned for an accurate measurement.

E.1 Parallax Error

Parallax error occurs when the line of sight of a person taking a measurement is at an incorrect angle relative to the instrument scale and the object being measured. As shown in Figure E.1, the viewer's line of sight must be parallel to the lines on the scale and perpendicular to the scale itself. Misalignment of the viewer's line of sight results in an inaccurate measurement due to parallax error.

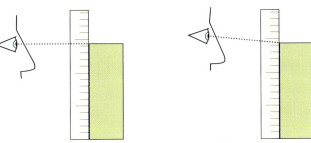

Figure E.1. In the sketch on the left, the measured object, measurement instrument, and viewer line of sight are correctly aligned. In the sketch on the right, the misalignment of the viewer causes parallax error in the measurement.

E.2 Measurements with a Meter Stick or Rule

1. For maximum accuracy, avoid using the end of a wooden rule. The end is usually subject to a lot of pounding and abrasion, which wears off or compresses the wood on the end.

2. As indicated in Figure E.2, arrange the rule against the object to be measured so the marks on the scale touch the object being measured. This helps minimize parallax error.

3. As indicated in Figure E.3, use a straight-edge to ensure that the end of a metal rule is accurately aligned with the edge of the object being measured.

incorrect

E.3   Liquid Measurements

When a liquid is placed in a container, the surface tension of the liquid causes the liquid to curve up or down at the walls of the container. In most liquids, including water, the surface of the liquid curves up at the container wall. (A well-known example of downward curvature is liquid mercury.) We concentrate here on the common upward curvature exhibited by water.

If the container is tall and narrow, as with a graduated cylinder, then the curving liquid at the edges gives the liquid surface an overall bowl shape. This bowl-shaped surface is called a *meniscus*. The correct way to read a volume of liquid is to read the liquid at the bottom of the meniscus. Figure E.4 illustrates the correct way to read a volume of liquid in a graduated cylinder by reading the liquid level at the bottom of the meniscus, avoiding paral-lax error.

correct

Figure E.2. Proper placement of a rule or meter stick.

E.4   Measurements with a Triple-Beam Balance

1. Calibrate the balance before making measurements. This is accomplished by turning the calibration knob under the pan until the scale's alignment marks are perfectly aligned when the pan is empty and the beam is at rest.

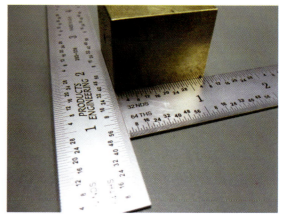

Figure E.3. Use of a straightedge for proper alignment of the end of the rule with the object being measured.

2. Make sure the 10-g and 100-g weights are locked into a notch on the beam. Otherwise, the measurement is not be correct.

3. As shown in Figure E.5, when adjusting the position of the gram weight, proper practice is to slide this weight with the tip of a pencil held below the beam instead

of with your finger. If done carefully, this technique allows the gram weight to be manipulated into position without disturbing the balance of the beam as the balance point is approached.

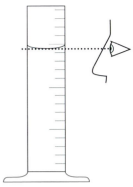

E.5   Measurements with an Analog Thermometer

1. Mercury thermometers are more accurate than spirit thermometers. However, if a mercury thermometer breaks, you have a real problem cleaning up and disposing of the spilled mercury. Thus, for student use, I recommend using only spirit thermometers.

2. When measuring temperatures, be sure to notice that the thermometers have a mark indicating the proper degree of immersion for the most accurate reading.

3. Thermometer accuracy can be severely compromised if gaps get into the liquid. Always store spirit thermometers vertically in an appropriate rack to help prevent gaps from getting into the liquid.

Figure E.4. Correctly reading a liquid volume in a graduated cylinder. The viewer's line of sight is at the bottom of the meniscus and perpendicular to the scale to avoid parallax error.

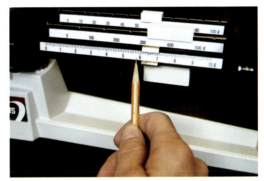

Figure E.5. Move the gram slider with a pencil to prevent disturbing the beam as the balance point is approached

APPENDIX F
References

CHAPTER 1

The quote from Sir Arthur Eddington is from Wikimedia Commons.

Information about Sir Humphry Davy and the Davy Safety Lamp is from *The Age of Wonder*, Richard Holmes, Pantheon, 2008.

CHAPTER 2

The definition for significant digits is quoted from *Trigonometry*, Charles McKeague and Mark Turner, 6th ed.

The quote in the box about Copernicus is from *On the Revolutions of the Heavenly Spheres*, Nicolaus Copernicus, Prometheus Books, 1995.

The quote in the box about Kepler is from *Harmonies of the World*, Johannes Kepler, Prometheus Books, 1995.

Galileo's recantation speech is found in *The Great Physicists from Galileo to Einstein*, George Gamow, Dover, 1988.

Some of the historical information about Copernicus, Rheticus, Osiander, and Tycho is from *A More Perfect Heaven*, Dava Sobel, Walker, 2011.

Some of the historical information about Tycho and Kepler is from *Kepler*, Max Caspar, Dover, 1993. Some of the details about Tycho are from wikipedia.org.

Details of Galileo's interaction with the Church and trial are from *Galileo's Mistake*, Wade Rowland, Arcade, 2003, and *God's Planet*, Owen Gingerich, Harvard, 2014.

Information about William and Caroline Herschel is from *The Age of Wonder*, Richard Holmes, Pantheon, 2008.

CHAPTER 3 AND CHAPTER 5

Statements about Newton's wording of the laws of motion are based on the revised Motte translation, *Sir Isaac Newton's Mathematical Principles*, University of California Press, 1947.

CHAPTER 6

Rutherford's comment about the artillery shell bouncing back from a piece of tissue paper is found on wikipedia.org, among other places.

Information about the strength of carbon nanotubes is from the Carbon Nanotube article at wikipedia.org.

CHAPTER 7

Most of the data in the box about the speed of air molecules are from Penn State College of Earth and Mineral Sciences website, http://www.ems.psu.edu/~bannon/moledyn.html.

Specific heat capacity and thermal conductivity data were obtained from engineeringtoolbox.com.

CHAPTER 9

The harmonic spectrum and frequency response images were captured on an iPad with an app called "n-track tuner."

CHAPTER 10

Much of the historical information about the history of electricity was taken from *The Great Physicists from Galileo to Einstein*, George Gamow, Dover 1988.

Information about Michael Faraday, including the quote about Michael Faraday's religious faith, is from the wikipedia.org article on Michael Faraday. Other information about Faraday is from *Creations of Fire*, Cathy Cobb and Harold Goldwhite, Plenum, 1995.

Information about Maxwell's color photograph is from the wikipedia.org article on Maxwell.

The reference to the pictures on Einstein's wall is from the wikipedia.org article on Michael Faraday.

CHAPTER 11

Information about the War of Currents, Thomas Edison, and George Westinghouse is from the wikipedia.org articles on same.

CHAPTER 12

Information about Nikola Tesla is from the wikipedia.org article on Tesla.

CHAPTER 13

A significant resource for this chapter was *Physics: Algebra/Trig*, 2e, Eugene Hecht, Brooks/Cole, 1998.

GENERAL

The science articles on wikipedia.org are very well maintained and reliable. These articles also specify original sources for the scientific information included in the article. The wikipedia.org articles were a resource for information in captions for licensed and public domain images, and for the Do You Know? boxes.

APPENDIX G
Image Credits

xvii. John D. Mays. 2. 1919_eclipse_positive.jpg via https://commons.wikimedia.org/wiki/File:1919_eclipse_positive.jpg. Author, F. W. Dyson, A. S. Eddington, and C. Davidson, public domain. 16. Wood,_lauren_(1).jpg via https://commons.wikimedia.org/wiki/File:Wood,_lauren_(1).jpg. Author: Bill Branson (Photographer), public domain. 17. On_a_USDA-ARS_test_plot_Utah_State_University_Research_assistant_pollinates_Snake_River_wheatgrass.jpg via https://commons.wikimedia.org/wiki/File:On_a_USDA-ARS_test_plot_Utah_State_University_Research_assistant_pollinates_Snake_River_wheatgrass.jpg. Author: USDA, licensed under CC-BY-SA-2.0. 19. Sir_Humphry_Davy,_Bt_by_Thomas_Phillips.jpg via https://upload.wikimedia.org/wikipedia/commons/8/8f/Sir_Humphry_Davy%2C_Bt_by_Thomas_Phillips.jpg. Author: Thomas Phillips, public domain. 19. Davy Safety Lamp: Davy_lamp.png via https://commons.wikimedia.org/wiki/File:Davy_lamp.png. Source: Bibliothek allgemeinen und praktischen Wissens für Militäranwärter Band III, 1905 / Deutsches Verlaghaus Bong & Co Berlin * Leipzig * Wien * Stuttgart, public domain. 20. John D. Mays. 24. Prototype_kilogram_replica.JPG via https://commons.wikimedia.org/wiki/File:Prototype_kilogram_replica.JPG. Author: Japs 88, licensed under CC-BY-SA-3.0. 27–34. 42. Aristotle_Altemps_Inv8575.jpg via http://commons.wikimedia.org/wiki/File:Aristotle_Altemps_Inv8575.jpg. Source: Ludovisi Collection, public domain. 43. Ptolemy_16century.jpg via https://commons.wikimedia.org/wiki/File:Ptolemy_16century.jpg. Source: Claudius Ptolemäus, Picture of 16th century book frontispiece, public domain. 45. Fairy_World_Spin1.JPG via https://commons.wikimedia.org/wiki/File:Fairy_World_Spin1.JPG. Author: Druyts.t, licensed under CC-BY-SA-4.0. 46. Eratosthenes.jpg via https://commons.wikimedia.org/wiki/File:Eratosthenes.jpg. Public domain. 50. Jan_Matejko-Astronomer_Copernicus-Conversation_with_God.jpg via https://commons.wikimedia.org/wiki/File:Jan_Matejko-Astronomer_Copernicus-Conversation_with_God.jpg. Artist: Jan Matejko, public domain. 51. Tycho_Brahe.JPG via https://commons.wikimedia.org/wiki/File:Tycho_Brahe.JPG. Author: Eduard Ender († 1883), public domain. 51. Uraniborgskiss_45.jpg via https://commons.wikimedia.org/wiki/File:Uraniborgskiss_45.jpg. Source: Tycho, Astronomiae instauratae mechanica (1598), public domain. 52. Johannes_Kepler_1610.jpg via https://commons.wikimedia.org/wiki/File:Johannes_Kepler_1610.jpg. Author: unknown, public domain. 55. Justus_Sustermans_-_Portrait_of_Galileo_Galilei,_1636.jpg via https://commons.wikimedia.org/wiki/File:Justus_Sustermans_-_Portrait_of_Galileo_Galilei,_1636.jpg. Justus Sustermans, public domain. 56. Pendule_de_Foucault.jpg via https://commons.wikimedia.org/wiki/File:Pendule_de_Foucault.jpg. Author: Arnaud 25, public domain. 58. GodfreyKneller-IsaacNewton-1689.jpg via https://commons.wikimedia.org/wiki/File:GodfreyKneller-IsaacNewton-1689.jpg. Sir Godfer Kneller, public domain. PSM_V09_D079_Herschel_40_foot_telescope_at_slough.jpg via https://commons.wikimedia.org/wiki/File:PSM_V09_D079_Herschel_40_foot_telescope_at_slough.jpg. Author: unknown, public domain. 59. Einstein1921_by_F_Schmutzer_2.jpg via https://commons.wikimedia.org/wiki/File:Einstein1921_by_F_Schmutzer_2.jpg. Author: Ferdinand Schmutzer, 1921, public domain. 59. Spacetime_curvature.png via http://commons.wikimedia.org/wiki/File:Spacetime_curvature.png. Author: Johnstone, licensed under CC-BY-SA-3.0. 64. Photograph by John D. Mays. 73. Earth_Western_Hemisphere.jpg via https://upload.wikimedia.org/wikipedia/commons/7/7b/Earth_Western_Hemisphere.jpg. Author: NASA, public domain. 73. FullMoon2010.jpg via https://commons.wikimedia.org/wiki/File:FullMoon2010.jpg. Author: Gregory H. Revera, licensed under CC-BY-SA-3.0. 78. STS120LaunchHiRes-edit1.jpg via https://commons.wikimedia.org/wiki/File:STS120LaunchHiRes-edit1.jpg. Author: NASA, public domain. 79. Photograph by John D. Mays. 83. John D. Mays. 84. CMS Higgs event.jpg via https://commons.wikimedia.org/wiki/File:CMS_Higgs-

domain. 167, 168. John D. Mays. 173. JPL_Spiderweb_Bolometer.jpg via https://commons.wikimedia.org/wiki/File:JPL_Spiderweb_Bolometer.jpg. Author: NASA/JPL-Caltech, public domain. 174. Defense.gov photo essay 120710-N-RY232-532.jpg via http://commons.wikimedia.org/wiki/File:Defense.gov_photo_essay_120710-N-RY232-532.jpg. Author: Petty Officer 2nd Class Julia A. Casper, public domain. 176. Blaise_pascal.jpg via http://commons.wikimedia.org/wiki/File:Blaise_pascal.jpg, public domain. 179. Evangelista_Torricelli2.jpg via https://commons.wikimedia.org/wiki/File:Evangelista_Torricelli2.jpg. Author: S.L. Pelaco, public domain. 182. Domenico-Fetti_Archimedes_1620.jpg via https://commons.wikimedia.org/wiki/File:Domenico-Fetti_Archimedes_1620.jpg. Artist: Domenico Fetti, public domain. 186. Archimede_bain.jpg via http://commons.wikimedia.org/wiki/File:Archimede_bain.jpg. Public domain. 188. Alexander_the_Great_diving_NOAA.jpg via http://commons.wikimedia.org/wiki/File:Alexander_the_Great_diving_NOAA.jpg, public domain. 189. Drum_(container).jpg via http://commons.wikimedia.org/wiki/File:Drum_(container).jpg. Author: Meggar at English Wikipedia, licensed under CC-BY-SA-3.0. 189. Rabe pump01.jpg via http://commons.wikimedia.org/wiki/File:Rabe_pump01.jpg. Author: Tomasz Kuran aka Meteor2017, licensed under CC-BY-SA-3.0. 190. Horn_Antenna-in_Holmdel,_New_Jersey.jpeg via https://commons.wikimedia.org/wiki/File:Horn_Antenna-in-Holmdel,_New_Jersey.jpeg. Author: NASA, public domain. 197–201. John D. Mays. 203. Taipei101.portrait.altonthompson.jpg via https://commons.wikimedia.org/wiki/File:Taipei101.portrait.altonthompson.jpg. Author: GREG, licensed under CC-BY-SA-3.0. 203. Taipei_101_Tuned_Mass_Damper_2010.jpg via https://commons.wikimedia.org/wiki/File:Taipei_101_Tuned_Mass_Damper_2010.jpg. Author: Armand du Plessis, licensed under CC-BY-SA-3.0. 203. Adapted from Taipei_101_Tuned_Mass_Damper_pl.png via https://commons.wikimedia.org/wiki/File:Taipei_101_Tuned_Mass_Damper_pl.png. Author: Bambosz, Someformofhuman, licensed under CC-BY-SA-3.0. 204. John D. Mays. 206. John D. Mays. 208. Schlieren_photograph_of_T-38_shock_waves.jpg via https://commons.wikimedia.org/wiki/File:Schlieren_photograph_of_T-38_shock_waves.jpg. Author: NASA/Dr. Leonard Weinstein, public domain. 214. Static_slide.jpg via https://commons.wikimedia.org/wiki/File:Static_slide.jpg. Author: Ken Bosma, licensed under CC-BY-SA 2.0. 215. William_Gilbert.jpg via https://commons.wikimedia.org/wiki/File:William_Gilbert.jpg. Author: Granger, public domain. 216. Leyden_jars,_maker_unknown,_c._1820-1860_-_DSC06563.JPG via https://commons.wikimedia.org/wiki/File:Leyden_jars,_maker_unknown,_c._1820-1860_-_DSC06563.JPG. Author: Daderot, public domain. 216. PSM_V05_D400_Joseph_Priestly.jpg via https://commons.wikimedia.org/wiki/File:PSM_V05_D400_Joseph_Priestley.jpg. Unknown, public domain. 217. Luigi_Galvani,_oil-painting.jpg via https://commons.wikimedia.org/wiki/File:Luigi_Galvani,_oil-painting.jpg. Artist unknown, public domain. 218. Alessandro_Volta.jpeg via https://commons.wikimedia.org/wiki/File:Alessandro Volta.jpeg. Source: http://www.anthroposophie.net/bibliothek/nawi/physik/volta/bib_volta.htm; public domain. 218. Pila_di_Volta.jpg via https://commons.wikimedia.org/wiki/File:Pila_di_Volta.jpg. Author: Luigi Chesa, licensed under CC-BY-SA-3.0. 219. Hans_Christian_Oersted_statue.jpg via https://commons.wikimedia.org/wiki/File:Hans_Christian_Oersted_statue.jpg. Author: Andrew Gray, licensed under CC-BY-SA-3.0. 219. Ampere_Andre_1825.jpg via https://commons.wikimedia.org/wiki/File:Ampere_Andre_1825.jpg. Author: Ambrose Tardieu, public domain. 220. M_Faraday_Th_Phillips_oil_1842.jpg via https://commons.wikimedia.org/wiki/File:M_Faraday_Th_Phillips_oil_1842.jpg. Author: Thomas Phillips, public domain. 220. James-clerk-maxwell3.jpg via https://commons.wikimedia.org/wiki/File:James-clerk-maxwell3.jpg. Source: Practical Physics, Millikan and Gale, 1920, public domain. 221. Tartan_Ribbon.jpg via https://commons.wikimedia.org/wiki/File:Tartan_Ribbon.jpg. James Clerk Maxwell (original photographic slides), public domain. 222. John D. Mays. 223. John D. Mays. 224. NeTube.jpg via https://commons.wikimedia.org/wiki/File:NeTube.jpg. Author: Pslawinski, licensed under CC-BY-SA-2.5. 226. John D. Mays. 229. Hw-newton.jpg via https://commons.wikimedia.org/wiki/File:Hw-newton.jpg. Source: H.F. Helmolt (ed.): History of the World. New York, 1901. Public domain. 229. Albert_Einstein_photo_1920.jpg via https://commons.wikimedia.org/wiki/File:Albert_Einstein_photo_1920.jpg. Photographer unknown, public domain. 229. Faraday_photograph_ii.jpg via https://upload.wikimedia.org/wikipedia/commons/6/62/Faraday_photograph_ii.jpg. Author: Maull & Polyblank, public domain. 229. James-clerk-maxwell_1.jpg via https://commons.wikimedia.org/wiki/File:James-clerk-maxwell_1.jpg. Public domain. 230. 4-fach-NAND-C10.JPG via https://commons.wikimedia.org/wiki/File:4-fach-NAND-C10.JPG. Author: Dgarte, licensed under CC-BY-SA-3.0. 237. Georg_Simon_Ohm3.jpg via https://commons.wikimedia.org/wiki/File:Georg_Simon_Ohm3.jpg. Public domain. 239–240. John D. Mays. 248. John D. Mays. 254. Gustav_Robert_Kirchhoff.jpg via https://commons.wikimedia.org/wiki/File:Gustav_Robert_Kirchhoff.jpg. Public domain. 264. Thomas_Edison2.jpg via https://commons.wikimedia.org/wiki/File:Thomas_Edison2.jpg. Louis Bachrach, Bachrach Studios, restored by Michel Vuijlsteke, public domain. 264. George_Westinghouse.jpg via https://commons.wikimedia.org/wiki/File:George_Westinghouse.jpg. Author: Joseph G. Gessford, public domain. 274. Superconducting_levitation_and_candle_on_a_magnet.JPG via https://commons.wikimedia.org/wiki/File:Superconducting_levitation_and_candle_on_a_magnet.JPG. Author: Julien Bobroff (user:Jubobroff), Frederic Bouquet (user:Fbouquet), LPS, Orsay, France, licensed under CC-BY-SA-3.0. 280–283. John D. Mays. 286. John D. Mays. 286. DBZ_trafo.jpg via https://commons.wikimedia.org/wiki/File:DBZ_trafo.jpg. Author: Zátonyi Sándor, (ifj.), licensed under CC-BY-SA-3.0. 286. Etzenricht_Transformer.JPG via http://commons.wikimedia.org/wiki/File:Etzenricht_

Index